METHODS IN MOLECULAR BIOLOGY™

Series Editor
John M. Walker
School of Life Sciences
University of Hertfordshire
Hatfield, Hertfordshire, AL10 9AB, UK

For other titles published in this series, go to
www.springer.com/series/7651

Embryonic Stem Cell Therapy for Osteo-Degenerative Diseases

Methods and Protocols

Edited by

Nicole I. zur Nieden

Department of Cell Biology & Neuroscience and Stem Cell Center,
University of California at Riverside, Riverside CA, USA

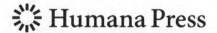 Humana Press

Editor
Nicole I. zur Nieden
Department of Cell Biology & Neuroscience
and Stem Cell Center
University of California at Riverside
Riverside, CA, USA
nicole.zurnieden@ucr.edu

ISSN 1064-3745   e-ISSN 1940-6029
ISBN 978-1-60761-961-1   e-ISBN 978-1-60761-962-8
DOI 10.1007/978-1-60761-962-8
Springer New York Dordrecht Heidelberg London

Library of Congress Control Number: 2010938795

Printed on acid-free paper

Humana Press is part of Springer Science+Business Media (www.springer.com)

Preface

Degenerative cartilage and bone diseases have challenged physicians for many decades due to the inherent inability of the tissue to repair itself when ailing. Patients afflicted not only suffer from physical pain, but their quality of life is decreased as these diseases often progress, going hand in hand with physical disability. Although conventional treatment regimens can ameliorate the condition, full recovery is never achieved. Here, tissue engineering using stem cells has brought a promising new approach to the treatment of such diseases.

Whereas adult stem cells, such as mesenchymal stem cells, are already tested in clinical trials and have brought some success in spinal disc and cartilage regeneration, this particular stem cell source may have some disadvantages in the context of certain other diseases. Mesenchymal stem cells senesce in culture, which ultimately means that maximum cell number reached after in vitro expansion might not suffice to fill and repair larger defects.

Embryonic stem cells (ESCs) on the other hand are stem cells of embryonic origin with an unlimited self-renewing capacity. Therefore, these cells represent an almost bottomless source for regenerative medicine and tissue engineering approaches. ESCs have been first derived from mice and just celebrated their 29th birthday. We have an adequate understanding of the processes that control pluripotency and are now diving deeper into explaining epigenetics and reprogramming. Although many groups have been studying their in vitro differentiation potential into cardiomyocytes and neurons, pancreatic beta-cells, and hepatocytes, the study of osteogenesis and chondrogenesis using ESCs as a model is a relatively young field. With the derivation of human and primate ESCs, however, the potential of ESCs is more intriguing than ever and more and more researchers enter this rising field.

With this textbook, I would like to set the ground stone on which those entering the field – cell biologists, molecular biologists, bioengineers, and clinicians – can build their mission to understand the mechanisms involved in differentiation of ESCs into skeletal cell types. To bring ESCs to clinic, much of this is needed and will not be the single undertaking of one laboratory. *Embryonic Stem Cell Therapy for Osteo-Degenerative Diseases: Methods and Protocols* provides detailed descriptions on how to expand ESCs from the most commonly used species ex vivo – mouse and human – in static culture as well as in controllable bioreactor processes. It summarizes the methods that may be used to differentiate the cells along the desired lineage of choice – be it osteoblasts, osteoclasts, or chondrocytes – and consequentially also offers analysis tools for the characterization of resulting cells and evaluation of differentiation effectiveness. Each chapter gives special care to address possible pitfalls and provides a troubleshooting guide.

Embryonic Stem Cell Therapy for Osteo-Degenerative Diseases: Methods and Protocols contains only protocols that are specific to ESCs and stem cell differentiation, and I refer the reader to *Bone Research Protocols*, edited by Helfrich and Ralston (1), for a detailed description for DNA/RNA isolation from bone tissue, specific considerations for embedding and sectioning bone tissue, and other issues concerning characterization of bone tissue the reader might want to refer to after using ESCs for in vivo transplantation.

The contributors of this book were chosen because they represent the experts in the field of stem cell self-renewal and the pioneers in skeletogenesis from ESCs. All have published in reputable, peer-reviewed journals. I sincerely hope that the methodology described in this book will be successful in your hands and that it allows you to carry out critical research needed in order to publish your own peer-reviewed articles and bring the field closer to the clinic. Such protocols merit ongoing consideration to explore the potential of progenitor cells for skeletal tissue repair within practical clinical guidelines that would allow for the widespread application of a successful strategy. Ultimately, this will advance the field and bring new knowledge on ESCs and skeletal development in general.

Tremendous gratitude is owed to Prof. John Walker at Humana Press, who gave me the opportunity to learn a great deal from my fellow authors and guided me through the process of editing, for providing me with the opportunity to add a new volume focusing on my personal scientific interest – pluripotent stem cells in skeletal regeneration – to the series, *Methods in Molecular Biology*.

Riverside, CA, USA ·   *Nicole I. zur Nieden*

Reference

1. Helfrich, M.H. and Ralston, S.H. (2003) Bone Research Protocols. Totowa, New Jersey, USA, p. 448

Contents

Preface .... *v*

Contributors .... *ix*

1 Embryonic Stem Cells for Osteo-Degenerative Diseases 1
  Nicole I. zur Nieden

2 Methods for Culturing Mouse and Human Embryonic Stem Cells 31
  Sabrina Lin and Prue Talbot

3 Serum-Free and Feeder-Free Culture Conditions for Human
  Embryonic Stem Cells 57
  Ludovic Vallier

4 Functional Assays for Human Embryonic Stem Cell Pluripotency 67
  Michael D. O'Connor, Melanie D. Kardel, and Connie J. Eaves

5 Using Cadherin Expression to Assess Spontaneous
  Differentiation of Embryonic Stem Cells 81
  Helen Spencer, Maria Keramari, and Christopher M. Ward

6 Generation of Human Embryonic Stem Cells Carrying Lineage
  Specific Reporters 95
  Parinya Noisa, Alai Urrutikoetxea-Uriguen, and Wei Cui

7 Manipulations of MicroRNA in Human Pluripotent Stem Cells
  and Their Derivatives 107
  *Stephanie N. Rushing, Anthony W. Herren, Deborah K. Lieu,
  and Ronald A. Li*

8 Large-Scale Expansion of Mouse Embryonic Stem Cells on Microcarriers 121
  *Ana Fernandes-Platzgummer, Maria Margarida Diogo,
  Cláudia Lobato da Silva, and Joaquim M.S. Cabral*

9 Embryoid Body Formation: Recent Advances in Automated
  Bioreactor Technology 135
  Susanne Trettner, Alexander Seeliger, and Nicole I. zur Nieden

10 Methods for Embryoid Body Formation: The Microwell Approach 151
   *Dawn P. Spelke, Daniel Ortmann, Ali Khademhosseini, Lino Ferreira,
   and Jeffrey M. Karp*

11   Human Embryonic Stem Cell-Derived Mesenchymal Progenitors:
      An Overview   163
      Peiman Hematti

12   Derivation of Mesenchymal Stem Cells from Human Embryonic
      Stem Cells   175
      Andre Choo and Sai Kiang Lim

13   Differentiation of Human Embryonic Stem Cells into Mesenchymal
      Stem Cells by the "Raclure" Method   183
      Emmanuel N. Olivier and Eric E. Bouhassira

14   Improved Media Compositions for the Differentiation
      of Embryonic Stem Cells into Osteoblasts and Chondrocytes   195
      Beatrice Kuske, Vuk Savkovic, and Nicole I. zur Nieden

15   Differentiation of Mouse Embryonic Stem Cells in Self-Assembling
      Peptide Scaffolds   217
      Núria Marí-Buyé and Carlos E. Semino

16   Methods for Investigation of Osteoclastogenesis Using Mouse
      Embryonic Stem Cells   239
      Motokazu Tsuneto, Toshiyuki Yamane, and Shin-Ichi Hayashi

17   Absorption-Based Assays for the Analysis of Osteogenic
      and Chondrogenic Yield.   255
      Lesley A. Davis, Anke Dienelt, and Nicole I. zur Nieden

18   Identification of Osteoclasts in Culture.   273
      *Nobuyuki Udagawa, Teruhito Yamashita, Yasuhiro Kobayashi,
      and Naoyuki Takahashi*

19   Analysis of Glycosaminoglycans in Stem Cell Glycomics   285
      *Boyangzi Li, Haiying Liu, Zhenqing Zhang, Hope E. Stansfield,
      Jonathan S. Dordick, and Robert J. Linhardt*

20   Drill Hole Defects: Induction, Imaging, and Analysis in the Rodent   301
      Andre Obenaus and Pedro Hayes

21   Measurement and Illustration of Immune Interaction After Stem
      Cell Transplantation   315
      Stephan Fricke

Index.   333

Contributors

ERIC E. BOUHASSIRA • *Departments Hematology and Cell Biology, Einstein Center for Human Embryonic Stem Cell Research Medicine, Albert Einstein College of Medicine, New York, NY, USA*

JOAQUIM M.S. CABRAL • *Centre for Biological and Chemical Engineering, Institute for Biotechnology and Bioengineering (IBB), Instituto Superior Técnico, Lisboa, Portugal*

ANDRE CHOO • *Bioprocessing Technology Institute, Agency for Science Technology and Research, Singapore*

WEI CUI • *Institute of Reproductive and Developmental Biology, Imperial College London, London, UK*

CLÁUDIA LOBATO DA SILVA • *Centre for Biological and Chemical Engineering, Institute for Biotechnology and Bioengineering (IBB), Instituto Superior Técnico, Lisboa, Portugal*

LESLEY A. DAVIS • *Department of Surgery, Cambridge Institute for Medical Research, Addenbrooke's Hospital, University of Cambridge, Cambridge, UK*

ANKE DIENELT • *Fraunhofer Institute for Cell Therapy and Immunology, Department of Cell Therapy, Applied Stem Cell Technology Unit, Leipzig, Germany*

MARIA MARGARIDA DIOGO • *Centre for Biological and Chemical Engineering, Institute for Biotechnology and Bioengineering (IBB), Instituto Superior Técnico, Lisboa, Portugal*

JONATHAN S. DORDICK • *Departments of Biology and Chemical and Biological Engineering, Center for Biotechnology and Interdisciplinary Studies, Rensselaer Polytechnic Institute, Troy, NY, USA*

CONNIE J. EAVES • *Terry Fox Laboratory, BC Cancer Agency, VancouverBC, Canada*

ANA FERNANDES-PLATZGUMMER • *Centre for Biological and Chemical Engineering, Institute for Biotechnology and Bioengineering (IBB), Instituto Superior Técnico, Lisboa, Portugal*

LINO FERREIRA • *Center of Neurosciences and Cell Biology, University of Coimbra, Coimbra, Portugal; Biocant- Center of Biotechnology Innovation Center, Cantanhede, Portugal*

STEPHAN FRICKE • *Fraunhofer Institute for Cell Therapy and Immunology, Leipzig, Germany*

SHIN-ICHI HAYASHI • *Division of Immunology, Department of Molecular and Cellular Biology, Faculty of Medicine, School of Life Science, Tottori University, Yonago, Japan*

PEDRO HAYES • *Non-Invasive Imaging Laboratory, Departments of Radiation Medicine, Pediatrics, and Radiology, Department of Biophysics and Bioengineering, School of Medicine, School of Science and Technology, Loma Linda University, Loma Linda, CA, USA*

PEIMAN HEMATTI • *Department of Medicine, School of Medicine and Public Health, University of Wisconsin*

PAUL P. CARBONE • *Comprehensive Cancer Center, University of Wisconsin Madison, Madison, WI, USA*

ANTHONY W. HERREN • *Department of Cell Biology and Human Anatomy, University of California at Davis School of Medicine, Davis, CA, USA*

MELANIE D. KARDEL • *Terry Fox Laboratory, BC Cancer Agency, Vancouver, BC, Canada*

JEFFREY M. KARP • *Department of Medicine, HST Center for Biomedical Engineering, Brigham and Women's Hospital, Harvard Medical School, Boston, MA, USA; Harvard–MIT Division of Health Sciences and Technology, Massachusetts Institute of Technology, Cambridge, MA, USA*

MARIA KERAMARI • *Faculty of Medical and Human Sciences, Centre for Molecular Medicine, The University of Manchester, Manchester, UK*

ALI KHADEMHOSSEINI • *Department of Medicine, HST Center for Biomedical Engineering, Brigham and Women's Hospital, Harvard Medical School, Boston, MA, USA*
Harvard–MIT Division of Health Sciences and Technology, Massachusetts Institute of Technology, Cambridge, MA, USA

YASUHIRO KOBAYASHI • *Institute for Oral Science, Matsumoto Dental University, Nagano, Japan*

BEATRICE KUSKE • *Department of Cell Biology & Neuroscience, University of California Riverside, Riverside, CA, USA*

BOYANGZI LI • *Department of Chemistry and Chemical Biology, Center for Biotechnology and Interdisciplinary Studies, Rensselaer Polytechnic Institute, Troy, NY, USA*

RONALD A. LI • *Department of Cell Biology and Human Anatomy, University of California at Davis School of Medicine, Davis, CA, USA*

DEBORAH K. LIEU • *Department of Cell Biology and Human Anatomy, University of California at Davis School of Medicine, Davis, CA, USA*

SABRINA LIN • *Department of Cell Biology & Neuroscience and Stem Cell Center, University of California Riverside, Riverside, CA, USA*

ROBERT J. LINHARDT • *Department of Chemistry and Chemical Biology and Department of Biology and Department of Chemical and Biological Engineering, Center for Biotechnology and Interdisciplinary Studies, Rensselaer Polytechnic Institute, Troy, NY, USA*

SAI KIANG LIM • *Institute of Medical Biology, Agency for Science Technology and Research, Singapore; Department of Surgery, Yong Loo Lin School of Medicine, National University of Singapore, Singapore*

HAIYING LIU • *Department of Biology, Center for Biotechnology and Interdisciplinary Studies, Rensselaer Polytechnic Institute, Troy, NY, USA*

NÚRIA MARÍ-BUYÉ • *Department of Bioengineering, Institut Químic de Sarrià, Universitat Ramon Llull, Barcelona, Spain*
Translational Centre for Regenerative Medicine (TRM), Universität Leipzig, Leipzig, Germany

PARINYA NOISA • *Institute of Reproductive and Developmental Biology, Imperial College London, London, UK*

ANDRE OBENAUS • *Non-Invasive Imaging Laboratory, Department of Biophysics and Bioengineering, Departments of Radiation Medicine, Pediatrics, and Radiology, School of Medicine, School of Science and Technology, Loma Linda University, Loma Linda, CA, USA*

MICHAEL D. O'CONNOR • *Terry Fox Laboratory, BC Cancer Agency, Vancouver, BC, Canada*

EMMANUEL N. OLIVIER • *Hematology and Cell Biology Departments, Einstein Center for Human Embryonic Stem Cell Research Medicine, Albert Einstein College of Medicine, New York, NY, USA*

DANIEL ORTMANN • *Department of Internal Medicine II–Cardiology, University of Ulm, Ulm, Germany; Department of Medicine, HST Center for Biomedical Engineering, Brigham and Women's Hospital, Harvard Medical School, Boston, MA, USA; Harvard–MIT Division of Health Sciences and Technology, Massachusetts Institute of Technology, Cambridge, MA, USA*

STEPHANIE N. RUSHING • *Department of Cell Biology and Human Anatomy, University of California at Davis School of Medicine, Davis, CA, USA*

VUK SAVKOVIC • *Fraunhofer Institute for Cell Therapy and Immunology, Department of Cell Therapy, Applied Stem Cell Technology Unit, Leipzig, Germany*

ALEXANDER SEELIGER • *Fraunhofer Institute for Cell Therapy and Immunology, Department of Cell Therapy, Applied Stem Cell Technology Unit, Leipzig, Germany*

CARLOS E. SEMINO • *Department of Bioengineering, Institut Químic de Sarrià, Universitat Ramon Llull, Barcelona, Spain; Translational Centre for Regenerative Medicine (TRM), Universität Leipzig, Leipzig, Germany Center for Biomedical Engineering, Massachusetts Institute of Technology, Cambridge, MA, USA*

DAWN P. SPELKE • *Department of Biological Engineering, Massachusetts Institute of Technology, Cambridge, MA, USA*

HELEN SPENCER • *Faculty of Medical and Human Sciences, Centre for Molecular Medicine, The University of Manchester, Manchester, UK*

HOPE E. STANSFIELD • *Departments of Chemistry and Chemical Biology and Biology, Center for Biotechnology and Interdisciplinary Studies, Rensselaer Polytechnic Institute, Troy, NY, USA*

NAOYUKI TAKAHASHI • *Institute for Oral Science, Matsumoto Dental University, Nagano, Japan*

PRUE TALBOT • *Department of Cell Biology & Neuroscience and Stem Cell Center, University of California Riverside, Riverside, CA, USA*

SUSANNE TRETTNER • *Fraunhofer Institute for Cell Therapy and Immunology, Department of Cell Therapy, Applied Stem Cell Technology Unit, Leipzig, Germany*

MOTOKAZU TSUNETO • *Max Planck Institute for Infection Biology,*
Berlin, Germany

NOBUYUKI UDAGAWA • *Department of Biochemistry,*
Matsumoto Dental University, Nagano, Japan

ALAI URRUTIKOETXEA-URIGUEN • *Institute of Reproductive and Developmental*
Biology, Imperial College London, London, UK

LUDOVIC VALLIER • *Department of Surgery, Cambridge Institute for Medical*
Research, Addenbrooke's Hospital, University of Cambridge,
Cambridge, UK

CHRISTOPHER M. WARD • *Faculty of Medical and Human Sciences, Centre for*
Molecular Medicine, The University of Manchester, Manchester, UK

TOSHIYUKI YAMANE • *Division of Genomics and Regenerative Biology, Department of*
Physiology and Regenerative Medicine, Mie University Graduate School of Medicine,
Institute of Medical Science, Tsu, Japan

TERUHITO YAMASHITA • *Institute for Oral Science, Matsumoto Dental University,*
Nagano, Japan

ZHENQING ZHANG • *Department of Chemistry and Chemical Biology,*
Center for Biotechnology and Interdisciplinary Studies,
Rensselaer Polytechnic Institute, Troy, NY, USA

NICOLE I. ZUR NIEDEN • *Fraunhofer Institute for Cell Therapy and Immunology,*
Department of Cell Therapy, Applied Stem Cell Technology Unit,
Leipzig, Germany
Department of Cell Biology & Neuroscience
and Stem Cell Center, University of California Riverside, Riverside, CA, USA

Chapter 1

Embryonic Stem Cells for Osteo-Degenerative Diseases

Nicole I. zur Nieden

Abstract

Current orthopedic practice to treat osteo-degenerative diseases, such as osteoporosis, calls for antiresorptive therapies and anabolic bone medications. In some cases, surgery, in which metal rods are inserted into the bones, brings symptomatic relief. As these treatments may ameliorate the symptoms, but cannot cure the underlying dysregulation of the bone, the orthopedic field seems ripe for regenerative therapies using transplantation of stem cells. Stem cells bring with them the promise of completely curing a disease state, as these are the cells that normally regenerate tissues in a healthy organism. This chapter assembles reports that have successfully used stem cells to generate osteoblasts, osteoclasts, and chondrocytes – the cells that can be found in healthy bone tissue – in culture, and review and collate studies about animal models that were employed to test the function of these in vitro "made" cells. A particular emphasis is placed on embryonic stem cells, the most versatile of all stem cells. Due to their pluripotency, embryonic stem cells represent the probably most challenging stem cells to bring into the clinic, and therefore, the associated problems are discussed to put into perspective where the field currently is and what we can expect for the future.

Key words: Embryonic stem cell, Transplantation, Teratoma, Immunological rejection, Osteo-degenerative disease, Skeletal injury

1. The Potential for Stem Cells in Therapeutic Applications

Degenerative bone diseases affect millions of people each year. Osteoporosis is probably the most common of such degenerative bone disorders and is particularly prevalent in the elderly population. With ten million patients presently in the United States only and increasing age of the world population, the overall numbers are expected to rise, challenging the already burdened health-care systems even more (1). The most common symptoms of not only osteoporosis, but also other degenerative bone diseases like Paget's disease, familial expansile osteolysis, expansile skeletal hyperphosphatasia, osteogenesis imperfecta, and osteopetrosis,

Nicole I. zur Nieden (ed.), *Embryonic Stem Cell Therapy for Osteo-Degenerative Diseases*, Methods in Molecular Biology, vol. 690, DOI 10.1007/978-1-60761-962-8_1, © Springer Science+Business Media, LLC 2011

are fractures, which either heal poorly by themselves or heal in a way that alters the microarchitecture of the bones leading to deformities and ultimately disability. Mostly, these symptoms are caused by an impaired interplay between the chondrocytes, osteoblasts, and osteoclasts, the cells that usually build the bones and take part in its continuous remodeling. In hard numbers, costs for osteoporosis-related fractures alone run at about $19 billion per year. Reason for even more concern is the life-threatening nature of most of these diseases. The current treatments of choice are antiresorptive medications, such as biphosphonates, or bone-forming medications. Sadly, however, both types of current treatments can ameliorate the symptoms, but they cannot cure the underlying disease. This, in fact, is true for degenerative diseases of other nature as well. Thus, novel therapeutic approaches, such as stem cell-based therapies, are being developed, which have the potential to completely cure the degenerative state as the stem cells would build the lost tissue. To date, adult stem cells have been successfully applied in clinical trials of ischemic heart disease (2), spinal cord lesions (3), Parkinson's disease (4), Huntington's disease (5), and diabetes (6). An overview of the degenerative diseases that stem cell treatment has been suggested for, including osteo-degenerative diseases, is given in Table 1.

In the case of joint injuries and diseases, adult stem cells with the potential to regenerate bone and cartilage, the mesenchymal stem cells (MSCs) cultured from human bone marrow, are already used in preclinical trials for treatment of osteogenesis imperfecta and nonunion bone fractures (7, 8).

Although these MSCs possess unique immunoregulatory features that suppress rejection of the grafted cells and show some success in treating osteoporosis, they are rarely found in adult tissues. Additionally, the ability of harvested cells to proliferate and secrete functional matrix diminishes with increasing age of the donor. Moreover, cells harvested from patients with certain osteo-degenerative deficiencies might harbor an overt or subtle genetic defect that could impair the ability of autologous cells to aid in tissue repair. Conversely, transplanting cells or tissues from allogeneic donors heightens the risk of immune rejection. Consequently, novel cell sources have to be identified. Therefore, new approaches to repair involving embryonic stem cells (ESCs) have emerged.

Embryonic stem cells, however, have not been used in clinical therapies, but the first clinical trial for the treatment of spinal cord lesions has just been approved. So far, ESCs have been successfully transplanted only into animals (8), which is discussed in more detail later in this chapter. The challenges that exist with the use of ESCs in clinical therapies include ethical concerns surrounding the derivation of the cells, tumorigenicity of the cells

Table 1
Collation of degenerative diseases suggested to be curable through the use of stem cells

Disease	Phenotype
Alzheimer's proteopathy, tauopathy	Loss of mental functions
Amyotrophic lateral sclerosis	Loss of motor functions (no motor signals)
Arthritis, Osteoarthritis-after trauma or aging, rheumatoid and psoriatic-autoimmune, septic-infection, gouty-uric crystals inflammation	
Atherosclerosis	Plugs in blood vessels
Autoimmune hepatitis	
Bovine spongiform encephalopathy, transmissible spongiform encephalopathies (TSEs)	
Cancer	
Charcot-Marie-Tooth disease	Demyelination of peripheral sensory paths due to defective myelin production
Diabetes	
Friedrichs ataxia	Spinal cord degeneration, movement disorganization, heart palpitations
Heart disease (cardiomyophatic/cardiovascular, congenital/hypertensive/inflammatory)	
Huntingtons disease	Congenital, neurodegeneration, motor and mental incapatitation
Inflammatory bowel disease	
Lewy body disease (alpha-synucleiopathy)	Accumulation in the nuclei of the neurons; elderly dementia, muscle stiffness
Ménière's disease	Change in fluid volume within the labyrinth, loss of hearing, tinnitus
Multiple sclerosis	Autoimmune demyelination of axons
Muscular dystrophy	Dystrophin deficiency in muscle fibers, muscle deterioration
Myasthenia gravis	Autoimmune damage of acetylcholine receptors at the neuromuscular junction, weakening of the muscle tonus of neck, face
Norris disease	Sight/hearing loss, mental challenge

(continued)

Table 1
(continued)

Disease	Phenotype
Osteoporosis	Reduction of bone mineral density and the amount of collagenous proteins
Parkinson's	Dopamine deficiency, loss of mental/motor functions
Prostatitis	Inflammation of the prostate gland
Scleroderma	Autoimmune thickening and tightening of the skin; serious damage to internal organs including the lungs, heart, kidneys, esophagus, and gastrointestinal tract
Shy–Drager syndrome	Multiple system atrophy due to orthostatic hypotension (excessive drop in blood pressure when standing up)
Spinal muscular atrophy	Congenital, motor neurons in spinal cord, weakening of the voluntary muscle movements, limp muscles

in vivo as well as the immune rejection of the cells following transplantation. The latter has recently been examined in several studies that have attempted to use somatic cell nuclear transfer and various methods to reprogram mature, fully differentiated cells to a pluripotent state. In the first method, a diploid nucleus from a patient's somatic cell is transferred into an enucleated egg, which then gives rise to cells that would possess very low host-graft rejection. In the second, pluripotency-associated genes are shuttled into somatic cells, i.e. fibroblasts or keratinocytes (9–11), and brought to expression before they are silenced, which is just enough to turn the differentiated cells into ESC-like cells with a pluripotent pheno- and genotype. Although there are many potential clinical applications of both embryonic and adult stem cells, this chapter discusses the usage of embryonic stem cells in skeletal tissue engineering in particular, briefly comparing them to other stem cells.

2. An Overview of Bone and Cartilage Tissue

2.1. Features of Mature Bone and Cartilage Tissue

In order to devise effective tissue engineering strategies, one must understand the target tissue as well as possible – that is, its composition and microarchitecture, as well as its molecular blueprint without which the particular function of the tissue would not be properly executable. The skeleton in vertebrates is composed primarily of bone and cartilage, which both secrete a characteristic and unique extracellular matrix (ECM). It is the ECM that gives these tissues their characteristic mechanical properties and lets them function as the scaffold for the body to give it strength, holding it in an upright position as well as supporting its movement. The ECM is full of collagens, which draw water into the tissue and therefore provide certain flexibility to the tissue to prevent breakage. Often, the presence or absence of specific collagens is therefore used as an analysis tool to identify osteoblasts and chondrocytes in culture. However, the type of collagen and the integration of other additional proteins into the matrix make for the specificity of cartilage and bone matrix.

The extracellular matrix of bone is inhabited by two main cell populations – osteoblasts and osteoclasts. Osteoblasts stem from the mesoderm or the neural crest of ectoderm; they are a source of matrix, cytokines, and activation factors for the hematopoietic osteoclasts. Both cell types interact and as a result of which bone undergoes cyclic modifications – osteoclasts resorb and reshape the deposited matrix that the osteoblasts produce. The bone ECM, in contrast to the cartilage ECM, associates an organic phase (noncollagen fibronectin, growth factors, and collagens) with a mineral phase (hydroxyapatite crystals). As osteoclasts are stimulated into resorption by the osteoblasts, this finely tuned perpetuation allows an adaptive modification of bone ECM.

The chondrocytes mainly secrete type II collagen (COL), particularly isoform COL IIB, which is the predominant collagen in adult cartilage (12). The collagen fibrils are organized such that a network is being established in which single fibrils are linked to each other in secondary, tertiary, and supramolecular structures (13). The orientation of the fibers and their thickness within the network depend on the plane that the fibers are located in, comparison to the articular surface (14). In addition to collagen, the main component of the cartilaginous matrix is proteoglycans, the most abundant being aggrecan with a percentile content of 80% (15).

In bone, collagen is prevalently represented by collagen type I, which also serves as a marker for preosteoblasts (16). Additional proteins that flag the osteoblastic character are osteopontin (17, 18), osteonectin, alkaline phosphatase, COL X (16), and the transcription factors Core binding factor alpha 1 (Cbfa1) and osterix (osx), which are discussed in more detail later (19–22). With increasing maturation, osteoblasts express bone sialoprotein and osteocalcin, which are considered to be truly exclusive markers for fully functioning osteoblasts.

Created by monocytic precursors of hematopoietic origin, the osteoclast is a bone-specific multinucleated specialized macrophage, which lives near the bone surface. Main control factors in osteoclastogenesis are colony-stimulating-factor 1 (CSF-1), also called M-CSF, and receptor activator of nuclear factor (NF)-kB ligand (RANKL), a cytokine from the tumor necrosis factor family, which are both involved in expansion of osteoclastic precursors and their maturation into osteoclasts (23–25). Markers of mature osteoclasts are tartrate-resistant acid phosphatase (TRAP), calcitonin receptor, vitronectin receptor, and cathepsin K, whereas the capacity of cells to resorb mineralized matrix is used as a functional marker.

2.2. Embryological Development of Bone and Cartilage

Already in the very early days of ESC culture, it became evident that their in vitro differentiation follows a timely coordinated succession of events (26). The sequence of genes expressed over time patterns embryological development. It is, therefore, critical to understand the embryological development of chondrocytes, osteoblasts, and osteoclasts, if we want to stage ESC differentiation in vitro. Osteoblasts and chondrocytes are nothing more than specialized mesenchymal cells. During embryological development, mesenchymal cells are located in the somites and the lateral plate mesoderm (27). In contrast, the osteoclasts stem from a myelomonocytic lineage and share a precursor with macrophages. These cells are, therefore, derived from a hematopoietic origin (28). The majority of bone tissue in the body is formed through a process termed endochondral bone formation, in which cartilage develops and is subsequently replaced by mineralized bone. However, some bones, particularly in the craniofacial region, are formed through intramembraneous ossification.

In endochondral ossification specifically, the chondrocytes play a critical role. At approximately embryonic stage 9.5 (E9.5), mesenchymal cells condense (29) into clusters of cells that subsequently differentiate into chondrocytes (E11.5), which in turn secrete extracellular matrix. It is these cartilage elements that provide a sort of blueprint, termed an anlagen, for future bone development. At approximately E13, the chondrocytes in the center of the condensations expand interstitially by continued matrix secretion and by differentiation into hypertrophic chondrocytes. Up to this point, COL IIB is excreted. Following differentiation of the chondrocytes, the cells begin to produce COL X, which subsequently calcifies into matrix that makes up the bone tissue. Overall, mineralization is initiated in the center of the anlagen and the periosteum, which is a thin layer of fibroblasts that coats the surface of the bone and forms around these central areas that contain the hypertrophic chondrocytes.

While undergoing apoptosis, the hypertrophic chondrocytes secrete vascular endothelial growth factor (VEGF), which induces vascularization of the bone matrix from the periosteum. Chondroclasts are able to migrate through the small channels that are formed during vascularization, and these cells digest the chondrocytic matrix as well as the hypertrophic chondrocytes. Ossification is then initiated by the osteoblast progenitors, which also migrate into the bone from the periosteum (30). These osteoprogenitors mainly secrete COL I, the main component of bone tissue (16), causing the composition of the extracellular matrix to change dramatically.

In contrast to this endochondral ossification, intramembraneous ossification is a much less complex process. Here, the mesenchymal tissue aggregates and condenses as it does in endochondral ossification, but the mesenchymal cells then differentiate directly into osteoblasts, omitting the development of cartilaginous anlagen. Interestingly, the majority of the mesenchymal cells in the craniofacial region originate from the neural crest region during early development and give rise to intramembraneous bone (27).

Cbfa1, a member of the runt family of transcription factors, also named runx2, is required for osteoblast differentiation, but is also expressed in certain nonosteoblastic cells without activating the differentiation process (19, 31). This may suggest that its activity is suppressed through a lineage-specific mechanism by cofactors, such as certain Smads (32). Cbfa1 was first identified as *Pebp2αA1* being a 513-amino acid protein that initiates in exon 2 of the Cbfa1 gene at the sequence MRIPV (33). Recent data, however, suggest that several isoforms are transcribed from the Cbfa1 gene. The expression is regulated by at least two distinct promoters that generate two N-terminal isoforms (34–36). *Pebp2αA1* is, therefore, now known as the type I isoform (20). The second major isoform (type II) was initially identified as *til-1* (37) and initiates in exon 1 at the sequence MASNS (34, 38). Cbfa1 type II is only 15 amino acids longer than the type I isoform. Whereas the type I transcript is constitutively expressed in nonosseous mesenchymal tissues and during osteoblast differentiation, only the expression of the type II transcript is regulated during osteoblast differentiation and can be induced by BMP-2 (39). Forced expression of both isoforms modulates transcription of skeletal genes (20, 40, 41), indicating that both proteins are functionally active in osteoblasts and hypertrophic chondrocytes.

It was suggested that Cbfa1 is somehow involved in osteoclastogenesis through regulation of osteoclast differentiation factor/osteoprotegerin ligand mRNA (42). Cbfa1 is also responsible for the transcriptional induction of the second osteoblast-specific transcription factor Osterix (osx) (22). Osx belongs to the Sp subgroup of the Krüppel family of transcription factors, containing three zinc finger DNA binding domains (43–45). Its human homologue has, therefore, been described as specificity protein 7 (Sp7) (46). In osx null mice, bone formation is completely absent; however, Cbfa1-expressing preosteoblasts can be found in the cartilaginous condensations of these mice during endochondral bone formation, suggesting that osx acts downstream of Cbfa1 (22, 47). In addition, evidence exists to suggest a negative feedback-loop from osx back to Cbfa1, since the level of Cbfa1 was decreased by the expression of osterix in MG-63 osteoblast cells (48).

Furthermore, the promoters of the bone-specific markers osteo-calcin, osteopontin, bone sialoprotein, alkaline phosphatase, and COL I (19, 22, 49) contain binding sites for Cbfa1 and osx (46, 50), suggesting Cbfa1 and osx as the two essential regulators of bone function.

BMP-2, BMP-4, BMP-6, and BMP-7 represent a family of growth factors that have been found to be involved in skeletal development; however, their exact role is not clear since most evidence of BMP involvement in osteogenesis has been indirect (28). Clinically, BMP-2 is mostly successfully used to enhance lumbar spinal fusions although bone resorption and thinning of the deposited bone has been described, as well as adverse side effects on vertebral bone (51, 52). During cranial bone forma-tion, FGF induces BMP-2, a process mediated by Cbfa1 (53). BMP-2 is then able to upregulate Sox-9 expression in a dose-dependent manner (54). In turn, Sox transcription factors are the transcriptional regulators that appear to prepare mesenchymal tis-sue to respond to BMP signaling (55). Sox transcription factors are high-mobility-group (HMG) domain transcription factors of which, Sox9, in particular, is involved in chondrocyte differentia-tion and formation of the extracellular matrix (56, 57). During embryogenesis, Sox9 is expressed in all cartilage primordia and cartilages, coincident with the expression of the collagen alpha1(II) gene (*Col2a1*). Sox9 as well as L-Sox5 and Sox6 bind to essential sequences in the *Col2a1* chondrocyte-specific enhancers and cooperatively activate the *aggrecan* gene (58, 59). Furthermore, the inactivation of Sox9 in limb buds before mesenchymal con-densations resulted in a complete absence of both cartilage and bone, due to inhibited cartilage proliferation and absence of Cbfa1 expression (60). Moreover, Sox9 is also needed to prevent conversion of proliferating chondrocytes into hypertrophic chon-drocytes (58).

During osteoclastogenesis, as many as 20 precursor cells can fuse into one such osteoclast (61). Two hematopoietic factors are necessary and sufficient for osteoclastogenesis, RANKL and M-CSF (24, 25). In addition to RANKL and M-CSF, 22 other gene loci have been identified so far, which either positively or negatively regulate osteoclastogenesis (61). The gene products that these encode for have only been identified for some, for example, the transcription factor microphthalmic, the growth fac-tor toothless, and osteoprotegerin. The latter has been shown to block the maturation of osteoclasts and is secreted by cells of the osteoblast lineage (62). Some of these genes, like M-CSF and PU.1 control the early stages of differentiation, more exactly the formation and survival of the precursor cell, whereas RANK, p50/p52 rel, and fos regulate the ability of the precursor to dif-ferentiate (63–66). Fetal liver kinase 1 (Flk-1) and SLC/tal-1, as crucial genes in generation of hematopoietic cells, and GATA-2, important for the generation of osteoclastic precursors, are used as additional osteoclastogenesis markers in the early stages of development (67). Towards maturation, the osteoclasts then produce tartrate-resistant acid phosphatase (TRAP) and cathepsin K in their activated state, which are both needed to digest the bone matrix (68, 69).

3. On the Way to the Clinic: ESCs, Differentiation, and Bone Tissue Engineering

3.1. Features of ESCs and Their Expansion In Vitro

In vivo, the embryo and the osteoblasts, osteoclasts, and chondrocytes develop from the inner cell mass of the blastocyst. These are the cells that are harvested and cultured in vitro as ESCs. As this is their normal in vivo fate, in vitro, ESCs have been shown to possess the ability to form any tissue type from all three germ layers (70, 71) including osteoblasts, osteoclasts, and chondrocytes (72–75).

The first ESCs were isolated from blastocysts of mice (76, 77). Since this first derivation, ESCs have also been isolated from various other rodents, such as hamster (78), rat (79), and rabbit (80, 81), and also farm animals (82). It was only in the late 1990s that the isolation of primate ESCs, including the rhesus and common marmoset monkey (83–85) and finally human ESCs (hESCs), was first described (86, 87).

ESCs can be propagated to an unlimited cell number, given their indefinite self-renewing capacity in vitro (88). First and foremost, however, they can be distinguished from adult stem cells by their capacity to differentiate into almost all of the cells of the mature body, a property known as pluripotency. Because of this particular feature, mouse ESCs (mESCs) have been successfully differentiated in vitro into a variety of specialized cells types of endodermal, mesodermal, and ectodermal origin (89–94), including osteoblasts (72) and chondrocytes (73, 95). Similarly, human ESCs have been used to generate insulin producing cells (96), cardiomyocytes (97), hematopoietic precursor cells (98), and neuronal precursor cells (87).

Although there does not seem to be a difference in the differentiation capability between ESCs from different species, the signals required to maintain their undifferentiated state and prevent spontaneous differentiation involves the activation or inhibition of different signaling cascades in vitro. Initially, cultivation of mESCs on mitotically inactivated mouse embryonic fibroblasts was required to maintain the undifferentiated, pluripotent state of mESCs (99). Today, a similar result can be achieved through supplementation of ESC cultures with a cytokine called leukemia inhibitory factor (LIF) (100–103), and in this case, the need to use feeder layers is avoided.

Interestingly, LIF is not sufficient to maintain expanding cultures of human or primate ESCs (86).

However, it seems to be the same transcription factor network in all species that downstream of signaling cascade activation or inhibition plays a critical role in maintaining the ESC state; one of these transcription factors is the POU-family transcription factor Oct-3/4 (104). Oct-3/4 expression has also been observed in oocytes, the inner cell mass, and the preimplantation embryo, speaking for its role in controlling differentiation events (105–108). However, changes in cell fate are not a yes or no response to Oct-3/4 expression. Rather, differentiation seems to be dependent on the expression level of Oct-3/4 (109). For example, while "normal" levels of Oct-3/4 maintain ESCs in a self-renewing state, a less than a two-fold increase in expression causes differentiation into endoderm and mesoderm. Likewise, a reduction to less than half of the "normal" expression level triggers trophoectoderm differentiation.

Regulation of the expression level of Oct-3/4 is not yet completely understood, but it has recently become clear that Oct-3/4 does not act alone. Expression of two other transcription factors, Sox2 and nanog, has been found in the inner cell mass and ESCs, suggesting that they may be essential for early embryonic development and ESC maintenance (110–112). Sox2 was specifically identified as a cooperative partner of Oct-3/4 in ESCs regulating their own expression in a circulatory manner (113–115). The Sox2–Oct-3/4 dimer also controls the transcriptional activation of their target nanog (116, 117) and together the Sox2–Oct-3/4–nanog complex potentially occupies the promoter regions of 353 genes in human ESCs, some for the purpose of repressing their transcription and some in order to activate their transcription (118).

Apart from transcriptional networks, higher-level transcriptional regulators, such as chromatin modulators, also control pluripotency. It has been suggested that ESCs are characterized by a bivalent chromatin state, where target genes are simultaneously enriched for repressive and activating trimethylation marks, making transcription permissive to the transcriptional machinery (119). Specifically, some lineage-specific genes carry repressive modifications on lysine 27 of histone 3 (H3K27me3) and two activating marks at lysines 9 and 4 (H3K9ac3 and H3K4me3) (120). The methyltransferase, which catalyzes the di- and trimethylation of histone 3 at lysine 27 is the polycomb group (PcG) protein PRC2 (121, 122). PRC2 promotes the recruitment of PRC1, another PcG protein, which maintains repression by mediating monoubiquitination of histone H2A at lysine 119, promoting chromatin condensation (123).

As a result, the expression of specific genes is prepared but halted by these opposing chromatin marks. This is critical in the light of the fact that for ESCs to be able to enter down a specific differentiation path, they need to be able to activate specific genes at a particular time of development. Differentiation is then accompanied by a loss of the bivalent state of these target genes (124).

3.2. Osteoblast, Osteoclast, and Chondrocyte Differentiation from Stem Cells

For the study of bone development, homeostasis, and bone tissue engineering, the ESC model has the advantage over other stem cell models that all three cell types inhabiting the bone ECM, the chondrocyte, osteoblast, and osteoclast, may be differentiated from these cells in vitro. This model consequently allows studying the interactions between these three cell types as well as matrix remodeling in vitro. This paragraph specifically reviews the methods that have been employed to differentiate the three different cell types from ESCs, some of which are presented in the later chapters of this book.

As has been explained above, both M-CSF and RANKL control osteoclast differentiation in vivo. However, RANKL alone is capable of inducing differentiation into osteoclasts from ESC culture (125), whereas M-CSF stimulates hematopoietic precursor expansion when applied early but inhibits the formation of mature osteoclasts later in culture (75). Osteoclasts can also be derived from ESCs by directly cultivating on plates (126) or on a monolayer of feeder cells, which can be either M-CSF-secreting bone-marrow-derived stromal cells (ST2) or new born calvaria stromal cells (OP9) supplemented with M-CSF (67, 75, 127, 128). Differentiation is performed in one step or stepwise – culturing the cells until the hematopoietic precursors are formed, followed by coculture. Cocultures vary in the differentiation stage of the cells at the time of coculturing and the time of stimulation (127). Here, dexamethasone and VD$_3$ are necessary cofactors for the osteoclast differentiation from ESCs in culture (129). Addition of ascorbic acid to the culture medium increases osteoclast precursors and promotes osteoclastogenesis further (126).

Moreover, the basic helix-loop-helix transcription factor SCL/tal-1 seems to be essential for osteoclastogenesis and preceding hematopoiesis. Mice lacking this transcription factor fail to develop hematopoietic cells in vivo (130), and no cells of hematopoietic origin can be detected in tal-1$^{-/-}$ ESCs in vitro (131). Forced expression of the transcription factor PU.1 in tal-1$^{-/-}$ ESCs, however, may rescue the development of the multinucleated osteoclasts. Nonetheless, the observed rescue is not sufficient to restore wild-type levels, and other cell types of the hematopoietic lineage are not generated (132). These results suggest that tal-1 may lie upstream of PU.1 regulating hematopoietic cell fate decisions in ESCs.

Aside from hematopoietic development, ESCs can be led to differentiate from pluripotency into mesoderm, mesenchyme, and subsequently cartilage, bone, or fatty tissue. The differentiation initiating factors are provided by neighboring cells in vivo. In culture, appropriate factors have to be supplied. Here, timed supplementation of the culture medium with growth factors and additives largely regulate cell fate with main changes between the early and late stages of osteogenic differentiation.

In the last 8 years, protocols have been established that allow ESC differentiation into bone and cartilage and their characteristic cell types. Our group has pioneered this field by developing methods to differentiate mESCs into osteoblasts and chondrocytes in culture (72, 73). Both methods involve removing pluripotency factors and allowing ESCs to form EBs as the first stage of differentiation. For example, the static suspension method of EB generation has been employed to differentiate osteoblasts from hESCs (133). Differentiation might also be begun starting from monolayer hESC colonies (134).

At differentiation day 3–4, ESCs express *T-Brachyury*, a gene typically expressed in the primitive streak, which develops by day 6.75 p.c. in vivo (135). The primitive streak contains cells with mesodermal character, a subpopulation of which will later become osteoblasts, osteoclasts, and chondrocytes. *T-Brachyury* expression is, therefore, widely used to characterize the output of differentiating mesendoderm (136, 137) and allows drawing conclusions as to the early differentiation events of osteogenesis, chondrogenesis, and osteoclastogenesis.

Then, specific differentiation factors are added to control differentiation and to direct the formation of osteoblasts. Typically, an alkaline environment is necessary to drive the cells to secrete a mineralized matrix, which is provided by the medium supplements beta-glycerophosphate and ascorbic acid (AA) (92, 138, 139). $1,25\text{-OH}_2$ Vitamin D_3 (VD_3) or dexamethasone acts as additional triggers for ESCs to activate the transcription of osteoblast-specific genes, including Cbfa1, COL I, osteocalcin, and bone sialoprotein. Alternatively, coculture systems or conditioned media can be employed to drive differentiation (140, 141). The complete process of osteogenesis from a pluripotent ESC to a mature osteoblast is quite similar between murine, human, and nonhuman primate ESCs; it transverses through four distinct osteogenic stages: proliferation (I), mesendoderm specification (II), matrix deposition and onset of alkaline phosphatase expression (III), and mineralization (IV) (142–145). The interrogation of signaling pathways with the help of microarray gene expression profiling has furthermore led to the conclusion that seven major signaling molecules are involved in regulating ESC-osteogenesis, such as retinoic acid, BMP-2, and Wnts (146).

The Wnt/beta-catenin canonical signaling pathway is thought to play a central role in osteogenesis not only in ESCs (147). Activation of this pathway promotes osteoblast differentiation from mesenchyme in vitro (148, 149), which can be attributed to the fact that beta-catenin (CatnB) is being stabilized and translocated to the nucleus, where it complexes with TCF and LEF family transcription factors and/or other cofactors in its function as a transcriptional activator and triggers transcription of a specific set of target genes (150–153). In osteoblasts specifically, CatnB and TCF-1 co-occupy the promoter of Cbfa1, potentially regulating not only Cbfa1 transcription, but also osteocalcin expression, which has known Cbfa1 binding sites in its enhancer region (50, 154, 155).

The key role of CatnB becomes clear considering that osteoblasts and osteoclasts interact through osteoprotegerin, an osteoclast-inhibiting factor secreted by osteoblasts, whose production is stimulated when the canonical Wnt/CatnB pathway is inhibited (156). The classically perceived role for CatnB in osteogenesis is currently being challenged by a number of publications. Whereas CatnB has been perceived as necessary for the development of osteoblasts, it now becomes clear that there are CatnB-dependent and independent phases of osteogenic development (142). In ESC osteogenesis specifically, CatnB expression levels and nuclear activity are regulated in a biphasic manner – it is decreased during early differentiation in vitro once the primitive streak-like cells have formed (146) and increased in the late stages of osteogenesis both in ESCs in vitro and in differentiating osteoblasts in vivo (157). Ultimately, its decrease at early differentiation stages is necessary for the osteogenic induction to occur.

The chondrogenic induction protocols differ either in the choice of media supplements and growth factors or in cell types used for cell culture, but generally employ external stimuli to activate chondrogenic gene sets (73, 158). For example, limb bud progenitor cells seem suitable to induce Sox9 and COL II expression as well as proteoglycan secretion in 60–80% of the cells (159).

Similar to the in vivo situation during embryonic development, osteogenesis in vitro can be direct, or the future osteoblasts can at first undergo a chondrocyte phase. An ESC-derived in vitro equivalent of the embryonal intramembraneous ossification would be the spontaneously differentiated mesenchymal precursors being pushed directly into the osteoblastic fate (73). In turn, the equivalent to embryonal endochondral ossification can also be found in ESC differentiations when they differentiate into mature chondrocytes, which undergo hypertrophy and still retain certain plasticity to give rise to osteocytes (73). Direct ossification seems to result in a higher degree of mineralization than the indirect (chondrocyte-mediated) differentiation.

Direct chondrocytic differentiation is mediated by bone morphogenic protein-2 (BMP-2) applied during day 3–5 of differentiation (95). Sox9 and scleraxis, cartilage-specific transcription factors, are upregulated at early stages of differentiation (60, 73, 160). Addition of BMP-2 increased the formation of cartilaginous matrix, indicated by gene expression of COL I, COL II, COL X and deposition of collagen, proteoglycans, and ECM proteins that correspond to the in vivo chondrocyte maturation (161). Such ESC-derived chondrocytes are able to undergo hypertrophy and calcify (162). In the ESC-derived chondrocytes that undergo hypertrophy and mineralize in the later stages of differentiation, osteoblast-specific genes are also upregulated towards the later stages of differentiation. This overlap of the chondrocyte-specific and osteoblast-specific gene expression suggests that ESC-derived osteoblasts may previously have undergone a chondrocyte stage. Even at the later stages (day 20) of the chondrocyte differentiation, a rescue of osteoblasts is possible by treatment with VD_3, AA, and β-glycerophosphate, and with a high degree of mineralization and full-scale expression of osteoblast markers and phenotype. These findings support the hypothesis that BMP-2 controls chondrogenesis and endochondral bone formation, whereas VD_3 induces intramembraneous ossification from the same mesenchymal progenitors.

The Sox triad (5, 6, 9) has been reported as inducers of a dramatic increase in chondrocyte marker gene expression (163). In an attempt to speed up the mESC differentiation into a chondrocytic phenotype, Kim and coworkers (164), have stably overexpressed human Sox9, a gene present during mesenchymal condensation and cartilage formation in embryos (165) and considered required for chondrogenesis. Without affecting the proportions of the cells in particular cell cycles, and without any exogenous stimulation, a very early increase in COL IIB has been reported over a modest level of COL IIA, suggesting an early shift between the splicing forms (73, 164). Sox9 acts in a triad together with Sox5 and Sox6, and its single overexpression resulted in very low levels of COL IIA, suggesting that additional cofactors are required for differentiation. Although still suboptimal, these studies could represent a large step forward in cartilage tissue engineering and will be elaborated further in Chapter 14.

More recent efforts veer towards the identification of progenitor populations with MSC character, which are minimally capable of differentiating into both osteoblasts and chondrocytes, if not also into adipocytes. Such MSC-like cells have been purified after coculture on OP9 stromal cells and isolation of CD73⁺ cells, a marker typically found on adult MSCs (166–168). Likewise, differentiating hESCs have been repeatedly passaged to isolate an expanding cell population with MSC character (169). In a similar approach, such cells were physically harvested as spontaneously differentiating cells in expanding hESC cultures (170, 171). A combination of both techniques, passaging and isolation, has also been described (172). The various techniques to harvest such progenitor cells are reviewed in more detail in Chapter 11.

3.3. Skeletal Tissue Engineering with ESCs

Pluripotency and self-renewal, the same two features that make ESCs such promising cells for clinical applications, also largely hinder their widespread use. Still problematic in transplantation settings is their potential to induce teratomas and the insufficient control of their proliferation and differentiation once transplanted.

It is likely that at least two factors influence the teratoma formation capacity of the transplanted cells – their maturation status as well as the microenvironment into which they are transplanted. Indeed, Brederlau and colleagues concluded from their study using hESCs differentiated into dopaminergic neurons and intra-striatally microinjected into immunosuppressed rat brains that ESCs lose their teratoma formation capacity with progressing differentiation and maturation in vitro (173). While 16-day predifferentiated hESCs formed teratomas in 100% of the cases ($n = 22$), none formed when the cells were left to differentiate in vitro for only 1 week longer ($n = 8$). We have been recently able to confirm this finding for cells differentiated down the osteogenic lineage and subcutaneously injected into SCID mice. Specifically, we found that triggering a specific differentiation program decreased their teratoma formation capacity even more. Whereas spontaneously differentiated control cells still formed teratomas in 16% of the cases when taken from day-10-old cultures, vitamin D_3-induced osteogenic cultures did not show any sign of teratoma formation at this point (174). By day 20 of differentiation, the control cultures had also lost their teratoma formation capacity.

The influence of the microenvironment on cell fate upon transplantation becomes clear when we consider the studies by Wakitani et al. on transplantation of undifferentiated murine ESCs into the knee joint with the purpose of treating articular cartilage defects. As one would expect, undifferentiated murine ESCs when injected into the knee joint of SCID mice form teratomas. They also irreparably destroyed the joint, although the incidence of newly formed cartilage was high in these teratomas (175), the rate at which the teratomas grew in the knee joint, however, was slower than upon subcutaneous injection, suggesting that the microenvironment in the knee joint is not as favorable for ESC proliferation. Surprisingly, if cells were injected into an inflammatory environment caused by a full-thickness osteochondral defect, the cells integrated and repaired the defect even in an allogeneic setting (176). Moreover, repair only occurs in mice in which the joints were weight-bearing, suggesting a role for mechanical loading during in vivo repair (177).

Transplantations with currently available human ESC lines, however, could only be allogeneic, which bring with them problems surrounding immune rejection. In similar studies carried out by two laboratories, undifferentiated murine ESCs unexpectedly did not form teratomas after injection into infarcted hearts

(178, 179). However, more stringent analyses revealed that this finding was only true for allogeneic transplantations and that teratomas were formed in all cases in syngeneic transplantations. It was furthermore confirmed that the rejection of the cells in the allogeneic setting was due to a $CD3^+$ T lymphocyte infiltration and invasion of $CD4^+$ and $CD8^+$ T cells (179). The immune reaction was most likely caused by the upregulation of MHC class I and II antigens on the surface of the cells, which were starting to differentiate in vivo (178).

The hope is to widely use the technique of reprogramming to dedifferentiate patient-specific somatic cells to a pluripotent state, creating a nonimmunogenic source of pluripotent cells. Before transplantation, these patient-specific induced pluripotent stem cells (iPSCs) could then be differentiated according to the protocols previously developed for the in vitro differentiation of ESCs, and appropriate precursors could be isolated as a transplantable source.

Bone tissue engineering, unlike stem cell transplantations into the myocardium, liver, or brain, however, requires the use of a scaffold to provide strength to the body while the cells are repairing the underlying defect. In order to achieve FDA approval for a tissue-engineered product, a shortcut would be to utilize biomaterials that are currently on the market and have former FDA approval. Not many scaffold materials have been characterized for their use with ESCs or iPSCs yet, but the few that have are reviewed in the following section and summarized in Table 2.

That hESCs are capable of forming new bone in vivo was first shown by Bielby and coworkers. When steered into the osteogenic fate with dexamethasone for 24 h in vitro, seeded onto Poly-(d, l)-lactic acid (PDLLA) scaffolds, and transplanted into the skin fold of SCID mice, no signs of teratoma formation were observed and neither was inflammation. However, new matrix was laid down, and vasculature was formed in close proximity to the bone front (144). In contrast, scaffolds seeded with cells that had not received the osteogenic trigger only formed fibrous tissue. Although it is known that subcutaneously injected ESCs form confined and enclosed teratomas without migrating into adjacent tissue, stricter controls are needed to examine for tumors away from the implant site. Furthermore, long-term follow-up analyses are required to properly and rigidly evaluate the tumor formation potential of transplanted cells and determine the risks associated with stem cell transplantation.

When using scaffolds, the fate of the cells largely depends on the material used. PDLLA was preferred in the presented study, as it is biodegradable and a polyester made from cornstarch or other renewable sources and has approval for use in humans as stents, sutures, and many more applications. Although a better

Table 2
List of studies that have used stem cells for bone and cartilage tissue engineering

Species	Lineage	Target site	Purification/ pretreatment	Recipient animal	Route of administration	Outcome measure	Time of assessment	Teratoma	References
M	U	Joint	None	Female 5-week-old SCID mice	Intra-articular joint injection in collagen	Teratoma formation H&E	1–6 Weeks	In 20% of cases ($n=5$)	(175)
M	U	Joint	None	Female 5-week-old SCID mice	Intra-articular joint injection in collagen	Teratoma formation H&E	8 Weeks	In 40% of cases ($n=5$)	(175)
M	U	OCD	None	Cyclosporine A immunosuppressed rats	Intra-articular joint injection	Survival Cartilage production	8 Weeks	None observed	(176)
M	U	OCD	None	SD rats immunosuppressed with cyclosporine A	Intra-articular joint injection	Histological grading scale	4 Weeks	In immobilized knee joints only ($n=9$)	(177)
M	Ch	Subcut	21 Day chondrogenic induction	Immune-deficient mice	Calcium phosphate ceramics	Histological staining Histomorphometry	3, 7, 14, 21 Days	None observed $n=6$	(180)
M	Ch	CD	21 Day chondrogenic induction	Immune-deficient rats	Calcium phosphate ceramics	Histological staining Histomorphometry	3, 7, 14, 21 Days	None observed $n=7$	(180)
H	Ch	Subcut	Coculture with chondrocytes	Athymic nude mice	PEGDA or PEG-RGD hydrogels	Histology GAG and Collagen content	12 and 24 Weeks	None observed $n=6$ each	(171)

(continued)

Table 2
(continued)

Species	Lineage	Target site	Purification/ pretreatment	Recipient animal	Route of administration	Outcome measure	Time of assessment	Teratoma	References
H	MSC	Subcut	Chondrocyte conditioned medium	Athymic mice	PEGDA hydrogel	Histology GAG and Collagen content	12 Weeks	$n = 6$	(182)
H	MSC	OCD	3 Day pellet culture	Athymic nude rats	Pellet culture no scaffold, fibrin glue seal	Histology Histomor- phometry	8 Weeks	$n = 6$	(182)
H	U		None	In vitro study	BSM-PGLA scaffold PGLA scaffold Collagraft®	Cytotoxicity Cell viability Adhesion	8 Days	n.a.	(180)
H	O	Subcut	Dexamethasone treatment d14 for 24 h	Immune-deficient SCID mice	PDLLA scaffold	Cell survival Vascularization Presence of mineralized cells	35 Days	None observed $n = 5$	(144)
H	O	Subcut	Coculture with primary bone derived cells	Immune-deficient BALB/c-nu, 7 weeks old, female	PGLA/ HA+fibrin glue	Histology PCR for bone markers Soft X-ray and calcium deposit	4 and 8 Weeks	None observed $n = 10$	(181)
M	O	Subcut	21 Day osteogenic induction	Immune-deficient mice	Calcium phosphate ceramics	Histological staining Histomor- phometry	21 Days	None observed $n = 6$	(180)

H	MSC	CD	ALP+	Immuno-compromised mice	Bio-OSS particles in fibrin glue	Radiographic analysis IHC for bone markers	6 Weeks	None observed	(169)
H	MSC	Subcut	10 Day osteogenic induction	6-Week-old athymic nude mice	(PLLA/PLGA) (1:1) with HA particles	Survival H&E staining	4 and 8 Weeks	None observed n=4	(182)

H human, M mouse, U undifferentiated, Ch chondrogenic, O osteogenic, MSC hESC-derived mesenchymal stem cell, subcut subcutaneous, CD cranial defect, OCD osteochondral defect in patellar groove, IHC immunohistochemistry choice for fast FDA approval, existing biomaterials may not be suitable for use with cells in general or ESCs in particular due to their physical and structural configuration. Indeed, Collagraft, a collagen–calcium phosphate ceramic graft material, and Poly(d, l)-lactide-co-glycolide (PLGA), that have both been used as scaffolds for bone tissue engineering, differ in their support of more basic cellular function, such as viability and adhesion (180). Only when mixed with a composite collagen isolated from bladder tissue or hydroxyapatite, PGLA promoted hESC adhesion and osteogenic differentiation, the latter also after subcutaneous engraftment (180–182).

In contrast, Arpornmaeklong proofed that the opposite is true and that indeed existing biomaterials may be suitable for ESC bone tissue engineering (169). They combined ALP+ osteogenic progenitors flow isolated from hESC-derived MSCs with Bio-Oss, a natural bone substitute made from bovine bone, that has already been used clinically in periodontal and maxillofacial surgery. The cell–granule mix was then inserted into a cranial defect. This hallmark study showed for the first time that human ESC-derived osteogenic progenitors can integrate and repair an injured bone without forming teratomas.

The only study published so far on cartilage tissue engineering using hESCs also employed existing biomaterials. For cartilage engineering, the scaffolds show different properties than for bone engineering and mostly are softer in nature. One such material that provides these properties is PEG-hydrogel. Poly(ethylene glycol)-diacrylate (PEGDA) or tyrosine–arginine–glycine–aspartate–serine (YRGDS)-modified PEGDA hydrogels (PEG-RGD) both supported cartilage formation, glycosaminoglycan deposition, and collagen secretion in vivo (182). While in this first study hESCs were predifferentiated into chondrogenically committed cells by coculture techniques with primarily isolated chondrocytes, later studies by the same group proofed that these materials can also help form cartilage when hESC-derived MSCs are used (182).

As has been explained in Subheading 2.2, bone formation occurs through two processes: intramembraneous ossification, in which bone is laid down from neural crest progenitors (183), and endochondral ossification, in which mesenchymal stem cells and chondrocytes derived from the primitive streak play a huge part. Indeed, ESCs have been shown to mineralize through both processes. Recently, cranial neural crest (CNC)-like cells have been derived from day-10-old differentiating human embryoid bodies by flow-sorting a frizzled-3 (Fzd3) and cadherin-11 double positive population (184). These two markers were chosen as they are expressed in the dorsal neural tube and in migrating neural crest cells, respectively (185–187). Similarly, Lee and colleagues used p75 and HNK1 to isolate CNC-like cells from neurally induced hESCs (188). Both flow-sorted putative CNC-like cells were responsive to dexamethasone. Additionally, they were capable of differentiating not only into chondrocytes but also the glia, neurons, and smooth muscles, cell fates typically associated with neural crest. The occurrence of the Fzd3$^+$/cad-11$^+$ CNC-like cells in the differentiating cell population, however, was very rare with 1.2%. Cad-11 single positive cells were more abundant (3%), suggesting that Cad-11 may mark other cells of mesodermal origin. This hypothesis is underlined by the findings of Kimura and colleagues, who have reported Cad-11 expression in the somites and limb buds of early mouse embryos (189). Neither the cad-11$^+$ the Fzd3$^+$/cad-11$^+$ nor the p75$^+$/HNK1$^+$ cells were transplanted into a cranial defect or implanted subcutaneously to verify their osteogenic potential in vivo.

Subcutaneous implantation of ceramic scaffolds seeded with mESCs and pretreated in vitro with osteogenic supplements did not show any formation of bone in vivo (190). The in vitro treatment regimen, however, seems to favor the development of neural crest like cells, since ESCs were treated with retinoic acid during early differentiation. Such treatment has previously been shown to favor neural crest differentiation from mESCs (191). When the cell-seeded scaffolds were, however, pretreated with a chondro-conducive medium, cartilage and bone formation was noted in subcutaneous implantation sites as well as in a cranial defect (190). Based on these results, it was suggested that ESCs effectively form bone through endochondral ossification.

It remains to be seen whether or not all progenitors with osteogenic potential derived from ESCs, such as the MSCs or the CNC-like cells, have the same repair potential in all possible treatment sites, such as the cranium, mandibular defects, and long bones, as well as in regenerating bones, which are injured or exhibit properties of degenerative bone diseases.

References

1. National Osteoporosis Foundation. http://www.nof.org/osteoporosis/diseasefacts.htm.

2. Kovacic, J.C., Muller, D.W., Harvey, R., and Graham, R.M. (2005) Update on the use of stem cells for cardiac disease. *Intern. Med. J.* **35(6)**, 348–356.

3. Feron, F., Perry, C., Cochrane, J., Licina, P., Nowitzke, A., Urquhart, S., et al. (2005) Autologous olfactory ensheathing cell transplantation in human spinal cord injury. *Brain* **128(Pt 12)**, 2951–2960.

4. Freed, C.R., Greene, P.E., Breeze, R.E., Tsai, W.Y., DuMouchel, W., Kao, R., et al. (2001) Transplantation of embryonic dopamine neurons for severe Parkinson's disease. *N. Engl. J. Med.* **344(10)**, 710–719.

5. Bachoud-Levi, A.C., Gaura, V., Brugieres, P., Lefaucheur, J.P., Boisse, M.F., Maison, P., et al. (2006) Effect of fetal neural transplants in patients with Huntington's disease 6 years after surgery: a long-term follow-up study. *Lancet Neurol.* **5(4)**, 303–309.

6. Farkas, G., and Karacsonyi, S. (1985) Clinical transplantation of fetal human pancreatic islets. *Biomed. Biochim. Acta* **44(1)**, 155–159.

7. Le Blanc, K., Gotherstrom, C., Ringden, O., Hassan, M., McMahon, R., Horwitz, E., et al. (2005). Fetal mesenchymal stem-cell

engraftment in bone after in utero transplantation in a patient with severe osteogenesis imperfecta. *Transplantation* **79(11)**, 1607–1614.

8. Tuch, B.E. (2006) Stem cells – a clinical update. *Aust. Fam. Physician* **35(9)**, 719–721.

9. Takahashi, K., and Yamanaka, S. (2006) Induction of pluripotent stem cells from mouse embryonic and adult fibroblast cultures by defined factors. *Cell* **126(4)**, 663–676.

10. Okita, K., Ichisaka, T., and Yamanaka, S. (2007) Generation of germline-competent induced pluripotent stem cells. *Nature* **448(7151)**, 313–317.

11. Aasen, T., Raya, A., Barrero, M.J., Garreta, E., Consiglio, A., Gonzalez, F., et al. (2009) Efficient and rapid generation of induced pluripotent stem cells from human keratinocytes. *Nat. Biotechnol.* **26(11)**, 1276–1284.

12. Ryan, M.C., and Sandell, L.J. (1990) Differential expression of a cysteine-rich domain in the NH2-terminal propeptide of type II (cartilage) procollagen. *J. Biol. Chem.* **265**, 10334–10339.

13. Eyre, D.R., Weis, M.A., and Wu, J.J. (2006) Articular cartilage collagen: an irreplaceable framework? *Eur. Cell Mater.* **12**, 57–63.

14. Eyre, D. (2004) Collagen of articular cartilage. *Arthritis Res.* **4(1)**, 30–35.

15. Heinegard, D., and Paulsson, M. (1984) Structure and metabolism of proteoglycans. In: Piez, K.A., Reddi, A.H. (eds). Extracellular matrix biochemistry. Elsevier, New York, p. 277.

16. Termine, J.D., and Robey, P.G. (1996) Bone matrix proteins and the mineralization process. In: Favus MJ (ed). Primer on the metabolic bone diseases and disorders of mineral metabolism. Lippincott-Raven, Philadelphia, pp. 24–28.

17. Aubin, J.E., Liu, F., Malaval, L., and Gupta, A.K. (1995) Osteoblast and chondroblast differentiation. *Bone* **17(2 Suppl)**, 77S–83S.

18. Davies JE (1996). In vitro modeling of the bone/implant interface. *Anat. Rec.* **245(2)**, 426–445.

19. Ducy, P., Zhang, R., Geoffroy, V., Ridall, A.L., and Karsenty, G. (1997) Osf2/Cbfa1: a transcriptional activator of osteoblast differentiation. *Cell* **89(5)**, 747–754.

20. Harada, H., Tagashira, S., Fujiwara, M., Ogawa, S., Katsumata, T., Yamaguchi, A., et al. (1999) Cbfa1 isoforms exert functional differences in osteoblast differentiation. *J. Biol. Chem.* **274(11)**, 6972–6978.

21. Kern, B., Shen, J., Starbuck, M., and Karsenty, G. (2001) Cbfa1 contributes to the osteoblast-specific expression of type I collagen genes. *J. Biol. Chem.* **276(10)**, 7101–7107.

22. Nakashima, K., Zhou, X., Kunkel, G., Zhang, Z., Deng, J.M., Behringer, R.R., et al. (2002) The novel zinc finger-containing transcription factor osterix is required for osteoblast differentiation and bone formation. *Cell* **108**, 17–29.

23. Tanaka, S., Takahashi, N., Udagawa, N., Tamura, T., Akatsu, T., Stanley, E.R., et al. (1993) Macrophage colony-stimulating factor is indispensable for both proliferation and differentiation of osteoclast progenitors. *J. Clin. Invest.* **91(1)**, 257–263.

24. Yasuda, H., Shima, N., Nakagawa, N., Yamaguchi, K., Kinosaki, M., Mochizuki, S., et al. (1998) Osteoclast differentiation factor is a ligand for osteoprotegerin/osteoclastogenesis-inhibitory factor and is identical to TRANCE/RANKL. *Proc. Natl. Acad. Sci. USA* **95(7)**, 3597–3602.

25. Lacey, D.L., Timms, E., Tan, H.L., Kelley, M.J., Dunstan, C.R., Burgess, T., et al. (1998) Osteoprotegerin ligand is a cytokine that regulates osteoclast differentiation and activation. *Cell* **93(2)**, 165–176.

26. Doetschmann, T.C., Eistetter, H., Katz, M., Schmidt, W., and Kemler, R. (1985) The in vitro development of blastocyst-derived embryonic stem cell lines: formation of visceral yolk sac, blood islands and myocardium. *J. Embryol. Exp. Morphol.* **87**, 27–45.

27. Olsen, B.R., Reginato, A.M., and Wang, W. (2000) Bone development. *Annu. Rev. Cell. Dev. Biol.* **16**, 191–220.

28. Karsenty, G., and Wagner, E.F. (2002) Reaching a genetic and molecular understanding of skeletal development. *Dev. Cell* **2(4)**, 389–406.

29. Yang, X., and Karsenty, G. (2002) Transcription factors in bone: developmental and pathological aspects. *Trends Mol. Med.* **8(7)**, 340–345.

30. Vu, T.H., Shipley, J.M., Bergers, G., Berger, J.E., Helms, J.A., Hanahan, D., et al. (1998) MMP-9/gelatinase B is a key regulator of growth plate angiogenesis and apoptosis of hypertrophic chondrocytes. *Cell* **93(3)**, 411–422.

31. Ducy, P., Starbuck, M., Priemel, M., Shen, J., Pinero, G., Geoffroy, V., et al. (1999) A Cbfa1-dependent genetic pathway controls

bone formation beyond embryonic development. *Genes Dev.* **13(8)**, 1025–1036.

32. Hjelmeland, A.B., Schilling, S.H., Guo, X., Quarles, D., and Wang, X.F. (2005) Loss of Smad3-mediated negative regulation of Runx2 activity leads to an alteration in cell fate determination. *Mol. Cell. Biol.* **25(21)**, 9460–9468.

33. Ogawa, E., Maruyama, M., Kagoshima, H., Inuzuka, M., Lu, J., Satake, M., et al. (1993) PEBP2/PEA2 represents a family of transcription factors homologous to the products of the Drosophila runt gene and the human AML1 gene. *Proc. Natl. Acad. Sci. USA* **90(14)**, 6859–6863.

34. Xiao, Z.S., Thomas, R., Hinson, T.K., and Quarles, L.D. (1998) Genomic structure and isoform expression of the mouse, rat and human Cbfa1/Osf2 transcription factor. *Gene* **214**, 187–197.

35. Pozner, A., Goldenberg, D., Negreanu, V., Le, S.Y., Elroy-Stein, O., Levanon, D., et al. (2000) Transcription-coupled translation control of AML1/RUNX1 is mediated by cap- and internal ribosome entry site-dependent mechanisms. *Mol. Cell. Biol.* **20**, 2297–2307.

36. Drissi, H., Luc, Q., Shakoori, R., Chuva, De Sousa Lopes, S., Choi, J.Y., et al. (2000) Transcriptional autoregulation of the bone related CBFA1/RUNX2 gene. *J. Cell. Physiol.* **184(3)**, 341–350.

37. Stewart, M., Terry, A., Hu, M., O'Hara, M., Blyth, K., Baxter, E., et al. (1997) Proviral insertions induce the expression of bone-specific isoforms of PEBP2alphaA (CBFA1): evidence for a new myc collaborating oncogene. *Proc. Natl. Acad. Sci. USA* **94(16)**, 8646–8651.

38. Thirunavukkarasu, K., Mahajan, M., McLarren, K.W., Stifani, S., and Karsenty, G. (1998) Two domains unique to osteoblast-specific transcription factor Osf2/Cbfa1 contribute to its transactivation function and its inability to heterodimerize with Cbfbeta. *Mol. Cell. Biol.* **18**, 4197–4208.

39. Banerjee, C., Javed, A., Choi, J.-Y., Green, J., Rosen, V., Van Wijnen, A.J., et al. (2001) Differential regulation of the two principal Runx2/Cbfa1 N-Terminal isoforms in response to bone morphogenetic protein-2 during development of the osteoblast phenotype. *Endocrinology* **142(9)**, 4026–4039

40. Javed, A., Guo, B., Hiebert, S., Choi, J.Y., Green, J., Zhao, S.C., et al. (2000). Groucho/TLE/R-esp proteins associate with the nuclear matrix and repress RUNX (CBF(alpha)/AML/PEBP2(alpha))

dependent activation of tissue-specific gene transcription. *J. Cell Sci.* **113(Pt 12)**, 2221–2231.

41. Javed, A., Barnes, G.L., Jasanya, B.O., Stein, J.L., Gerstenfeld, L., Lian, J.B., et al. (2001) runt homology domain transcription factors (Runx, Cbfa, and AML) mediate repression of the bone sialoprotein promoter: evidence for promoter context-dependent activity of Cbfa proteins. *Mol. Cell. Biol.* **21(8)**, 2891–2905.

42. Gao, Y.H., Shinki, T., Yuasa, T., Kataoka-Enomoto, H., Komori, T., Suda, T., et al. (1998) Potential role of cbfa1, an essential transcriptional factor for osteoblast differentiation, in osteoclastogenesis: regulation of mRNA expression of osteoclast differentiation factor (ODF). *Biochem. Biophys. Res. Commun.* **252(3)**, 697–702.

43. Philipsen, S., and Suske, G.A. (1999) Tale of three fingers: the family of mammalian Sp/XKLF transcription factors. *Nucleic Acids Res.* **27**, 2991–3000.

44. Black, A.R., Black, J.D., and Azizkhan-Clifford, J. (2001) Sp1 and kruppel-like factor family of transcription factors in cell growth regulation and cancer. *J. Cell. Physiol.* **188**, 143–160.

45. Göllner, H., Dani, C., Phillips, B., Philipsen, S., and Suske, G. (2001) Impaired ossification in mice lacking the transcription factor Sp3. *Mech. Dev.* **106**, 77–83.

46. Milona, M.A., Gough, J.E., and Edgar, A.J. (2003) Expression of alternatively spliced isoforms of human Sp7 in osteoblast-like cells. *BMC Genomics* **4(1)**, 43.

47. Nishio, Y., Dong, Y., Paris, M., O'Keefe, R.J., Schwarz, E.M., and Drissi, H. (2006) Runx2-mediated regulation of the zinc finger Osterix/Sp7 gene. *Gene* **372**, 62–70.

48. Maehata, Y., Takamizawa, S., Ozawa, S., Kato, Y., Sato, S., Kubota, E., et al (2006) Both direct and collagen-mediated signals are required for active vitamin D3-elicited differentiation of human osteoblastic cells: roles of osterix, an osteoblast-related transcription factor. *Matrix Biol.* **25(1)**, 47–58.

49. Shen, Q., and Christakos, S. (2005) The vitamin D receptor, Runx2, and the Notch signaling pathway cooperate in the transcriptional regulation of osteopontin. *J. Biol. Chem.* **280(49)**, 40589–40598.

50. Xiao, Z.S., Hinson, T.K., and Quarles, L.D. (1999) Cbfa1 isoform overexpression upregulates osteocalcin gene expression in non-osteoblastic and pre-osteoblastic cells. *J. Cell. Biochem.* **74(4)**, 596–605.

51. McClellan, J.W., Mulconrey, D.S., Forbes, R.J., and Fullmer, N. (2006) Vertebral bone resorption after transforaminal lumbar interbody fusion with bone morphogenetic protein (rhBMP-2). *J. Spinal Disord. Tech.* **19(7)**, 483–486.

52. Vaidya, R., Weir, R., Sethi, A., Meisterling, S., Hakeos, W., and Wybo, C.D. (2007) Interbody fusion with allograft and rhBMP-2 leads to consistent fusion but early subsidence. *J. Bone Joint Surg. Br.* **89(3)**, 342–345.

53. Choi, K.Y., Kim, H.J., Lee, M.H., Kwon, T.G., Nah, H.D., Furuichi, T., et al. (2005) Runx2 regulates FGF2-induced Bmp2 expression during cranial bone development. *Dev. Dyn.* **233(1)**, 115–121.

54. Zehentner, B.K., Dony, C., and Burtscher, H. (1999) The transcription factor Sox9 is involved in BMP-2 signaling. *J. Bone Miner. Res.* **14(10)**, 1734–1741.

55. Chimal-Monroy, J., Rodriguez-Leon, J., Montero, J.A., Ganan, Y., Macias, D., Merino, R., et al. (2003) Analysis of the molecular cascade responsible for mesodermal limb chondrogenesis: Sox genes and BMP signaling. *Dev. Biol.* **257(2)**, 292–301.

56. Lefebvre, V., and de Crombrugghe, B. (1998) Toward understanding SOX9 function in chondrocyte differentiation. *Matrix Biol.* **16(9)**, 529–540.

57. Smits, P., Li, P., Mandel, J., Zhang, Z., Deng, J.M., Behringer, R.R., et al. (2001) The transcription factors L-Sox5 and Sox6 are essential for cartilage formation. *Dev. Cell* **1(2)**, 277–290.

58. Akiyama, H., Chaboissier, M.C., Martin, J.F., Schedl, A., and de Crombrugghe, B. (2002) The transcription factor Sox9 has essential roles in successive steps of the chondrocyte differentiation pathway and is required for expression of Sox5 and Sox6. *Genes Dev.* **16(21)**, 2813–2828.

59. Lefebvre, V., Behringer, R.R., and de Crombrugghe, B. (2001) L-Sox5, Sox6 and Sox9 control essential steps of the chondrocyte differentiation pathway. *Osteoarthritis Cartilage* **9(Suppl A)**, S69–S75.

60. Bi, W., Deng, J.M., Zhang, Z., Behringer, R.R., and de Crombrugghe, B. (1999) Sox9 is required for cartilage formation. *Nat. Genet.* **22(1)**, 85–89.

61. Boyle, W.J., Simonet, W.S., and Lacey, D.L. (2003) Osteoclast differentiation and activation. *Nature* **423(6937)**, 337–342.

62. Lanzi, R., Losa, M., Villa, I., Gatti, E., Sirtori, M., Dal Fiume, C., et al. (2003) GH replacement therapy increases plasma osteoprotegerin levels in GH-deficient adults. *Eur. J. Endocrinol.* **148(2)**, 185–191.

63. Kwon, O.H., Lee, C.K., Lee, Y.I., Paik, S.G., and Lee, H.J. (2005) The hematopoietic transcription factor PU.1 regulates RANK gene expression in myeloid progenitors. *Biochem. Biophys. Res. Commun.* **335(2)**, 437–446.

64. Xing, L., Bushnell, T.P., Carlson, L., Tai, Z., Tondravi, M., Siebenlist, U., et al. (2002) NF-kappaB p50 and p52 expression is not required for RANK-expressing osteoclast progenitor formation but is essential for RANK- and cytokine-mediated osteoclastogenesis. *J. Bone Miner. Res.* **17(7)**, 1200–1210.

65. Kuroki, Y., Shiozawa, S., Sugimoto, T., Kanatani, M., Kaji, H., Miyachi, A., et al. (1994) Constitutive c-fos expression in osteoblastic MC3T3-E1 cells stimulates osteoclast maturation and osteoclastic bone resorption. *Clin. Exp. Immunol.* **95(3)**, 536–539.

66. Grigoriadis, A.E., Wang, Z.Q., Cecchini, M.G., Hofstetter, W., Felix, R., Fleisch, H.A., et al. (1994) c-Fos: a key regulator of osteoclast-macrophage lineage determination and bone remodeling. *Science.* **266(5184)**, 443–448.

67. Yamane, T., Kunisada, T., Yamazaki, H., Nakano, T., Orkin, S.H., and Hayashi, S.I. (2000) Sequential requirements for SCL/tal-1, GATA-2, macrophage colony-stimulating factor, and osteoclast differentiation factor/osteoprotegerin ligand in osteoclast development. *Exp. Hematol.* **28(7)**, 833–840.

68. Lindunger, A., MacKay, C.A., Ek-Rylander, B., Andersson, G., and Marks, S.C. Jr. (1990) Histochemistry and biochemistry of tartrate-resistant acid phosphatase (TRAP) and tartrate-resistant acid adenosine triphosphatase (TrATPase) in bone, bone marrow and spleen: implications for osteoclast ontogeny. *Bone Miner.* **10(2)**, 109–119.

69. Bossard, M.J., Tomaszek, T.A., Thompson, S.K., Amegadzie, B.Y., Hanning, C.R., Jones, C., et al. (1996) Proteolytic activity of human osteoclast cathepsin K. Expression, purification, activation, and substrate identification. *J. Biol. Chem.* **271(21)**, 12517–12524.

70. Itskovitz-Eldor, J., Schuldiner, M., Karsenti, D., Eden, A., Yanuka, O., Amit, M., et al. (2000) Differentiation of human embryonic stem cells into embryoid bodies compromising the three embryonic germ layers. *Mol. Med.* **6(2)**, 88–95.

71. Gerecht-Nir, S., Cohen, S., and Itskovitz-Eldor, J. (2004) Bioreactor cultivation enhances the efficiency of human embryoid body (hEB) formation and differentiation. *Biotechnol. Bioeng.* **86(5)**, 493–502.

72. zur Nieden, N.I., Kempka, G., and Ahr, H.J. (2003). In vitro differentiation of embryonic stem cells into mineralized osteoblasts. *Differentiation* **71(1)**, 18–27.

73. zur Nieden, N.I., Kempka, G., Rancourt, D.E., and Ahr, H.J. (2005) Induction of chondro-, osteo- and adipogenesis in embryonic stem cells by bone morphogenetic protein-2: effect of cofactors on differentiating lineages. *BMC Dev. Biol.* **5(1)**, 1.

74. zur Nieden, N.I., Cormier, J.T., Rancourt, D.E., and Kallos, M.S. (2007) Embryonic stem cells remain highly pluripotent following long term expansion as aggregates in suspension bioreactors. *J. Biotechnol.* **129(3)**, 421–432.

75. Yamane, T., Kunisada, T., Yamazaki, H., Era, T., Nakano, T., and Hayashi, S.I. (1997) Development of osteoclasts from embryonic stem cells through a pathway that is c-fms but not c-kit dependent. *Blood* **90(9)**, 3516–35123.

76. Evans, M.J., and Kaufman, M.H. (1981) Establishment in culture of pluripotential cells from mouse embryos. *Nature* **292(5819)**, 154–156.

77. Martin, G.R. (1981) Isolation of a pluripotent cell line from early mouse embryos cultured in medium conditioned by teratocarcinoma stem cells. *Proc. Natl. Acad. Sci. USA* **78(12)**, 7634–7638.

78. Doetschman, T., Williams, P., and Maeda, N. (1998) Establishment of hamster blastocyst-derived embryonic stem (ES) cells. *Dev. Biol.* **127(1)**, 224–227.

79. Iannaccone, P.M., Taborn, G.U., Garton, R.L., Caplice, M.D., and Brenin, D.R. (1994) Pluripotent embryonic stem cells from the rat are capable of producing chimeras. *Dev. Biol.* **163(1)**, 288–292.

80. Giles, J.R., Yang, X., Mark, W., and Foote, R.H. (1993) Pluripotency of cultured rabbit inner cell mass cells detected by isozyme analysis and eye pigmentation of fetuses following injection into blastocysts or morulae. *Mol. Reprod. Dev.* **36(2)**, 130–138.

81. Graves, K.H., and Moreadith, R.W. (1993) Derivation and characterization of putative pluripotential embryonic stem cells from preimplantation rabbit embryos. *Mol. Reprod. Dev.* 36(**4**), 424–433.

82. Sims, M., and First, N.L. (1994) Production of calves by transfer of nuclei from cultured inner cell mass cells. *Proc. Natl. Acad. Sci. USA* **91(13)**, 6143–6147.

83. Thomson, J.A., Kalishman, J., Golos, T.G., Durning, M., Harris, C.P., Becker, R.A., et al. (1995) Isolation of a primate embryonic stem cell line. *Proc. Natl. Acad. Sci. USA* **92(17)**, 7844–7848.

84. Thomson, J.A., Kalishman, J., Golos, T.G., Durning, M., Harris, C.P., and Hearn, J.P. (1996) Pluripotent cell lines derived from common marmoset (*Callithrix jacchus*) blastocysts. *Biol. Reprod.* **55(2)**, 254–259.

85. Sasaki, E., Hanazawa, K., Kurita, R., Akatsuka, A., Yoshizaki, T., Ishii, H., et al. (2005) Establishment of novel embryonic stem cell lines derived from the common marmoset (*Callithrix jacchus*). *Stem Cells* **23(9)**, 1304–1313.

86. Thomson, J.A., Itskovitz-Eldor, J., Shapiro, S.S., Waknitz, M.A., Swiergiel, J.J., Marshall, V.S., et al. (1998) Embryonic stem cell lines derived from human blastocysts. *Science* **282(5391)**, 1145–1147.

87. Reubinoff, B.E., Pera, M.F., Fong, C.Y., Trounson, A., and Bongso, A. (2000) Embryonic stem cell lines from human blastocysts: somatic differentiation in vitro. *Nat. Biotechnol.* **18(4)**, 399–404.

88. Amit, M., and Itskovitz-Eldor, J. (2002) Derivation and spontaneous differentiation of human embryonic stem cells. *J. Anat.* **200(Pt 3)**, 225–232.

89. Maltsev, V.A., Rohwedel, J., Hescheler, J., and Wobus, A.M. (1993). Embryonic stem cells differentiate in vitro into cardiomyocytes representing sinusnodal, atrial and ventricular cell types. *Mech. Dev.* **44(1)**, 41–50.

90. Klug, M.G., Soonpaa, M.H., Koh, G.Y., and Field, L.J. (1996) Genetically selected cardiomyocytes from differentiating embryonic stem cells form stable intracardiac grafts. *J. Clin. Invest.* **98(1)**, 216–224.

91. Soria, B., Roche, E., Berná, G., León-Quinto, T., Reig, J.A., and Martín, F. (2000) Insulin-secreting cells derived from embryonic stem cells normalize glycemia in streptozotocin-induced diabetic mice. *Diabetes* **49(2)**, 157–162.

92. Hamazaki, T., Iiboshi, Y., Oka, M., Papst, P.J., Meacham, A.M., Zon, L.I., et al. (2001) Hepatic maturation in differentiating embryonic stem cells in vitro. *FEBS Lett.* **497(1)**, 15–19.

93. Okabe, S., Forsberg-Nilsson, K., Spiro, A.C., Segal, M., and McKay, R.D. (1996) Development of neuronal precursor cells and functional postmitotic neurons from embryonic stem cells in vitro. *Mech. Dev.* **59(1)**, 89–102.

94. Keller, G., Kennedy, M., Papayannopoulou, T., and Wiles, M.V. (1993) Hematopoietic commitment during embryonic stem cell differentiation in culture. *Mol. Cell. Biol.* **13(1)**, 473–486.

95. Kramer, J., Hegert, C., Guan, K., Wobus, A.M., Müller, P.K., and Rohwedel, J. (2000) Embryonic stem cell-derived chondrogenic differentiation in vitro: activation by BMP-2 and BMP-4. *Mech. Dev.* **92(2)**, 193–205.

96. Assady, S., Maor, G., Amit, M., Itskovitz-Eldor, J., Skorecki, K. L., and Tzukerman, M. (2001) Insulin production by human embryonic stem cells. *Diabetes* **50(8)**, 1691–1697.

97. Kehat, I., Kenyagin-Karsenti, D., Snir, M., Segev, H., Amit, M., Gepstein, A., et al. (2001) Human embryonic stem cells can differentiate into myocytes with structural and functional properties of cardiomyocytes. *J. Clin. Invest.* **108(3)**, 407–414.

98. Kaufman, D.S., Hanson, E.T., Lewis, R.L., Auerbach, R., and Thomson, J.A. (2001) Hematopoietic colony-forming cells derived from human embryonic stem cells. *Proc. Natl. Acad. Sci. USA* **98(19)**, 10716–10721.

99. Smith, A.G. (2001) Embryo-derived stem cells: of mice and men. *Annu. Rev. Cell Dev. Biol.* **17**, 435–462.

100. Smith, A.G., Heath, J.K., Donaldson, D.D., Wong, G.G., Moreau, J., Stahl, M., et al. (1988) Inhibition of pluripotential embryonic stem cell differentiation by purified polypeptides. *Nature* **336(6200)**, 688–690.

101. Williams, R.L., Hilton, D.J., Pease, S., Willson, T.A., Stewart, C.L., Gearing, D.P., et al. (1988) Myeloid leukaemia inhibitory factor maintains the developmental potential of embryonic stem cells. *Nature* **336(6200)**, 684–687.

102. Nichols, J., Evans, E.P., and Smith, A.G. (1990) Establishment of germ-line-competent embryonic stem (ES) cells using differentiation inhibiting activity. *Development* **110(4)**, 1341–1348.

103. Metcalf, D. (1990) The induction and inhibition of differentiation in normal and leukaemic cells. *Philos. Trans. R. Soc. Lond. B. Biol. Sci.* **327(1239)**, 99–109.

104. Nichols, J., Zevnik, B., Anastassiadis, K., Niwa, H., Klewe-Nebenius, D., Chambers, I., et al. (1998) Formation of pluripotent stem cells in the mammalian embryo depends on the POU transcription factor Oct4. *Cell* **95(3)**, 379–391.

105. Palmieri, S.L., Peter, W., Hess, H., and Schöler, H.R. (1994) Oct-4 transcription factor is differentially expressed in the mouse embryo during establishment of the first two extraembryonic cell lineages involved in implantation. *Dev. Biol.* **166(1)**, 259–267.

106. Rosner M.H., Vigano M.A., Ozato K., Timmons P.M., Poirier F., Rigby P.W., et al. (1990) A POU-domain transcription factor in early stem cells and germ cells of the mammalian embryo. *Nature* **345(6277)**, 686–692.

107. Schöler, H.R., Dressler, G.R., Balling, R., Rohdewohld, H., and Gruss, P. (1990) Oct-4: a germline-specific transcription factor mapping to the mouse t-complex. *EMBO J.* **9(7)**, 2185–2195.

108. Schöler, H.R., Ruppert, S., Suzuki, N., Chowdhury, K., and Gruss, P. (1990) New type of POU domain in germ line-specific protein Oct-4. *Nature* **344(6265)**, 435–439.

109. Niwa, H., Miyazaki, J., and Smith, A.G. (2000) Quantitative expression of Oct-3/4 defines differentiation, dedifferentiation or self-renewal of ES cells. *Nat. Genet.* **24(4)**, 372–376.

110. Avilion, A.A., Nicolis, S.K., Pevny, L.H., Perez, L., Vivian, N., and Lovell-Badge, R. (2003) Multipotent cell lineages in early mouse development depend on SOX2 function. *Genes Dev.* **17(1)**, 126–140.

111. Mitsui, K., Tokuzawa, Y., Itoh, H., Segawa, K., Murakami, M., Takahashi, K., et al. (2003). The homeoprotein Nanog is required for maintenance of pluripotency in mouse epiblast and ES cells. *Cell* **113(5)**, 631–642.

112. Chambers, I., Colby, D., Robertson, M., Nichols, J., Lee, S., Tweedie, S., et al. (2003) Functional expression cloning of Nanog, a pluripotency sustaining factor in embryonic stem cells. *Cell* **113(5)**, 643–655.

113. Tomioka, M., Nishimoto, M., Miyagi, S., Katayanagi, T., Fukui, N., Niwa, H., et al. (2002). Identification of Sox-2 regulatory region which is under the control of Oct-3/4-Sox-2 complex. *Nucleic Acids Res.* **30(14)**, 3202–3213.

114. Chew, J.-L., Loh, Y.-H., Zhang, W., Chen, X., Tam, W.-L., Yeap, L.-S., et al. (2005) Reciprocal transcriptional regulation of Pou5f1 and Sox2 via the Oct4/Sox2 complex in embryonic stem cells. *Mol. Cell. Biol.* **25(14)**, 6031–6046.

115. Okumura-Nakanishi, S., Saito, M., Niwa, H., and Ishikawa, F. (2005) Oct-3/4 and Sox2 regulate Oct-3/4 gene in embryonic stem cells. *J. Biol. Chem.* **280(7)**, 5307–5317.

116. Kuroda, T., Tada, M., Kubota, H., Kimura, H., Hatano, S.Y., Suemori, H., et al. (2005). Octamer and Sox elements are required for transcriptional cis regulation of Nanog gene expression. *Mol. Cell. Biol.* **25**(6), 2475–2485

117. Rodda, D.J., Chew, J.L., Lim, L.H., Loh, Y.H., Wang, B., Ng, H.H., et al. (2005) Transcriptional regulation of nanog by OCT4 and SOX2. *J. Biol. Chem.* **280**(26), 24731–24737.

118. Boyer, L.A., Lee, T.I., Cole, M.F., Johnstone, S.E., Levine, S.S., Zucker, J.P., et al. (2005) Core transcriptional regulatory circuitry in human embryonic stem cells. *Cell.* **122**(6), 947–956.

119. Bernstein, B.E., Kamal, M., Lindblad-Toh, K., Bekiranov, S., Bailey, D.K., Huebert, D.J., et al. (2005). Genomic maps and comparative analysis of histone modifications in human and mouse. *Cell* **120**(2), 169–181.

120. Azuara, V., Perry, P., Sauer, S., Spivakov, M., Jørgensen, H.F., John, R.M., et al. (2006) Chromatin signatures of pluripotent cell lines. *Nat. Cell Biol.* **8**(5), 532–538.

121. Kirmizis, A., Bartley, S.M., Kuzmichev, A., Margueron, R., Reinberg, D., Green, R., et al. (2004) Silencing of human polycomb target genes is associated with methylation of histone H3 Lys 27. *Genes Dev.* **18**(13), 1592–1605.

122. Boyer, L.A., Plath, K., Zeitlinger, J., Brambrink, T., Medeiros, L.A., Lee T.I., et al. (2006) Polycomb complexes repress developmental regulators in murine embryonic stem cells. *Nature* **441**(7091), 349–353.

123. Wang, H., Wang, L., Erdjument-Bromage, H., Vidal, M., Tempst, P., Jones, R.S., et al. (2004) Role of histone H2A ubiquitination in Polycomb silencing. *Nature* **431**(7010), 873–878.

124. Schwartz, Y.B., and Pirrotta, V. (2007) Polycomb silencing mechanisms and the management of genomic programmes. *Nat. Rev. Genet.* **8**(1), 9–22.

125. Wittrant, Y., Theoleyre, S., Couillaud, S., Dunstan, C., Heymann, D., and Rédini, F. (2004) Relevance of an in vitro osteoclastogenesis system to study receptor activator of NF-kB ligand and osteoprotegerin biological activities. *Exp. Cell Res.* **293**(2), 292–301.

126. Tsuneto, M., Yamazaki, H., Yoshino, M., Yamada, T., and Hayashi, S. (2005) Ascorbic acid promotes osteoclastogenesis from embryonic stem cells. *Biochem. Biophys. Res. Commun.* **335**(4), 1239–1246.

127. Okuyama, H., Tsuneto, M., Yamane, T., Yamazaki, H., and Hayashi, S. (2003) Discrete types of osteoclast precursors can be generated from embryonic stem cells. *Stem Cells* **21**(6), 670–680.

128. Hemmi, H., Okuyama, H., Yamane, T., Nishikawa, S., Nakano, T., Yamazaki, H., et al. (2001) Temporal and spatial localization of osteoclasts in colonies from embryonic stem cells. *Biochem. Biophys. Res. Commun.* **280**(2), 526–534.

129. Hayashi, S., Yamane, T., Miyamoto, A., Hemmi, H., Tagaya, H., Tanio, Y., et al. (1998) Commitment and differentiation of stem cells to the osteoclast lineage. *Biochem. Cell Biol.* **76**(6), 911–922.

130. Porcher, C., Swat, W., Rockwell, K., Fujiwara, Y., Alt, F.W., and Orkin, S.H. (1996) The T cell leukemia oncoprotein SCL/tal-1 is essential for development of all hematopoietic lineages. *Cell* **86**(1), 47–57.

131. Elefanty, A.G., Robb, L., Birner, R., and Begley, C.G. (1997) Hematopoietic-specific genes are not induced during in vitro differentiation of scl-null embryonic stem cells. *Blood* **90**(4), 1435–1447.

132. Tsuneto, M., Tominaga, A., Yamazaki, H., Yoshino, M., Orkin, S.H., and Hayashi, S. (2005). Enforced expression of PU.1 rescues osteoclastogenesis from embryonic stem cells lacking Tal-1. *Stem Cells* **23**(1), 134–143.

133. Cao, T., Heng, B.C., Ye, C.P., Liu, H., Toh, W.S., Robson, P., et al. (2005) Osteogenic differentiation within intact human embryoid bodies result in a marked increase in osteocalcin secretion after 12 days of in vitro culture, and formation of distinct nodule-structure. *Tissue Cell.* **37**(4), 325–34.

134. Yao, S., Chen, S., Clark, J., Hao, E., Beattie, G.M., Hayek, A., et al. (2006) Long-term self-renewal and directed differentiation of human embryonic stem cells in chemically defined conditions. *Proc. Natl. Acad. Sci. USA* **103**(18), 6907–6912.

135. Beddington, R.S., Rashbass, P., Wilson, V. (1992) Brachyury – a gene affecting mouse gastrulation and early organogenesis. *Dev. Suppl.* 157–165.

136. Gadue, P., Huber, T.L., Paddison, P.J., Keller, G.M. (2006) Wnt and TGF-beta signaling are required for the induction of an in vitro model of primitive streak formation using embryonic stem cells. *Proc. Natl. Acad. Sci. USA* **103**(45), 16806–16811.

137. Nakanishi, M., Kurisaki, A., Hayashi, Y., Warashina, M., Ishiura, S., Kusuda-Furue, M., et al. (2009) Directed induction of anterior

and posterior primitive streak by Wnt from embryonic stem cells cultured in a chemically defined serum-free medium. *FASEB J.* **23(1)**, 114–122.

138. Phillips, B.W., Belmonte, N., Vernochet, C., Ailhaud, G., and Dani, C. (2001) Compactin enhances osteogenesis in murine embryonic stem cells. *Biochem. Biophys. Res. Commun.* **284(2)**, 478–484.

139. Buttery LD, Bourne S, Xynos JD, Wood H, Hughes FJ, Hughes SP, et al. (2001) Differentiation of osteoblasts and in vitro bone formation from murine embryonic stem cells. *Tissue Eng.* **7(1)**, 89–99.

140. Ahn, S.E., Kim, S., Park, K.H., Moon, S.H., Lee, H.J., Kim, G.J., et al. (2006) Primary bone-derived cells induce osteogenic differentiation without exogenous factors in human embryonic stem cells. *Biochem. Biophys. Res. Commun.* **340**, 403–408.

141. Hwang, Y.S., Randle, W.L., Bielby, R.C., Polak, J.M., and Mantalaris, A. (2006) Enhanced derivation of osteogenic cells from murine embryonic stem cells after treatment with hepG2-conditioned medium and modulation of the embryoid body formation period: application to skeletal tissue engineering. *Tissue Eng.* **12**, 1381–1392.

142. Davis, L.A., and zur Nieden, N.I. (2008) Mesodermal fate decisions of a stem cell: the Wnt switch. *Cell. Mol. Life Sci.* **65(17)**, 2658–2674.

143. Sottile, V., Thomson, A., and McWhir, J. (2003) In vitro osteogenic differentiation of human ES cells. *Cloning Stem Cells* **5**, 149–155.

144. Bielby, R.C., Boccaccini, A.R., Polak, J.M., and Buttery, L.D. (2004) In vitro differentiation and in vivo mineralization of osteogenic cells derived from human embryonic stem cells. *Tissue Eng.* **10(9–10)**, 1518–1525.

145. Yamashita, A., Takada, T., Narita, J., Yamamoto, G., and Torii, R. (2005) Osteoblastic differentiation of monkey embryonic stem cells in vitro. *Cloning Stem Cells* **7**, 232–237.

146. zur Nieden, N.I., Price, F.D., Davis, L.A., Everitt, R.E., Rancourt, D.E. (2007) Gene profiling on mixed embryonic stem cell populations reveals a biphasic role for beta-catenin in osteogenic differentiation. *Mol. Endocrinol.* **21(3)**, 674–685.

147. Krishnan, V., Bryant, H.U., and Macdougald, O.A. (2006) Regulation of bone mass by Wnt signaling. *J. Clin. Invest.* **116(5)**, 1202–1209.

148. Gong, Y., Slee, R.B., Fukai, N., Rawadi, G., Roman-Roman, S., Reginato, A.M., et al. (2001) LDL receptor-related protein 5 (LRP5) affects bone accrual and eye development. *Cell* **107(4)**, 513–523.

149. Bain, G., Muller, T., Wang, X., and Papkoff, J. (2003) Activated beta-catenin induces osteoblast differentiation of C3H10T1/2 cells and participates in BMP2 mediated signal transduction. *Biochem. Biophys. Res. Commun.* **301(1)**, 84–91.

150. Tolwinski, N.S., and Wieschaus, E. (2004) A nuclear escort for beta-catenin. *Nat. Cell Biol.* **6(7)**, 579–580.

151. van de Wetering, M., Cavallo, R., Dooijes, D., van Beest, M., van Es, J., Loureiro, J., et al. (1997) Armadillo coactivates transcription driven by the product of the Drosophila segment polarity gene dTCF. *Cell* **88(6)**, 789–799.

152. Behrens, J., von Kries, J.P., Kuhl, M., Bruhn, L., Wedlich, D., Grosschedl, R., et al. (1996) Functional interaction of beta-catenin with the transcription factor LEF-1. *Nature* **382(6592)**, 638–642.

153. Topol, L., Jiang, X., Choi, H., Garrett-Beal, L., Carolan, P.J., and Yang, Y. (2003) Wnt-5a inhibits the canonical Wnt pathway by promoting GSK-3-independent beta-catenin degradation. *J. Cell. Biol.* **162(5)**, 899–908.

154. Gaur, T., Lengner, C.J., Hovhannisyan, H., Bhat, R.A., Bodine, P.V., Komm, B.S., et al. (2005) Canonical WNT signaling promotes osteogenesis by directly stimulating Runx2 gene expression. *J. Biol. Chem.* **280(39)**, 33132–33140.

155. Schinke, T., and Karsenty, G. (1999) Characterization of Osf1, an osteoblast-specific transcription factor binding to a critical cis-acting element in the mouse Osteocalcin promoters. *J. Biol. Chem.* **274(42)**, 30182–30189.

156. Kolpakova, E., and Olsen, B.R. (2005) Wnt/beta-catenin – a canonical tale of cell-fate choice in the vertebrate skeleton. *Dev. Cell.* **8(5)**, 626–627.

157. Day, T.F., Guo, X., Garrett-Beal, L., and Yang, Y. (2005) Wnt/beta-catenin signaling in mesenchymal progenitors controls osteoblast and chondrocyte differentiation during vertebrate skeletogenesis. *Dev. Cell.* **8(5)**, 739–750.

158. Vats, A., Bielby, R.C., Tolley, N., Dickinson, S.C., Boccaccini, A.R., Hollander, A.P., et al. (2006) Chondrogenic differentiation of human embryonic stem cells: the effect of the micro-environment. *Tissue Eng.* **12(6)**, 1687–1697.

159. Sui, Y., Clarke, T., and Khillan, J.S. (2003) Limb bud progenitor cells induce

differentiation of pluripotent embryonic stem cells into chondrogenic lineage. *Differentiation* **71**, 578–585.

160. Cserjesi, P., Brown, D., Ligon, K.L., Lyons, G.E., Copeland, N.G., Gilbert, D.J., et al. (1995) Scleraxis: a basic helix-loop-helix protein that prefigures skeletal formation during mouse embryogenesis. *Development* **121(4)**, 1099–1110.

161. Toh, W.S., Yang, Z., Liu, H., Heng, B.C., Lee, E.H., and Cao, T. (2007) Effects of culture conditions and bone morphogenetic protein 2 on extent of chondrogenesis from human embryonic stem cells. *Stem Cells* **25(4)**, 950–960.

162. Hegert, C., Kramer, J., Hargus, G., Muller, J., Guan, K., Wobus, A.M., et al. (2002) Differentiation plasticity of chondrocytes derived from mouse embryonic stem cells. *J. Cell Sci.* **115(Pt 23)**, 4617–4628.

163. Ikeda, T., Kamekura, S., Mabuchi, A., Kou, I., Seki, S., Takato, T., et al. (2004) The combination of SOX5, SOX6, and SOX9 (the SOX trio) provides signals sufficient for induction of permanent cartilage. *Arthritis Rheum.* **50(11)**, 3561–3573.

164. Kim, I.S., Otto, F., Zabel, B., and Mundlos, S. (1999) Regulation of chondrocyte differentiation by Cbfa1. *Mech. Dev.* **80**, 159–170.

165. Wright, E., Hargrave, M.R., Christiansen, J., Cooper, L., Kun, J., Evans, T., et al. (1995) The Sry-related gene Sox9 is expressed during chondrogenesis in mouse embryos. *Nat. Genet.* **9**, 15–20.

166. Barberi, T., Willis, L.M., Socci, N.D., and Studer, L. (2005). Derivation of multipotent mesenchymal precursors from human embryonic stem cells. *PLoS Med.* **2(6)**, e161.

167. Trivedi, P., and Hematti, P. (2007) Simultaneous generation of CD34+ primitive hematopoietic cells and CD73+ mesenchymal stem cells from human embryonic stem cells cocultured with murine OP9 stromal cells. *Exp. Hematol.* **35(1)**, 146–154.

168. Trivedi, P., and Hematti, P. (2008) Derivation and immunological characterization of mesenchymal stromal cells from human embryonic stem cells. *Exp. Hematol.* **36(3)**, 350–359.

169. Arpornmaeklong, P., Brown, S.E., Wang, Z., Krebsbach, P.H. (2009) Phenotypic characterization, osteoblastic differentiation, and bone regeneration capacity of human embryonic stem cell-derived mesenchymal stem cells. *Stem Cells Dev.* **18(7)**, 955–968.

170. Olivier, E.N., Rybicki, A.C., and Bouhassira, E.E. (2006) Differentiation of human embryonic stem cells into bipotent mesenchymal stem cells. *Stem Cells* **24(8)**, 1914–1922.

171. Hwang, N.S., Varghese, S., Lee, H.J., Zhang, Z., Ye, Z., Bae, J., et al. (2008) In vivo commitment and functional tissue regeneration using human embryonic stem cell-derived mesenchymal cells. *Proc. Natl. Acad. Sci. USA* **105(52)**, 20641–20646.

172. Lian, Q., Lye, E., Suan Yeo, K., Khia Way Tan, E., Salto-Tellez, M., Liu, T.M., et al. (2007) Derivation of clinically compliant MSCs from CD105+, CD24- differentiated human ESCs. *Stem Cells* **25(2)**, 425–436.

173. Brederlau, A., Correia, A.S., Anisimov, S.V., Elmi, M., Paul, G., Roybon, L., et al. (2006) Transplantation of human embryonic stem cell-derived cells to a rat model of Parkinson's disease: effect of in vitro differentiation on graft survival and teratoma formation. *Stem Cells* **24(6)**, 1433–1440.

174. Taiani, J., Krawetz, R.J., zur Nieden, N.I., Wu, E.Y., Kallos, M.S., Matyas, J.R., Rancourt, D.E. (2010) Reduced differentiation efficiency of murine embryonic stem cells in stirred suspension bioreactors. *Stem Cells Dev.* **19(7)**, 989–998.

175. Wakitani, S., Takaoka, K., Hattori, T., Miyazawa, N., Iwanaga, T., Takeda, S., et al. (2003) Embryonic stem cells injected into the mouse knee joint form teratomas and subsequently destroy the joint. *Rheumatology* **42(1)**, 162–165.

176. Wakitani, S., Aoki, H., Harada, Y., Sonobe, M., Morita, Y., Mu, Y., et al. (2004) Embryonic stem cells form articular cartilage, not teratomas, in osteochondral defects of rat joints. *Cell. Transplant.* **13(4)**, 331–336.

177. Nakajima, M., Wakitani, S., Harada, Y., Tanigami, A., and Tomita, N. (2007) In vivo mechanical condition plays an important role for appearance of cartilage tissue in ES cell transplanted joint. *J. Orthop. Res.* **26(1)**, 10–17.

178. Nussbaum, J., Minami, E., Laflamme, M.A., Virag, J.A., Ware, C.B., Masino, A., et al. (2007) Transplantation of undifferentiated murine embryonic stem cells in the heart: teratoma formation and immune response. *FASEB J.* **21(7)**, 1345–1357.

179. Swijnenburg, R.J., Tanaka, M., Vogel, H., Baker, J., Kofidis, T., Gunawan, F., et al. (2005) Embryonic stem cell immunogenicity increases upon differentiation after transplantation into ischemic myocardium. *Circulation* **112(9 Suppl)**, I166–I172.

180. Lee, S.J., Lim, G.J., Lee, J.-W., Atala, A., and Yoo, J.J. (2006) In vitro evaluation of a poly(lactide-co-glycolide)–collagen composite scaffold for bone regeneration. *Biomaterials* **27(18)**, 3466–3472.

181. Kim, S., Kim, S.S., Lee, S.H., Eun Ahn, S., Gwak, S.J., Song, J.H., et al. (2008) In vivo bone formation from human embryonic stem cell-derived osteogenic cells in poly(d,l-lactic-co-glycolic acid)/hydroxyapatite composite scaffolds. *Biomaterials* **29(8)**, 1043–1053.

182. Hwang, N.S., Varghese, S., and Elisseeff, J. (2008) Derivation of chondrogenically-committed cells from human embryonic cells for cartilage tissue regeneration. *PLoS One* **3(6)**, e2498.

183. Webster, W.S., Johnston, M.C., Lammer, E.J., and Sulik, K.K. (1986) Isotretinoin embryopathy and the cranial neural crest: an in vivo and in vitro study. *J. Craniofac. Genet. Dev. Biol.* **6(3)**, 211–222.

184. Zhou, Y., and Snead, M.L. (2008) Derivation of cranial neural crest-like cells from human embryonic stem cells. *Biochem. Biophys. Res. Commun.* **376(3)**, 542–547.

185. Borello, U., Buffa, V., Sonnino, C., Melchionna, R., Vivarelli, E., and Cossu, G. (1999) Differential expression of the Wnt putative receptors Frizzled during mouse somitogenesis. *Mech. Dev.* **89(1–2)**, 173–177.

186. Kimura, Y., Matsunami, H., Takeichi, M. (1996) Expression of cadherin-11 delineates boundaries, neuromeres, and nuclei in the developing mouse brain. *Dev. Dyn.* **206(4)**, 455–462.

187. Hoffmann, I., and Balling, R. (1995) Cloning and expression analysis of a novel mesodermally expressed cadherin. *Dev. Biol.* **169(1)**, 337–346.

188. Lee, G., Kim, H., Elkabetz, Y., Al Shamy, G., Panagiotakos, G., Barberi, T., et al. (2007) Isolation and directed differentiation of neural crest stem cells derived from human embryonic stem cells. *Nat. Biotechnol.* **25(12)**, 1468–1475.

189. Kimura, Y., Matsunami, H., Inoue, T., Shimamura, K., Uchida, N., Ueno, T., et al. (1995) Cadherin-11 expressed in association with mesenchymal morphogenesis in the head, somite, and limb bud of early mouse embryos. *Dev. Biol.* **169(1)**, 347–358.

190. Jukes, J.M., Both, S.K., Leusink, A., Sterk, L.M., van Blitterswijk, C.A., and de Boer, J. (2008) Endochondral bone tissue engineering using embryonic stem cells. *Proc. Natl. Acad. Sci. USA* **105(19)**, 6840–6845.

191. Billon, N., Iannarelli, P., Monteiro, M.C., Glavieux-Pardanaud, C., Richardson, W.D., Kessaris, N., et al. (2007) The generation of adipocytes by the neural crest. *Development* **134(12)**, 2283–2292.

Chapter 2

Methods for Culturing Mouse and Human Embryonic Stem Cells

Sabrina Lin and Prue Talbot

Abstract

Mouse embryonic stem cells (mESCs) were first derived and cultured almost 30 years ago and ever since have been valuable tools for creating knockout mice and for studying early mammalian development. More recently (1998), human embryonic stem cells (hESCs) have been derived from blastocysts, and numerous methods have evolved to culture hESCs in vitro in both complex and defined media. hESCs are especially important at this time as they could potentially be used to treat degenerative diseases and to access the toxicity of new drugs and environmental chemicals. For both human and mouse ESCs, fibroblast feeder layers are often used at some phase in the culturing protocol. The feeders – often mouse embryonic fibroblasts (mEFs) – provide a substrate that increases plating efficiency, helps maintain pluripotency, and facilitates survival and growth of the stem cells. Various protocols for culturing embryonic stem cells from both species are available with newer trends moving toward feeder-free and serum-free culture. The purpose of this chapter is to provide basic protocol information on the isolation of mouse embryonic fibroblasts and establishment of feeder layers, the culture of mESCs on both mEFs and on gelatin in serum-containing medium, and the culture of hESCs in defined media on both mEFs (hESC culture medium) and Matrigel (mTeSR). These basic protocols are intended for researchers wanting to develop stem cell research in their labs. These protocols have been tested in our laboratory and work well. They can be modified and adapted for any relevant user's particular purpose.

Key words: Embryonic fibroblasts, Mouse embryonic stem cells, Human embryonic stem cells, Culture methods, Feeder layer

1. Introduction

The first derivation of mouse embryonic stem cells (mESCs) was reported independently in 1981 by laboratories in the USA (1) and in England (2). Success in deriving and culturing mESCs grew from prior experience with teratocarcinoma cells, which required a feeder layer for survival and growth, and the use of fibroblast feeder layers became a standard for the derivation and

Nicole I. zur Nieden (ed.), *Embryonic Stem Cell Therapy for Osteo-Degenerative Diseases*, Methods in Molecular Biology, vol. 690, DOI 10.1007/978-1-60761-962-8_2, © Springer Science+Business Media, LLC 2011

growth of mESCs (3). Feeder layers, which are still often used today with mESCs, have been valuable in providing conditions that support survival, proliferation, and the maintenance of pluripotency in stem cell populations. Since their initial isolation in 1981, numerous labs derived new mESC lines and used them to study mammalian development and to create knockout mice (4). Various protocols are now available for culturing mESCs, and to some extent, specific protocols work best with lines derived from specific strains of mice (3).

In 1998, Dr. Thomson's laboratory at the University of Wisconsin reported, for the first time, that embryonic stem cells can be derived from human blastocysts and propagated in vitro, opening the possibility of creating pluripotent cell lines with the potential to treat and cure numerous human diseases (5). Initially, culturing protocols for human embryonic stem cells (hESCs) involved the use of feeder layers as substrates and media containing serum and animal proteins. However, feeder layers add complexity to stem cell cultures and have the potential to introduce animal viruses and unwanted immunogens into the stem cell populations, which would preclude the use of hESCs grown on feeders in therapeutic applications. Subsequently, various protocols have been developed for feeder-free culturing of hESCs. The first of these replaced the feeders with Matrigel (6) and replaced serum with medium conditioned by murine embryonic fibroblasts (mEFs) (7). While Matrigel is readily available and easy to use, it is not strictly defined, may vary from lot to lot, and contains animal proteins. There has, thus, been interest in refining the substrate, and other options, such as laminin and fibronectin, have been successful (7, 8).

Media conditioned by fibroblasts or media containing serum have been valuable in initial work on embryonic stem cells, but there is a trend to eliminate these undefined components, especially from hESC culture protocols. Serum is highly variable from lot to lot and in some instances may promote differentiation rather than pluripotency of embryonic stem cells. Likewise, media conditioned by fibroblasts is not defined and contributes animal proteins to the culture milieu. New defined media formulations (containing some animal proteins) that work well with embryonic stem cells, such as Knockout Serum Replacement (SR or KSR) (Invitrogen) and mTeSR (StemCell Technologies), have been developed recently (9, 10) and are commercially available at affordable prices. More refined xeno-free media can also be purchased at a higher cost for hESC work demanding animal-protein-free media. Recently, three-dimensional culture systems, which would be beneficial for differentiation of hESCs, have also been introduced (11–13). Culture methods for ESCs are continually evolving, and protocols can be expected to improve and become more accessible in the future.

While many different protocols have been developed for both mESCs and hESCs and while the exact choice of media and culture conditions will be determined by the needs of individual investigators and purpose of their work, we describe in this review the fundamental methods needed for isolating mouse embryonic fibroblasts (mEFs) that can be used as feeder layers for both mouse and human ESCs. We also present a standard method for culturing mESCs in a serum-containing medium. It is possible to grow mESCs in serum-free knockout medium, and this would be an alternative strategy that might be preferable depending on the research goal. Finally, we present methods for growing hESCs on both mEFs and Matrigel in defined conditions using hESC medium in which serum is replaced with KSR or a new medium, mTeSR, developed by Ludwig and Thomson (9). We have had good success with hESC culture medium and mTeSR medium, both of which can be used with induced pluripotent stem cells (iPSCs) as well as hESCs. This set of protocols is intended for those who have not worked with ESCs before and need a starting point for accomplishing basic steps that would lead to setting up embryonic stem cell experiments in their laboratory. We present protocols as we often perform them in our laboratory. Amounts and sizes of preparations can be scaled up or down as needed. Once these methods have been mastered, numerous other techniques can be used in conjunction with ESCs, or more advanced methods of culturing can be added to this starting repertoire.

2. Materials

2.1. General Considerations

All methods described in this chapter need to be done using strict sterile technique. For more information on sterile technique, see the reviews by Phelan and by Cote (14, 15). For more information on contamination, how to prevent it, and how to deal with it when it occurs, see the excellent technical bulletin by Ryan (16). Testing for Mycoplasma infection of cultures should be done routinely (16). We recommend that all methods be done in a clean room, which is accessed only by users of the room. Individuals entering the clean room should step onto a sticky mat and then put on lab coats, booties, and masks, all of which are always kept in the clean room (disposable sticky mats, lab coats, latex gloves, face masks, and booties can be purchased from internet sources at considerable savings).

Floors should be cleaned at least once a week by users. Friday is a good time for this so that the clean rooms are ready for use at the beginning of the next week. All items that are removed from the sterile hoods (e.g. centrifuge tubes) should be sprayed with 75% ethanol, which is allowed to evaporate before returning them to the hood. Controlling contamination will be easier if air entering the rooms is HEPA-filtered. The need for fastidious sterile technique is especially important for hESCs, which are usually cultured without antibiotics.

2.2. Culturing Mouse Embryonic Fibroblasts

2.2.1. Isolating mEFs

1. Ethanol spray bottle (75%). All items going into the sterile hoods should be sprayed with ethanol, which will help kill microorganisms upon evaporation.

2. Stereoscopic (dissecting) microscope.

3. Pipet-Aid (i.e. Drummond).

4. Sterile plastic pipettes (10 and 25 mL, individually wrapped).

5. Trypsin/Ethylenediaminetetraacetic acid (EDTA): 40 mL of 0.25% trypsin is used for isolating mEFs, while 0.05% trypsin/EDTA solution is used for passaging cells.

6. Sterile stir bar and stir plate. Sterilize by allowing 75% ethanol to evaporate from their surfaces.

7. Autoclaved 100-mL Erlenmeyer flask (for use with one to two mice for isolating mEFs).

8. Autoclaved aluminum foil (small piece to cover Erlenmeyer flask).

9. Sterile deionized water (dH$_2$O).

10. Phosphate-buffered saline (PBS) without Mg^{2+} and Ca^{2+} (2,600 mL dH$_2$O, 8.3 mg NaH$_2$PO$_4$·H$_2$O, 28.4 mg Na$_2$HPO$_4$, pH = 7.4). Autoclave and store at 4°C.

11. 0.2% gelatin (i.e. Sigma, tissue culture grade) in PBS.

12. T75-cm^2 tissue culture flasks: When isolating mEFs, generally 4–5 T75-cm^2 flasks are needed for each pregnant mouse. When passaging mEFs or mESCs, we normally use T25-cm^2 flasks.

13. One or two pregnant female mice: for isolating mEFs, mice should be used between day 12.5 and 13.5 of pregnancy (day 1 = first day after mating) (see Note 1).

14. Autoclaved paper towels.

15. Sterile dissecting tools (scissors, two pairs of forceps, scalpel). Sterilize by autoclaving.

16. Sterile 60-mm diameter petri dishes (i.e. Falcon, Fisher Scientific).

17. DNase (100 μg/mL) (i.e. Sigma, 10 mg/mL at 10,000 total units).

18. Heat-inactivated and filtered fetal bovine serum (FBS): thaw FBS in a 37°C water bath (if frozen) until serum is liquid. Gently shake and then incubate at 56°C with whole volume of serum immersed in water bath for 30 min (do not immerse cap).

Swirl every 5–10 min and when done, cool in an ice bath. Do not overheat. Serum should be filtered (0.2-μm filter), aliquoted, and stored frozen at –20°C until needed.

19. Stericup filter flask (0.22 μm, 150 mL or larger depending on volume of medium being made).

20. mEF medium: 500 mL of high-glucose Dulbecco's Modified Eagle's Medium (DMEM), 6 mL of 200 mM L-glutamine (Invitrogen), 6 mL of penicillin/streptomycin (5,000 units of penicillin + 5 mg/mL streptomycin), and 55 mL of heat-inactivated FBS. After combining ingredients, sterilize using a 0.22-μm filter flask and store at 4°C for up to 2 weeks. Medium can also be made in smaller batches.

21. 100-μm cell strainer (BD Falcon, 100 μm nylon).

22. Sterile 50-mL conical tubes.

23. Inverted microscope (phase contrast or Hoffman modulation).

2.2.2. Freezing and Passaging mEFs

1. Ethanol spray bottle (75%). All items going into the sterile hoods should be sprayed with ethanol, which will help kill microorganisms upon evaporation.

2. Cryovials (1.8 mL).

3. mEF freezing medium: 20% dimethyl sulfoxide (DMSO) and 80% mEF medium (see Note 2).

4. 0.25% trypsin/EDTA (Invitrogen).

5. mEF medium (see above).

6. Sterile 50-mL conical tubes.

7. T25-cm² Nunc flasks (Fisher).

8. 0.2% Gelatin (i.e. Sigma, tissue culture grade) in PBS (see Note 3).

9. Sterile 15-mL conical tubes.

10. Pipettes (P1000, P200, P20) and sterile tips appropriate for each pipette.

2.2.3. Preparing mEF Feeder Layers for Subsequent Culture of mESCs

1. mEFs Isolated from these animals or mEFs purchased from commercial sources (e.g. ATCC, StemCell Technologies) which will be used to make feeder layers for mESCs and hESCs.

2. 0.2% Gelatin (i.e. Sigma, tissue culture grade) in PBS.

3. Ethanol spray bottle (75%). All items going into the sterile hoods should be sprayed with ethanol, which will help kill microorganisms upon evaporation.

4. mEF medium (see above).

5. Sterile 50-mL conical tubes.

6. T25-cm² Nunc flasks (Fisher).

7. Sterile 15-mL conical tubes.

8. Pipettes (P1000, P200, P20) and sterile tips appropriate for each pipette.

9. Radiation source that can produce 8,000 rads.

10. Leukemia inhibitory factor (LIF, Millipore), see Note 4.

11. Low LIF mESC medium: 125 mL DMEM, 22.5 mL FBS (various sources), 1.5 mL of 100 mM sodium pyruvate (Invitrogen), 1.5 mL of 100× nonessential amino acids (ATCC, catalog no. 30-2116), 1.5 mL of 200 mM L-glutamine, 750 μL penicillin/streptomycin (5,000 units of penicillin + 5mg/mL streptomycin), 1 μL of 2-mercaptoethanol (99%, tissue culture grade, Sigma, Catalog no. M7522), and 0.4 μL of LIF from the stock bottle (10^6 units/mL).

12. Mitomycin C: reconstitute mitomycin C in PBS (without Mg and Ca) and 5% DMSO to reach a final concentration of 1 mg/mL (e.g. 0.1 mL DMSO + 1.9 mL PBS + 2 mg mitomycin C). Mix ingredients carefully until all of mitomycin C powder has dissolved (see Notes 5 and 6).

13. mEF medium + mitomycin C: for each T25-cm^2 flask of mEFs, 4 mL of mitomycin C containing medium is used. The final concentration of mitomycin C is 10 μg/mL (e.g. for every 10 mL of mEF medium, add 100 μL of aliquoted mitomycin C solution).

14. Waste beaker (500 mL).

15. 0.05% trypsin/EDTA solution.

16. Sterile 15-mL conical tubes.

2.3. Culturing Murine Embryonic Stem Cells

1. Mitotically inactivated mEF feeder layers growing in T25-cm^2 flasks.

2. Murine embryonic stem cells (e.g. the D3 line from ATCC).

3. Low LIF mESC medium (see above).

4. Ethanol spray bottle (75%).

5. Sterile 15-mL conical tubes.

6. Pipettes (P1000, P200, P20) and sterile tips appropriate for each pipette.

7. mESC freezing medium: mESC medium plus10% FBS, and 10% DMSO. For one T25-cm^2 flask, 4 mL of medium should be sufficient. (The amount can be scaled up if more than one flask is used or if a larger flask is used).

8. 0.2% gelatin (i.e. Sigma, tissue culture grade) in PBS.

9. PBS without Mg^{2+} and Ca^{2+} (see above).

10. 0.05% trypsin/EDTA solution.

11. High LIF mESC medium: as described above for low LIF mESC medium, except for LIF, make a small amount of working solution by diluting stock solution (10^6 units/mL) tenfold, and add 1 μL of working solution for each mL of high LIF mESC medium (e.g. use 10 μL of working solution for 10 mL of mESC medium).

2.4. Culturing Human Embryonic Stem Cells

1. 0.05% trypsin/EDTA solution.

2. Gelatin-coated T25-cm^2 tissue culture flasks and six-well plates (see Subheading 3.1.1, steps 5 and 23).

3. hESCs (e.g. H9 line from WiCell). This could be a frozen vial or live culture.

4. Human basic fibroblast growth factor (hbFGF, i.e. Peprotech): Stock solution is made by dissolving 10 μg of hbFGF in 5 mL of PBS with 0.1% BSA (Fraction V). hbFGF is very sticky, so when using, prewet all pipette tips, tubes, and the filter with PBS + 0.1% BSA. Stock solution should be aliquoted (250 μL) and stored at –20°C (short term) or at –80°C (long term). Once thawed, aliquots can be stored at 4°C for a month.

5. hESC medium: 400 mL DMEM/F-12, 100 mL Knockout Serum Replacement (KSR, Invitrogen), 5 mL nonessential amino acids (NEAA, Invitrogen, catalog no. 11140-050), 5 mL L-glutamine-2-mercaptoethanol mix (7 μL 2-mercaptoethanol plus 5 mL of 200 mM L-glutamine), and 1 mL of hbFGF (10 μg/5 mL of PBS with 01% BSA) to a final concentration of 4 ng/mL. Filter through a 0.22-μm filter and store at 4°C (it is stable for 2 weeks). It can be made in smaller batches.

6. Ethanol spray bottle (75%).

7. Sterile 15-mL conical tubes.

8. Matrigel™.

9. DMEM/F-12.

10. mTeSR culture medium kit (includes basal medium plus supplements) (StemCell Technologies). Medium is made by combining 400 mL of mTeSR basal medium with 100 mL of mTeSR supplement. Complete medium can be made in smaller aliquots. Complete medium is stable at 4°C for 2 weeks.

11. PBS without Mg^{2+} and Ca^{2+} (see above).

12. Accutase (eBioscience) or collagenase IV or trypsin/EDTA (0.05% trypsin, 0.53 mM EDTA): Collagenase IV is made by warming DMEM/F12 in a 37°C water bath, then dissolving 0.01 mg of collagenase per mL of DMEM/F12. The solution is passed through a 0.22-μm filter before use.

13. Glass beads (3 mm diameter, i.e. Fisher Scientific). To prepare beads, place a bottle of glass beads in a beaker (beaker should be 1/3 to 1/2 full). Rinse the beads three times with dH_2O, then cover the beads with 10 N HCl and soak overnight at room temperature. Add an equal volume of 10 N NaOH (carefully) to neutralize HCl and then run dH_2O over the beads overnight. The next day, remove all water from the beaker and wash four times with dH_2O. Dry off the glass beads with an autoclaved paper towel or in a drying oven and put the beads back in beaker and cover with aluminum foil. Autoclave the beaker/beads and glass test tubes with plastic snap on caps. Allow the beads to dry overnight and then aliquot the beads into the glass test tubes. Autoclave the test tubes with the beads and allow them to dry overnight, after which they will be ready for use in passaging hESCs.

14. Sterilized inverted light microscope: Sterilize with 75% ethanol and UV light in hood.

15. Sterile scalpel.

16. Sterile 10-mL Pasteur pipettes.

---

3. Methods

3.1. Culturing mEFs

3.1.1. Procedure for Isolating mEFs

1. Place 75% ethanol-sterilized dissection microscope in the sterile hood.

2. Wipe down all bench tops that will be used with 75% ethanol.

3. Place 40 mL of 0.25% trypsin and a sterile stir bar in the Erlenmeyer flask, cap the flask with autoclaved aluminum foil, and preheat trypsin solution in a 37°C water bath.

4. Preheat 40 mL of mEF medium in a 50-mL conical tube in the 37°C water bath.

5. Coat T75-cm² culture flasks with 6 mL of 0.2% gelatin (see Note 7) and incubate at 37°C for a minimum of 15 min or until needed (see Note 3).

6. Sacrifice female mice (one at a time) using CO_2 gas.

7. Place sacrificed female mouse on her back on autoclaved paper towels.

8. Spray the mouse with 75% ethanol.

9. Open the peritoneal cavity by making a Y-incision (Fig. 1).

10. Dissect out embryos from the uterus and remove all tissue surrounding each embryo using sterile forceps.

11. Carefully transfer dissected embryos into 60-mm petri dishes containing sterile PBS.

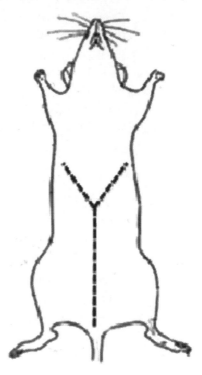

Fig. 1. Diagram showing where to make incisions on the ventral surface of a mouse during isolation of mEFs.

12. Rinse embryos in clean PBS twice and place them in new dish with fresh PBS for dissection.

13. Remove the head and internal organs from the embryos carefully with sterile forceps and scissors.

14. Place all dissected embryos in a fresh 60-mm dish of PBS and cut tissue into approximately 1 mm pieces with the scalpel.

15. Transfer all tissue pieces into the sterile Erlenmeyer flask containing 40 mL of preheated (37°C) 0.25% trypsin/EDTA solution and sterile stir bar.

16. Stir cell solution for 40 min at room temperature on a stirring plate.

17. Observe solution periodically. Add 200 µL of 100 µg/mL DNase if solution appears viscous and clumpy (add additional DNase in 200 µL increments if necessary).

18. After 40 min, add 40 mL of prewarmed (37°C) mEF medium and swirl the solution gently.

19. Strain the solution through a 100 µm cell strainer into a 50-mL conical tube in the sterile hood. Repeat this procedure twice. Use a new strainer each time.

20. Centrifuge the cell suspension in the 50-mL conical tube at $270 \times g$ for 4 min.

21. Decant supernatant into a fresh sterile 50-mL conical tube (save the supernatant in case there are not enough cells for plating).

22. Resuspend the pellet using fresh mEF medium (1 mL for each T75-cm² flask, e.g. if using four flasks, break the pellet with 4 mL of medium).

23. Aspirate excess gelatin out of T75-cm² flasks and replace it with 8 mL/flask of mEF medium.

24. Add 1 mL of cell suspension to each T75-cm² flask and rock the flask back and forth gently to distribute cells evenly across the bottom of the flask.

25. Observe cells with the inverted microscope to make sure that they look normal, for example, round with smooth surfaces and not apoptotic (see Note 8).

26. Incubate cells in the 37°C incubator.

27. Change medium after 24 h of plating and thereafter on alternate days.

28. Allow cells to reach 90–95% confluency (which should take 3–4 days) before passaging or freezing (Fig. 2b) (see Notes 9 and 10).

3.1.2. Freezing and Storing Passage 1 mEFs

1. Label cryovials with cell line, passage number, date, and initials (see Note 11).

2. Aspirate mEF medium and wash the cells twice with 5 mL of room temperature PBS.

3. Aspirate PBS and add 5 mL of 0.05% trypsin/EDTA to each T75-cm² flask and incubate at 37°C for 1 min.

4. Remove flasks from incubator and gently tap the sides of each flask. Do not leave cells in trypsin for more than 5 min (see Note 12).

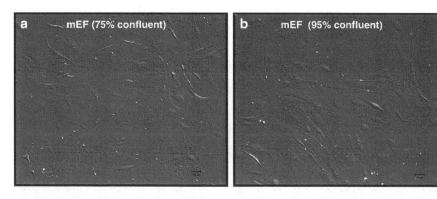

Fig. 2. mEFs that have been plated and inactivated using mitomycin C at 75% confluency (**a**) and 95% confluency (**b**). These cells are now ready to use as feeder layers for culturing mESCs (**b**) or hESCs, respectively (**a**). Note that the density of the cells is greater when they are prepared for mESCs.

5. Add 10 mL of mEF medium to each flask and mix gently. Amount of mEF medium added should be double the amount of trypsin solution that is used.

6. Aspirate the cell suspension from each flask and transfer it to one sterile 50-mL conical tube.

7. Centrifuge the cells at $270 \times g$ for 4 min and then decant off the supernatant.

8. Break the pellet with 5 mL of freezing medium and then add the remaining 5 mL of the freezing medium and mix by gently inverting the tube (e.g. for five T75-cm² flasks, the total volume of freezing medium will be 50 mL).

9. Add 1 mL of the cell suspension in freezing medium to each sterile cryovial.

10. Transfer the cryovials to a –80°C freezer and leave it overnight.

11. Remove cryovials from freezer, place on dry ice, and transfer to liquid nitrogen for long-term storage (see Note 13).

3.1.3. Preparing and Freezing Passage 2 mEFs

1. It is convenient to grow and freeze a number of passage 2 vials of mEFs for eventual use as feeder layers (see Notes 14 and 15).

2. Coat 3–4 T25-cm² flasks with 0.2% gelatin and incubate for at least 15 min.

3. Get a passage 1 vial of mEFs from the liquid nitrogen storage and bring the vial to a 37°C water bath on dry ice (see Note 16).

4. Thaw the cryovial in the water bath for no longer than 1.5 min or until a small ice crystal is left (see Note 17).

5. Spray the cryovial with 75% alcohol and transfer it to the sterile hood. Allow the alcohol to evaporate off before opening the vial.

6. Place 5 mL of fresh mEF medium into a 15-mL conical tube.

7. Transfer thawed cells into the conical tube using the P1000 pipette.

8. Rinse cryovial with 1 mL of mEF medium and transfer to the conical vial.

9. Cap the tube and centrifuge for 4 min at $270 \times g$. Spray the tube with 75% ethanol before transferring it to the sterile hood.

10. Decant supernatant and break the pellet with 3–4 mL of fresh mEF medium. Use 1 mL of mEF medium per T25-cm² flask.

11. Aspirate gelatin from the T25-cm² flasks.

12. Add 3 mL of fresh mEF medium into each T25-cm² flask.

13. Plate 1 mL of mEFs into each T25-cm² flask, rock flasks gently to distribute cells, and incubate at 37°C. Observe flasks after 20 min to verify that cells are attaching.

14. Change the medium 24 h after plating and then change the medium on alternate days.

15. When 90–95% confluency is reached, a flask may be used to prepare the feeder layer (Fig. 2b). Cells in the remaining flasks can be frozen as has been done for passage 1 cells (Subheading 3.1.2).

3.1.4. Preparing mEF Feeder Layers for Subsequent Culture of mESCs

1. Coat T25-cm² flasks with 0.2% gelatin and incubate for at least 15 min (use two flasks for each vial of mEFs).

2. Get stock vial of passage 2 mEFs from liquid nitrogen storage or use a T25-cm² flask containing a live culture of passage 2 mEFs (see Subheading 3.1.3 and Notes 14–16).

3. If frozen, thaw cells in the 37°C water bath for 1.5 min or until only a small crystal of ice remains. Keep cap above water during thawing.

4. Before transferring to the sterile hood, spray vial with 75% ethanol.

5. In a sterile hood, transfer thawed cells (1 mL) to a 15-mL conical tube containing 5 mL of mEF medium using a P1000 pipette. Rinse cyrovial with 1 mL of mEF medium and add this to the conical tube.

6. Centrifuge at $270 \times g$ for 4 min to create a loose pellet of cells.

7. Decant supernatant from the conical tube and gently break the pellet with 1–2 mL of mEF medium per pellet.

8. Aspirate excess gelatin from the T25-cm² flasks.

9. Pipette 0.5–1.0 mL of mEF medium containing the cell suspension into each of two T25-cm² flasks containing 3 mL of fresh medium and transfer flasks to the incubator.

10. After about 15–20 min, check to be sure that mEFs are attaching to the bottom of the flasks, using an inverted microscope. Then, observe the flasks daily for cell growth, normal appearance, and absence of contamination.

11. Cells will double about every 24 h. For use with mESCs, let mEFs grow until they are about 90–95% confluent.

12. mEF medium should be changed every other day.

3.1.5. Mitotic Inactivation of mEFs Using Irradiation

1. To prevent mEFs from dividing when stem cells are plated on them, they must be inactivated with either irradiation (for example cesium exposure) or mitomycin C (see Note 18).

2. To irradiate mEFs, place flasks in a cesium source irradiator and irradiate at 8,000 rads for approximately 2 h. It may be necessary to empirically determine the length of exposure and dosage for your particular instrument.

3. Remove flasks promptly and return them to the cell culture room.

4. Upon return to the lab, immediately replace mEF medium with 4 mL of room temperature low LIF mESC medium and incubate at 37°C for 30–60 min, after which mEFs can be used as feeder layers (see Notes 19 and 20).

3.1.6. Mitotic Inactivation of mEFs Using Mitomycin C

1. If an irradiator is not available, mEFs can be inactivated using mitomycin C.

2. Label a 50-mL conical vial to dispose of any waste containing mitomycin C.

3. Aspirate mEF medium out of T25-cm² flasks containing mEFs, add 4 mL of mEF medium + mitomycin C into each flask, and incubate at 37°C for 2–2.5 h.

4. For each T25-cm² flask of mitomycin-treated mEFs, coat one T25-cm² flask with gelatin and incubate for at least 15 min.

5. After treatment, aspirate the mitomycin C-containing mEF medium out of the T25-cm² flask (see Note 5).

6. Wash mitomycin C-treated mEFs with 5 mL of PBS for three times and discard the wash into the waste bottle containing mitomycin C medium.

7. Add 2 mL of 0.05% trypsin/EDTA to each T25-cm² flask and incubate at 37°C for 1 min.

8. Remove the flasks from the incubator and gently tap the sides of the flasks to loosen the cells. Check the flasks using the inverted microscope periodically to make sure that mEFs detach from the bottom of the flask. Do not leave mEFs in trypsin for more than 5 min (see Note 12).

9. Add 4 mL of mEF medium to each flask to inactivate the trypsin.

10. Transfer the 6 mL of cell solution from each flask into a single sterile 15-mL conical tube and cap the tube.

11. Centrifuge the cell solution at $270 \times g$ for 4 min. Spray conical tube with 75% ethanol before returning it to the sterile hood.

12. Decant the supernatant and gently break pellet with 1–2 mL of fresh mEF medium by pipetting repeatedly with a P1000 pipette. Use 1 mL of mEF medium per flask being plated.

13. Aspirate excess gelatin in T25-cm² flasks and add 3 mL of mEF medium to each flask.

Table 1
Schedule for mEF preparation

Day of week	Action
Monday	Plate passage 2 mEFs on gelatin coated T25-cm² flasks
Tuesday	Change medium for newly plated mEFs
Wednesday	Observe mEFs using a inverted light microscope for confluency and contamination
Thursday	Mitomycin C-treat or irradiate mEFs. Irradiated mEFs can be used on Thursday for plating mESCs
Friday	Mitomycin C-inactivated mEFs can be used for plating ESCs; change medium if not used

14. Plate mitomycin C-treated mEFs on two fresh 0.2% gelatin-coated T25-cm² flasks. Observe plated cells after 20 min (see Note 21).

15. mEFs freshly treated with mitomycin C can be used as feeders after overnight incubation (see Note 22).

16. mEF medium should be changed 24 h after mitomycin C treatment and every other day thereafter.

3.1.7. Schedule for Preparing mEFs

1. It is a good idea to set up a schedule for harvesting mEFs so that you will have them ready when needed for your stem cells. An example of a possible schedule is given in Table 1.

2. All mitotically inactivated mEFs should be fed on alternative days, and they are good for 2 weeks.

3.2. Culturing mESCs

3.2.1. Thawing and Expanding mESCs on mEFs

1. Prepare mEF feeder layer as described in Subheadings 3.1.4–3.1.6.

2. Aspirate mEF medium off of mEFs and add fresh low LIF mESC medium (4 mL to a T25-cm² flask) and then incubate for at least 30–60 min (see Note 23).

3. Get a vial of frozen mESCs from liquid nitrogen storage and transport it on dry ice to the cell culture room.

4. Immediately put the frozen vial in the 37°C water bath without submerging the cap and thaw for no more than 90 s or until a small ice crystal is left.

5. Remove vial and spray down with 75% ethanol before placing it in the sterile hood.

6. Put 5 mL of fresh low LIF mESC medium in a 15-mL conical tube.

7. Transfer the mESCs into the conical tube using the P1000 pipette and then wash the vial with 1 mL of low LIF mESC medium and add wash to the tube. Cap the tube before removing it from the hood.

8. Spin down cells in centrifuge at $270 \times g$ for 4 min and then spray the conical tube with 75% ethanol before returning it to the hood.

9. Decant the supernatant into the waste beaker and add 1 mL of low LIF mESC medium to the tube. Pipette gently with the P1000 pipette to break the pellet. Be sure that the cells are fully dispersed, but do not pipette too hard or else the cells may be damaged.

10. Set the cells aside and aspirate the low LIF mESC medium out of the T25-cm² flask and replace it with fresh low LIF mESC medium (4 mL).

11. Add 1 mL of mESC suspension to the T25-cm² flask using a P1000 and then rock the solution in the flask back and forth gently to evenly distribute the cells. Observe using the inverted microscope and be sure that the cells appear normal and are not apoptotic.

12. Incubate at 37°C and check the next day for evidence of mESC attachment.

13. Observe cultures every day to be sure that there is no evidence of contamination or differentiation (see Note 24).

14. Change the mESC medium every day (replace 5 mL of old medium with 5 mL of fresh low LIF mESC medium).

15. For vials that are freshly thawed, it will generally take 3–4 days for mESCs to become 50–60% confluent (adjacent colonies should not be touching each other).

16. Figure 3a and b show examples of pluripotent mESCs growing in colonies. These colonies are round, three-dimensional, and have defined edges. Colonies in Fig. 3b have been labeled to show alkaline phosphatase activity, a marker for pluripotency. Figure 3c and d, in contrast, show mESC colonies in which differentiation has begun. Colonies are flatter, and cells (arrow) have begun to migrate out of a colony.

3.2.2. Passaging and Freezing mESCs Using mEFs

1. Aspirate mEF medium from a new flask of mitotically inactivated mEFs, add 4 mL of low LIF mESC medium to the flask, and incubate at 37°C for 30–60 min.

2. Coat a 60-mm tissue culture dish with 0.2% gelatin for at least 15 min and incubate at 37°C until ready for use.

3. Aspirate the low LIF mESC medium out of the T25-cm² flask containing mESCs.

4. Add 5 mL of PBS to wash any serum out of the flask.

5. Aspirate out PBS and add 2 mL of 0.05% trypsin/EDTA to the flask with mESCs, then incubate for 1 min at 37°C (see Note 12).

6. Remove the flask from the incubator and gently tap its sides to loosen the cells. Check using the inverted microscope to be sure that the cells have detached.

7. Add 4 mL of low LIF mESC medium to the flask with mESCs and rock back and forth (flask contains a total of 6 mL).

8. Aspirate out 1 mL of solution containing mESCs using a P1000 pipette and transfer to a 15-mL conical tube and cap tube.

9. Aspirate out the remaining 5 mL of solution containing mESCs and transfer to a second 15-mL conical tube and cap tube.

10. Centrifuge the two conical tubes for 4 min at $270 \times g$. Spray down with 75% ethanol before placing tubes back in the hood.

11. Decant medium from conical tube containing 5 mL and gently break the pellet with 4 mL of freezing medium.

12. Add 1 mL of cell suspension in freezing medium per cyrovial (four vials total for one T25-cm² flask).

13. Transfer to a –80°C freezer for 24 h and then transfer to liquid nitrogen (use dry ice to carry vials from freezer to liquid-nitrogen tank).

14. Decant the medium from conical tube containing 1 mL of solution.

15. Break the pellet gently with 1 mL of low LIF mESC medium and set aside in the hood.

16. Aspirate out excess gelatin from a 60-mm culture dish and add 3 mL of fresh low LIF mESC medium.

17. Add 1 mL of the mESC suspension to the 60-mm culture dish using a P1000 pipette and mix solution carefully with the pipette.

18. Place the 60-mm culture dish in the incubator and wait 20–30 min to allow mEFs to attach (see Note 25).

19. After 20–30 min, aspirate the cell suspension containing mESCs from the 60-mm dish and transfer to a 15-mL conical tube and cap tube (see Note 26).

20. Centrifuge the cell suspension for 4 min at $270 \times g$.

21. While centrifuging, change the medium in the T25-cm² flask containing mEFs by replacing 4 mL of mESC medium with an equal amount of low LIF mESC medium. Spray the conical tube before returning it to the hood.

22. Decant the medium from the conical tube and break the pellet using 1 mL of low LIF mESC medium.

23. Transfer all of the cell suspension to a T25-cm^2 flask containing mEFs in fresh low LIF medium and rock the flask back and forth to distribute the cells.

24. Incubate the flasks at 37°C.

25. Routinely check and change medium everyday as described in Subheading 3.2.1.

3.2.3. Plating mESCs on Gelatin

1. Add 3 mL of 0.2% gelatin to T25-cm^2 flasks and incubate at 37°C for at least 15 min (one T25-cm^2 flask of mESCs can be passaged to three new flasks).

2. Decant medium in the T25-cm^2 flask containing mESCs and wash the flask using 5 mL of PBS.

3. Aspirate PBS and add 2 mL of 0.05% trypsin/EDTA to the T25-cm^2 flask and incubate for 1 min at 37°C.

4. Remove the flask from the incubator and the tap sides of the flask gently. Check cells with the inverted light microscope to see if all mESCs have detached.

5. Once all mESCs have detached (which should be no longer than 5 min), add 4 mL of high LIF mESC medium to neutralize the trypsin (see Notes 12 and 27).

6. Aspirate cell suspension and transfer to a sterile 15-mL conical tube and cap tube.

7. Centrifuge mESCs at $270 \times g$ for 4 min.

8. While centrifuging, remove gelatin-coated flasks from incubator and decant excess gelatin in sterile hood.

9. Add 3 mL of high LIF mESC medium to the flask and set in hood (see Note 27).

10. Decant supernatant and break pellet gently with 3 mL of high LIF mESC medium.

11. Add 1 mL of cell suspension to each flask and rock flasks back and forth for even distribution. Observe cells using an inverted light microscope for normal cell morphology (round, no apoptosis).

12. Incubate mESCs at 37°C and change medium every day.

13. mESCs should be ready for passaging once the flask reaches 70–75% confluency (Fig. 2a).

14. Figure 3e shows pluripotent mESCs plated on gelatin. The colony is flat and has the characteristic cobblestone appearance of mESCs on gelatin. Figure 3f shows mESCs that are starting to differentiate on gelatin. Cells have begun migrating out of the colony.

15. Once a frozen vial of mESCs has been expanded and passaged to produce sufficient frozen stock, experiments can be set up with the mESCs.

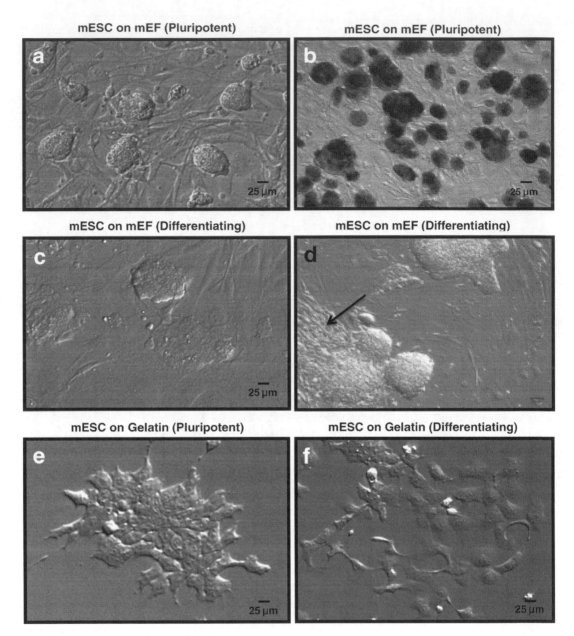

Fig. 3. Well-formed colonies of mESCs (**a**, **b**) growing on mEFs. The edges of the colonies are sharp, and colonies are round and three-dimensional, indicative of pluripotency. The colonies in (**b**) appear *dark* because they have been stained for alkaline phosphatase activity which shows the sharply defined edges of the colonies and indicates their pluripotent nature. mESC colonies that have begun differentiating on mEFs are shown in (**c**) and (**d**). In (**c**), the colonies have become flat, an early indicator of differentiation, while in (**d**), the colony at the lower left has lost its sharp edges, and differentiating cells are moving away from the colony (*arrow*). mESC colonies appear cobblestoned when grown on gelatin (**e**). When mESC differentiate on gelatin, they move away from the colony (**f**).

3.3. Culturing Human Embryonic Stem Cells

3.3.1. Prepping mEFs for hESC Culture Using Mitomycin C

1. Plate a vial of frozen mEFs (passage 2) in a gelatin-coated T25-cm² flask (see Subheading 3.1.1 and Note 14).

2. Allow the mEFs to grow to 95–100% confluency (Fig. 2b).

3. Mitomycin C-treat the T25-cm² flask of mEFs (see Subheading 3.1.6).

4. After treatment, wash the flask twice with PBS and then remove the mEFs using trypsin/EDTA (see Subheading 3.1.6).

5. Centrifuge the cell suspension, remove the supernatant, and break pellet with fresh mEF medium as described in Subheading 3.1.6.

6. One T25-cm² flask of mitomycin C-treated mEFs can be replated into four to five wells of a six-well plate (this will give about 70% confluency in each well) (Fig. 2a).

7. Alternatively, the mitomycin C-treated mEFs can be placed back into a T25-cm² flask if desired.

8. Put the plate or flask in the incubator for 24 h (check after 20 min to be sure that cells begin to attach).

9. After 24 h, the mEFs are ready to be used for plating hESCs.

3.3.2. Prepping mEFs for hESC Culture Using Irradiation

1. Plate two vials of frozen passage 2 mEFs in a gelatin-coated six-well plate. Two vials contain enough cells for six wells.

2. Allow mEFs to grow to 70–75% confluency in the incubator (Fig. 2a).

3. Irradiate as described in Subheading 3.1.5.

3.3.3. Thawing and Plating hESCs on mEFs

1. For the mitomycin C-treated or the irradiated mEFs, replace mEF medium with hESC medium and incubate for 30–60 min at 37°C. Use 1 mL for each well of a six-well plate or 3 mL into a T25-cm² flask.

2. Obtain a frozen vial of hESCs from liquid-nitrogen storage (see Notes 16 and 28). One frozen vial of hESCs is usually plated in one well of a six-well plate or in one T25-cm² flask.

3. Thaw the vial in 37°C water bath for no more than 1.5 min or until a small crystal is left (do not submerge the cap). Spray vial with 75% ethanol before transferring to sterile hood.

4. Put 5 mL of fresh hESC medium into a 15-mL conical tube.

5. Add hESCs to the conical tube dropwise by running drops down the sides of the tube.

6. Centrifuge the tube at $200 \times g$ for 3 min at room temperature. Spray the tube with 75% ethanol before returning it to the sterile hood.

7. Decant the supernatant and gently break the pellet using 1 mL of hESC medium.

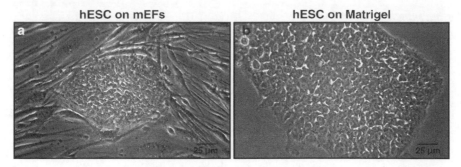

hESC on mEFs   **hESC on Matrigel**

Fig. 4. Examples of hESCs growing on mEFs (**a**) and Matrigel (**b**). In both cases, colonies appear flatter than the mESC colonies. hESCs grow between the mEFS, while mESCs grow on mEFs. Note that the edges of the colonies are sharply defined.

8. Replace the hESC medium in the six-well plate (1 mL) or flask (3 mL) with fresh hESC medium.

9. Add the 1 mL of hESC suspension dropwise into one well of a six-well plate or one T25-cm² flask.

10. Routinely check and change medium every day.

11. Figure 4a shows hESCs growing on mEFs. The hESCs actually grow in between the mEFs and form a colony. The colony is flat and has a cobblestone appearance and well-defined edges.

3.3.4. Preparing Matrigel Plates for hESCs

1. Thaw a 5-mL bottle of stock Matrigel in the refrigerator overnight. Do not thaw at room temperature because it may clump and not be usable.

2. Add 5 mL of DMEM/F12 (at 4°C) to the Matrigel bottle and mix thoroughly by pipetting up and down gently (avoid formation of bubbles at the surface of the solution).

3. Aliquot 500 μL of Matrigel solution to twenty 15-mL conical tubes. Aliquots that will not be used should be stored at –20°C. When using a frozen aliquot, it should be thawed for 1–2 h in the refrigerator (4°C) before use.

4. When making Matrigel plates, add 7 mL of DMEM/F12 (at 4°C) to a 500-μL aliquot of Matrigel prepared as described above and pipette gently without making bubbles.

5. Add 1 mL of Matrigel working solution to each well in a six-well plate and rock to evenly distribute the solution.

6. Leave the plate at room temperature for 1–2 h or leave it in the refrigerator overnight (see Note 29).

3.3.5. Thawing and Plating hESCs on Matrigel

1. One frozen vial of hESCs is usually plated in one well of a six-well plate or in one T25-cm² flask.

2. Thaw a vial of hESCs at 37°C for 1.5 min in a water bath or until a small ice crystal is left.

3. Add 5 mL of mTeSR to a 15-mL conical tube.

4. Transfer the cells in the thawed vial to the conical tube by adding cells dropwise to the sides of the tube and cap the tube.

5. Centrifuge for 3 min at $200 \times g$ and spray the tube with 75% ethanol before placing it in the sterile hood.

6. Decant the supernatant from the tube and gently break the pellet using 1 mL of mTeSR medium. Do not pipette too many times; cells should remain in colonies.

7. Aspirate out excess Matrigel from the six-well plate and add 1 mL of fresh mTeSR to each Matrigel-coated well.

8. Add 1 mL of cell suspension to the Matrigel-coated well dropwise, keeping drops near the surface of the well.

9. Incubate the plate at 37°C overnight.

10. Change the mTeSR medium every day.

11. Passage when hESCs are 70–75% confluent.

12. Figure 4b shows a hESC colony plated on Matrigel. The colony is flat with a cobblestone appearance. The cells are tightly joined to each other in a colony.

3.3.6. Passaging hESC on Either mEFs or Matrigel-Coated Dishes with Glass Beads

1. If plating hESC on mEFs, remove mEF medium and replace with 1 mL of hESC medium and incubate at 37°C for 30–60 min before using. If plating on Matrigel, prepare Matrigel-coated dishes 1–2 h before they are needed.

2. Wash hESCs with 1 mL of PBS per well and then aspirate out PBS.

3. Add 1 mL of Accutase or collagenase IV to each well and incubate at 37°C for 1 min (see Note 30).

4. Remove plate from incubator and observe with an inverted microscope. Edges of the colonies will start to curl.

5. Add 10–12 sterile glass beads to each well (see Notes 28 and 29).

6. Rock plate back and forth gently. Observe the plate from the bottom. hESC colonies should become detached and float in solution.

7. Do not leave hESC in Accutase or collagenase for more than 3 min.

8. If plating on mEFs, add 2 mL of hESC medium to the well. If plating on Matrigel, add 2 mL of mTeSR medium to the well.

9. Aspirate the cell suspension and transfer to a 15-mL conical tube and cap the tube.

10. Centrifuge the cell suspension at $200 \times g$ for 3 min. Spray the conical tube with 75% ethanol before placing in the hood.

11. Decant supernatant and gently break the pellet using either hESC medium (for mEF plating) or mTeSR medium (for

Matrigel plating). If two wells of hESCs will be passaged into six new wells, then break pellet with 3 mL of medium, or 500 µL per well.

12. Do not pipette too many times to prevent colonies from breaking up (hESCs survive better in colonies and not as single cells).

13. If plating onto mEFs, aspirate hESC medium from the wells and add 1 mL of fresh medium. If plating on Matrigel, aspirate out Matrigel from the wells and add 1 mL of mTeSR medium.

14. Replate hESCs by adding 500 µL of cell suspension dropwise to each well. Add drops right above the surface of the solution. Rock the plate in all four directions gently.

15. Clumps of hESCs should be observed using the inverted microscope.

16. Incubate the cells at 37°C overnight and change the medium (hESC medium or mTeSR) every day.

17. Passage again when colonies reach 70–75% confluency.

18. Once a frozen vial of hESCs has been expanded and passaged to produce sufficient frozen stock, experiments can be set up with the hESCs.

3.3.7. "Cut and Paste" Passaging

1. Instead of using the beads, the "cut and paste" method of passaging may be used. For this alternative, place a sterilized inverted light microscope in the sterile culture hood.

2. Place a six-well plate of hESCs on the stage of the microscope and open the lid.

3. Observe hESC colonies and choose those that look pluripotent and healthy (Fig. 4a, b).

4. Use a scalpel to gently cut the colonies into quadrants (Fig. 5a).

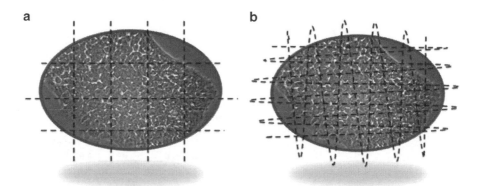

Fig. 5. Diagrams showing how to score colonies for passaging with the cut-and-paste (**a**) and the mechanical (**b**) method.

5. Gently scrape off the desired portions of the colonies.

6. Once all the pieces have been mechanically removed from the plate, transfer the solution in each well to a 15-mL conical tube.

7. Centrifuge the cell suspension and break pellet with appropriate media (mTeSR if plating on Matrigel and hESC medium if plating on mEFs).

8. Replate hESC colonies as described in Subheading 3.3.6, steps 8–18.

3.3.8. Passaging hESCs Mechanically

1. Instead of using the beads, the mechanical method can also be used to passage cells. For this alternative, place the hESC plate in the sterile hood.

2. Add 1 mL of Accutase or collagenase IV to each hESC well and incubate for 1 min at 37°C.

3. Return the plate to the sterile hood.

4. Unwrap a 10-mL Pasteur pipette in sterile hood. Do not allow the tip to touch any surfaces.

5. Scrape the colonies off from the bottom of the wells in the directions shown in Fig. 5b. You should see colonies floating in medium.

6. Add 2 mL of hESC medium or mTeSR medium into each well.

7. Aspirate all cell suspensions into one 15 mL conical tube and cap tube.

8. Centrifuge the cell suspension at $200 \times g$ for 3 min.

9. Break pellet and replate hESCs on Matrigel or mEFs as described in Subheading 3.3.6, steps 8–18.

4. Notes

1. Although mEFs from various strains of mice have been used successfully in stem cell culture, we have found that mEFs from NIH Swiss mice work very well with both mESCs and hESCs.

2. Make 10 mL of freezing medium per T75-cm^2 flask of cells that will be frozen (e.g. for five flasks, make 50 mL of freezing medium).

3. When plating mEFs or mESCs, T25-cm^2 and T75-cm^2 flasks can be used with or without gelatin coating. We have found that mEFs and mESCs stick better and are healthier with a gelatin coating, and therefore we routinely coat flasks for mEF isolation or passaging.

4. LIF is expensive. It can be made in small batches so as not to waste it. The stock bottle is stored at 4°C. Check expiration date of each bottle.

5. Mitomycin C is a potent chemical that inhibits cell division. It should be handled carefully under a hood while wearing gloves. Be sure to discard mitomycin C waste in the appropriate toxic waste bin.

6. Aliquots of mitomycin C stock solution can be stored at −20°C for 6 months. The unconstituted powder is stable at −20°C for 1 year.

7. Cells from about two to three embryos can be plated on one T75-cm² flask.

8. Observe cells with the inverted microscope at 20 min after plating. If mEFs are healthy, they will have begun attaching by this time.

9. It normally takes the mEFs about 3–4 days to become 90% confluent. If it takes longer than this, the cells may be unhealthy.

10. We refer to these cells as passage 1 mEFs (some labs call this stage passage 0).

11. One 90–95% confluent T75-cm² flask of passage 1 mEFs will usually yield 10–12 frozen vials.

12. It is important not to overtrypsinize the cells. To determine if cells have completely dislodged, you can examine the flask carefully with naked eye or look at it with the inverted microscope. Do not leave mESCs in trypsin for more than 4–5 min. Staying in trypsin too long will reduce plating efficiency.

13. Although it is customary to store mEFs in liquid nitrogen, we have also stored mEFs at −80°C for up to 6 months with success.

14. Passages 3 or 4 are the best for use as feeder layers. We have found that younger or older passages do not work well.

15. Our frozen vials of mEFs normally contain $3–5 \times 10^6$ cells/vial.

16. It is important to keep cells frozen until you thaw them in the water bath. If necessary, they can be transferred from the liquid-nitrogen storage site to the water bath on dry ice.

17. Do not submerge the cap.

18. The use of a cesium source machine may require special training at your institution. Try to receive training and gain access to the cesium source before beginning your work.

19. Irradiation can also be done when cells are in 15-mL conical tubes before plating on flasks as feeders.

20. Once the mEFs have been mitotically inactivated, they are only good for 10 days.

21. mEFs should begin to attach by 15 min after plating.

22. Irradiation of cells in flasks has the advantage that cells are ready to use after the relatively short irradiation period. Mitomycin C-treated cells have the disadvantage of needing to be removed from their flasks and replated which requires an extra night of culture before they can be used. We have used both methods with success. If mitomycin C-treated mEFs have less than 70% confluency when replated, the mitomycin C treatment may have damaged some of the mEFs.

23. FBS is highly variable from batch to batch. Even embryonic-stem-cell-qualified batches may not support pluripotency well in ESC populations. It is important to screen batches of FBS to ascertain which are suitable for your work. Often companies will hold specific lots of FBS in reserve for you if you commit to purchase them in the future.

24. Be sure to also observe the morphology of the mEFs. If mEFs start to die or detach, mESCs will not be properly supported.

25. Check dish after 15 min to see if most mEFs have already attached; if so, the mEFs are healthy.

26. Do this step one time if passaging mESCs or two times if preparing mESCs for an experiment.

27. High LIF medium should be used when plating mESC on gelatin to keep them from differentiating.

28. A number of lines of hESCs are available for culture. Some lines require their own particular protocols. The protocol described here works well with the WiCell H9 line.

29. Matrigel plates are good for 2 weeks if refrigerated, although fresh plates are preferable.

30. Accutase will give mainly single cells, while collagenase IV will give mainly colonies.

Acknowledgments

We are grateful for support from the California Tobacco-Related Disease Research Program, the UCR Office of Research, the Honors Program, and the Academic Senate at UCR which helped fund establishment of these protocols in our lab. S. L. was supported in part by a Deans' Fellowship from the Graduate Division. We also thank Connie Martin for her helpful suggestions and comments on this manuscript.

References

1. Martin, G. R. (1981) Isolation of a pluripotent cell line from early mouse embryos cultured in medium conditioned by teratocarcinoma stem cells. *Proc. Natl. Acad. Sci. USA* **78**, 7634–7638.

2. Evans, M. J., and Kaufman, M. H. (1981) Establishment in culture of pluripotential cells from mouse embryos. *Nature* **292**, 154–156.

3. Tremml, G., Singer, M., and Malavarca, R. (2008) Culture of mouse embryonic stem cells. *Curr. Protoc. Stem Cell Biol.* Chapter 1, Unit 1C 4.

4. Downing, G. J., and Battey, J. F., Jr. (2004) Technical assessment of the first 20 years of research using mouse embryonic stem cell lines. *Stem Cells* **22**, 1168–1180.

5. Thomson, J. A., Itskovitz-Eldor, J., Shapiro, S. S., Waknitz, M. A., Swiergiel, J. J., Marshall, V. S., et al. (1998) Embryonic stem cell lines derived from human blastocysts, *Science* **282**, 1145–1147.

6. Kleinman, H. K., and Martin, G. R. (2005) Matrigel: basement membrane matrix with biological activity. *Semin. Cancer Biol.* **15**, 378–386.

7. Xu, C., Inokuma, M. S., Denham, J., Golds, K., Kundu, P., Gold, J. D., et al. (2001) Feeder-free growth of undifferentiated human embryonic stem cells. *Nat. Biotechnol.* **19**, 971–974.

8. Amit, M., Shariki, C., Margulets, V., and Itskovitz-Eldor, J. (2004) Feeder layer- and serum-free culture of human embryonic stem cells. *Biol. Reprod.* **70**, 837–845.

9. Ludwig, T., Bergendahl, V., Levenstein, M. E., Yu, J., Probasco, M. D., and Thomson, J., A. (2007) Defined, feeder-independent medium for human embryonic stem cell culture. *Curr. Protoc. Stem Cell Biol.* Chapter 1, Unit 1C 2.

10. Ludwig, T. E., Levenstein, M. E., Jones, J. M., Berggren, W. T., Mitchen, E. R., Frane, J. L., et al. (2006) Derivation of human embryonic stem cells in defined conditions. *Nat. Biotechnol.* **24**, 185–187.

11. Hanjaya-Putra, D., and Gerecht, S. (2009) Vascular engineering using human embryonic stem cells. *Biotechnol. Prog.* **25**, 2–9.

12. Hewitt, K. J., Shamis, Y., Carlson, M. W., Aberdam, E., Aberdam, D., and Garlick, J. (2009) Three-dimensional epithelial tissues generated from human embryonic stem cells. *Tissue Eng. Part A* **15**(11), 3417–3426.

13. Wang, X., and Ye, K. (2009) Three-dimensional differentiation of embryonic stem cells into islet-like insulin-producing clusters. *Tissue Eng. Part A* **15**(8), 1941–1952.

14. Cote, R. J. (2001) Aseptic technique for cell culture. *Curr. Protoc. Cell Biol.* Chapter 1, Unit 1.3.

15. Phelan, M. C. (2006) Techniques for mammalian cell tissue culture. *Curr. Protoc. Hum. Genet.* Appendix 3, Appendix 3G.

16. Ryan, J. Understanding and Managing Cell Contamination. http://catalog2.corning.com/Lifesciences/en-US/TDL/techInfo.aspx?categoryname=cell_culture|Contamination.

Chapter 3

Serum-Free and Feeder-Free Culture Conditions for Human Embryonic Stem Cells

Ludovic Vallier

Abstract

Human embryonic stem cells (hESCs) are pluripotent cells derived from the embryo at the blastocyst stage. Their embryonic origin confers upon them the capacity to proliferate indefinitely in vitro while maintaining the capacity to differentiate into a large variety of cell types. Based on these properties of self-renewal and pluripotency, hESCs represent a unique source to generate a large quantity of certain specialized cell types with clinical interest for transplantation based therapy. However, hESCs are usually grown in culture conditions using fetal bovine serum and mouse embryonic fibroblasts, two components that are not compatible with clinical applications. Consequently, the possibility to expand hESCs in serum-free and in feeder-free culture conditions is becoming a major challenge to deliver the clinical promises of hESCs. Here, we describe the basic principles of growing hESCs in a chemically defined medium (CDM) devoid of serum and feeders.

Key words: Human embryonic stem cells, Serum-free, Feeder-free, Pluripotency

1. Introduction

The pluripotent status of human embryonic stem cells (hESCs) confers upon them the capacity to differentiate into a large variety of cell types. This unique property also implies that hESCs have an unstable phenotype and that their environment can easily induce their differentiation. This phenotypic instability represents a major challenge in maintaining hESCs in vitro, and a large number of culture systems have been developed to solve this issue. The first hESC line derived in 1998 by J.A. Thomson and colleagues was initially grown on mouse embryonic fibroblasts and in a medium containing serum (1). These cultures conditions have been progressively improved by replacing FBS with an artificial serum (serum replacement) (2), which decreased the problem of

Nicole I. zur Nieden (ed.), *Embryonic Stem Cell Therapy for Osteo-Degenerative Diseases*, Methods in Molecular Biology, vol. 690, DOI 10.1007/978-1-60761-962-8_3, © Springer Science+Business Media, LLC 2011

batch variations encountered with different lots of FBS. However, the serum replacer formulation is not defined or publicly available, and it includes factors such as BMP, which can influence hESC self-renewal and differentiation (3). Another progress in culture systems in growing hESCs was the use of conditioned media by MEFs supplemented with basic fibroblast growth factor (bFGF) (4, 5). Although this approach avoided a direct contact between hESCs and the feeder cells and improved the phenotypic stability of hESCs, it was based on media containing serum and thus unknown factors.

The first attempt to develop chemically defined conditions to grow hESCs were based on studies trying to uncover the signaling pathways maintaining their pluripotency (3, 6–10). Indeed, these studies demonstrated that Activin/Nodal/TGFβ signaling and FGF signaling pathways were respectively necessary for pluripotency and self-renewal of hESCs. Later studies also showed that Activin A and/or bFGF were sufficient to grow hESCs in the absence of serum and/or feeder cells (10–12). Based on these observations, several culture systems (Table 1) were developed to grow hESCs over a prolonged period of time in defined conditions (11–16).

In this chapter, we describe a culture system (CDM A + F) efficient for growing H9 (WiCell, Madison, WI) and hSF-6 (UCSF, San Francisco, CA) hESC lines (Fig. 1), two lines easily available and commonly used for basic studies. This method provides the knowledge necessary to grow hESCs in the absence of serum and feeder cells. However, different hESCs might have different requirements, and three key factors that might influence behavior of hESCs grown in defined conditions need to be considered and are reviewed in the following sections.

1.1. Growth Factors

Activin/TGFβ and bFGF are necessary for pluripotency and self-renewal of hESCs (8, 10). However, additional growth factors can have a positive effect on hESCs including Wnt/beta-catenin, which can easily be activated by adding lithium chloride to the culture media as it inhibits the enzyme that usually tags beta-catenin for proteasomal degradation. Nevertheless, the function of Wnt signaling in hESC pluripotency remains controversial (7, 17, 18), and its positive effect on hESC growth might depend on other factors provided in the culture medium.

1.2. Bovine Serum Albumin

All the media currently available to grow hESCs contain serum albumin of animal or human origin. The latter source of serum albumin is rarely used because of its cost and the important variability between different lots. Indeed, it is highly recommended to batch-screen bovine serum albumin (BSA) since it can affect the proliferation and the adhesion of hESCs. A BSA lot is usually validated when it can be used to grow hESCs more than five passages without decrease in proliferation or adhesion.

Table 1
Summary of culture systems currently available to grow hESCs in the absence of serum and feeder cells

Medium	Growth factors	ECM	Basal medium	Additives	Splitting methods	Cell lines	References
CDM A+F	Activin bFGF	Fibronectin or F3S	IMDM/F12	Bovine or human SA, insulin, transferrin, lipids	Collagenase IV or dispase or accutase	H9, H1, hSF-6	(6, 12)
TesR1	TGFβ bFGF	Matrigel	DMEM/F12	Bovine or human SA, LiCl, γ-aminobutyric acid, pipecolic acid lipids	Collagenase IV	H1, H7, H9, H14	(11)
N2/B27 -CDM	bFGF	Matrigel	DMEM/F12	Bovine SA, N2 supplement, B27 supplement	Dispase	H1, hSF-6	(14)
HESCO	Wnt3a/BAFF bFGF	Matrigel or fibronectin	DMEM/F12	Bovine SA, insulin, transferrin, lipids	Collagenase IV or trypsin–EDTA	H9, BGO1	(15)
hESF9	bFGF	Collagen I	DMEM/F12	Bovine SA, L-ascorbic acid 2-phosphate, heparin sulfate, sodium selenite	Collagenase IV or trypsin–EDTA	HUES-1, Shef1, Shef4, Shef5, H7	(16)

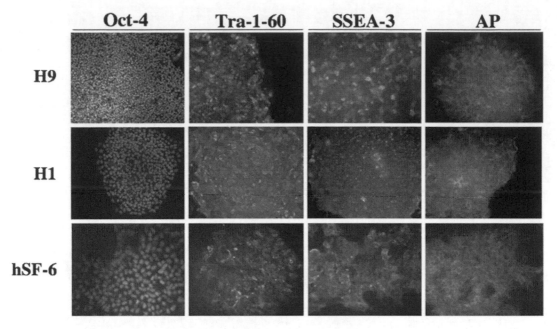

Fig. 1. Human ESCs grown in CDM supplemented with Activin and FGF maintain their pluripotent status. H9, H1, and hSF-6 hESCs were grown respectively for 35, 11, and 20 passages, and the expression of the pluripotency markers Oct-4, Tra-1-60, SSEA-3, and alkaline phosphatase (AP) was analyzed using immunostaining.

1.3. Extracellular Matrix

Problem of adhesion will immediately lead to cell death or differentiation of hESCs, especially if colonies start to form three-dimensional structures (or embryoid body-like structure). Human ESCs need to be grown as flat and compact colonies. Indeed, single hESCs excluded from the colonies differentiate into "stroma cells." This background of differentiation can produce growth factors, such as FGF or insulin growth factor, that can strongly interfere with the behavior of hESCs, especially during their differentiation into specific cell types (19). Absence of feeder cells increases this problem, since one of their key functions is to maintain hESC organization into compact colonies. Therefore, hESCs need specific extracellular matrix, not only to attach properly on plastic but also to maintain their pluripotent status. The usual purified ECM proteins, such as laminin, give mixed results and are consequently often replaced with Matrigel. However, Matrigel is prepared from murine Engelbreth-Holm-Swarm tumor cells and contains unknown factors capable of interfering with experimental outcomes. Fibronectin and vitronectin give the best results with the culture system described below (CDM A + F). However, the cost of these proteins precludes their use for large-scale experiments and routine cell culture. For these kinds of applications, plastic plates precoated with medium containing FBS (see Subheading 3.3) are recommended. This low-cost approach is very robust with most of the hESC cell lines. Importantly, we never observed differences between FBS precoated plates and fibronectin coated plate, suggesting that this method does not interfere with pluripotency or differentiation of hESCs.

In summary, we present a simple and robust method to grow hESCs in serum- and feeder-free conditions. Importantly, this method can be easily adapted to different hESC lines, and it can be modified to avoid any components of animal origin.

---

2. Materials

2.1. Precoating Plates

1. Plastic plates (Costar).

2. Phosphate-buffered saline (PBS): Combine 137 mM NaCl, 2.7 mM KCl, and 10 mM sodium phosphate dibasic (Na_2HPO_4) in ddH_2O. Adjust pH to 7.4 and autoclave at 121°C for 60 min. Store at 4°C.

3. Fibronectin (Chemicon) at a concentration of 10 mg/mL: the liquid is sterile-filtered and should be maintained at 4°C for up to 6 months from the date of receipt.

4. Gelatin from porcine skin: dilute to 0.1% in embryo tested water (Sigma).

5. Serum-containing media: prepare serum-containing media using 500 mL of Advanced Dulbecco's Modified Eagle's Medium (DMEM, Invitrogen) and 10 mL of fetal bovine serum (FBS, Biosera).

2.2. Human ESC Culture Media and Formulations

1. Chemically defined hESC medium (CDM): Mix 50% Iscove's Modified Eagle's Medium (IMDM) with 50% F12 medium. Add BSA to a final concentration of 5 mg/mL, 1% (v/v) lipid (100×, Invitrogen), 450 mM monothioglycerol, Insulin (10 mg/mL stock solution, Roche) to a final concentration of 7 µg/mL, and transferrin (Roche, stock solution 30 mg/mL) to a final concentration of 15 µg/mL (see Note 1).

2. Collagenase solution: mix 500 mL of DMEM (Invitrogen) with 0.5 g of collagenase IV. This makes a 1 mg/mL collagenase solution.

3. Activin (RnD Systems): prepare a stock solution of 10 µg/mL in 1× PBS with 0.1% BSA.

4. Basic fibroblast growth factor (bFGF, RnD Systems): prepare a 4 µg/mL stock solution in 10 mM Tris-HCL, pH 7.4 containing 0.1% BSA. Store at −20°C.

5. CDM A + F: CDM as above plus Activin A (10 ng/mL) and bFGF (12 ng/mL).

6. Dispase 1 mg/mL.

2.3. Karyotyping

1. Colcemid solution (i.e. Sigma) at a concentration of 0.1 mg/mL in CDM.

2. Potassium chloride (KCl) at a concentration of 0.037 M.

3. Trypsin, 0.25% (1×) with ethylenediaminetetraacetic acid (EDTA) 4Na, liquid (Invitrogen).

4. Methanol and acetic acid (3:1).

---

3. Methods

There are a large number of methods currently available to grow hESCs in the absence of serum and feeder cells (Table 1). The diversity of these culture systems reflects on the variability between different hESC lines (see Note 2) (20, 21).

3.1. Precoating Plates

3.1.1. Precoating Plates with Human Fibronectin

1. Add 1.5 mL of PBS per well of a 6-well plate. Then, add 15 µL of fibronectin (10 mg/mL stock solution) per well. For a 12-well plate, add 0.5 mL of PBS per well and 5 µL of fibronectin.

2. Incubate at 4°C overnight or 30 min to 1 h at 37°C. Then, store the plates at 37°C for a week or more.

3.1.2. Precoating Plates with Fetal Bovine Serum-Containing Medium

1. Precoat plates with 0.1% gelatin for 15–60 min at room temperature.

2. Discard gelatin and add medium containing 10% FBS for 24 h to 7 days at 37°C (see Note 3).

3.2. Transferring hESCs from Feeder Layers

1. Wash the cells once in PBS and then add 1.5 mL of collagenase IV 1 mg/mL per well of a 6-well plate (see Note 4). Incubate for 15 min at 37°C. Scrape the colonies with a 5-mL pipette and then dissociate them into clumps by pipetting gently three to four times. Compared to passaging of hESCs onto feeder layers, try to generate relatively big clumps (Fig. 2) (see Note 5).

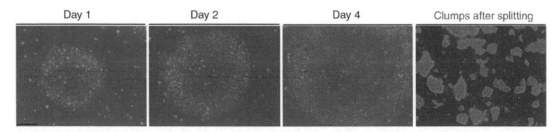

| Day 1 | Day 2 | Day 4 | Clumps after splitting |

Fig. 2. Human ESC colony grown in defined culture conditions. H9 hESCs were grown for 4 days in CDM A + F (Day1, Day 2, Day4) and then split using collagenase. The resulting clumps (*left panel*) were transferred in a new well for expansion. Scale bar 200 µm.

2. Wash the clumps in 3 mL of CDM (without Activin or bFGF).

3. Wash the fibronectin/FBS-coated plate once with PBS.

4. Plate the clumps at low density (100–300 clumps/well) on a fibronectin-coated 6-well plate in CDM A+F (see Note 6).

5. Leave 5–7 days in 5% CO_2 at 37°C until the colonies reach a very large size (4–10 times bigger than colony grown on feeder cells, see Notes 7 and 8).

3.3. Splitting in Chemically Defined Conditions Using Collagenase

1. Wash colonies once with PBS and then incubate in 1.5 mL of collagenase IV 1 mg/mL for 10 min at room temperature.

2. Scrape the colonies with a 5-mL pipette and transfer into a 15-mL tube containing 3 mL of CDM.

3. Dissociate the cells into clumps by pipetting gently 1–2 times [big clumps are definitively better (Fig. 2)].

4. Centrifuge the clumps at $200 \times g$ for 3 min and then plate the dissociated colonies on 3–4 new fibronectin/FBS-coated plates. Low-density culture is better.

5. Replace medium with fresh CDM containing 10 ng/mL Activin A and 12 ng/mL bFGF (CDM A+F) 24 h after splitting. Change the medium every day until the cells reach confluency (see Note 9).

3.4. Splitting in Chemically Defined Conditions Using Dispase

1. Wash colonies once with PBS and then incubate in 1.5 mL of dispase 1 mg/mL for 30 min at 37°C.

2. Scrape the colonies with a 5-mL pipette and transfer into a 15-mL tube containing 3 mL of CDM.

3. Dissociate the cells into clumps by pipetting gently once (big clumps are definitively better).

4. Centrifuge the clumps at $200 \times g$ for 3 min.

5. Add 3 mL of CDM medium and centrifuge the clumps at $200 \times g$ for 3 min.

6. Plate the dissociated colonies on 3–4 new fibronectin/FBS-coated plates. Low-density culture is better.

7. Replace medium with fresh CDM A+F 24 h after splitting. Then, change the medium every day until the cells reach confluency (see Notes 10 and 11).

3.5. Karyotyping

1. Incubate hESC in normal growth medium plus 0.1 mg/mL Colcemid solution (Sigma) for 3.5 h at 37°C (Incubation in Colcemid time might vary with cell lines and their proliferation rate).

2. Pour off supernatant and keep for step 5.

3. Wash cells with PBS and then add 1 mL trypsin/EDTA for 30 s.

4. Dissociate the colonies until the cells are in single suspension.

5. Wash with MEF medium to inhibit trypsin, add supernatant from step 2, and centrifuge for 4 min at $1,500 \times g$.

6. Wash pellet twice with PBS, each spin for 4 min at $1,500 \times g$.

7. Add 2 mL of KCl for 10 min at room temperature.

8. Add 2.5 mL methanol and acetic acid mixture and then centrifuge for 4 min at $1,500 \times g$.

9. Wash three times in methanol and acetic acid mixture, each time spinning at $1,500 \times g$ for 4 min.

10. Resuspend pellet in 0.5 mL of methanol and acetic acid mixture (3:1).

11. Leave at $-20°C$ for at least 2 h.

12. Drop onto slides and then stain for G banding analysis.

4. Notes

1. Alternatively, human serum albumin (HSA, Sigma) may be used at a final concentration of 1 mg/mL to substitute BSA. BSA is a key component of the CDM, and it needs to be batch-tested. We have now used around ten different batches of BSA, and the behavior of the cells (meaning proliferation and adhesion) has been slightly different for each batch. So far, the best substitute for BSA is polyvinyl alcohol (PVA). H9 grow well in CDM/PVA, and they stay undifferentiated at least for 15 passages (85% Tra-1-60, 60% SSEA-3, 85% SSEA-4).

2. H9, H1, and hSF-6 hESC have been grown successfully in CDM supplemented with Activin and bFGF (12). However, each of those lines shows some variability. H9 and hSF-6 cells have been grown respectively for 85 passages and 40 passages in CDM A+F without signs of differentiation (Fig. 1). However, hSF-6 proliferate less quickly than H9 cells, and thus, they require more time to reach confluency. In addition, hSF-6 need to form bigger colonies before they are split, otherwise they lose their organization upon adhesion and then differentiate into stroma cells. Finally, H1 cells have been grown for 15 passages in CDM A+F on FBS-coated plates, but they produce stroma cells, and they strongly differentiate on fibronectin-coated plates. These observations underline the differences between hESC lines and the necessity to adapt culture conditions to each hESC line.

3. Use 0.5 mL serum-containing media per well of a 12-well plate, 1.5 mL per well of a 6-well plate, and 6 mL per 100-mm dish.

4. Transferring hESCs from feeder cell and serum-containing culture condition into defined culture system does not erase background of differentiation. Therefore, it is essential to start with a homogeneous population of hESCs, which homogeneously express pluripotency markers (also Note 6).

5. The method used to detach hESC colonies and to dissociate the resulting clumps represents a key part of the culture system described above. Indeed, small-size clumps will generate colonies that lose their organization and thus generate stroma cells, while large-size clumps will form colonies with three-dimensional structure resulting in differentiation. Importantly, the ideal clumps size (Fig. 2) might vary between different hESC lines, and several assays might be required to find the right level of dissociation. Collagenase IV is commonly used to split hESCs in defined culture conditions. However, dispase (1 mg/mL) can be used especially when important cell death is observed after splitting with collagenase.

6. We typically plate one confluent ø 60-mm dish of hESCs grown on feeder cells in 6–8 wells of a 6-well plate.

7. The cells can have some problems to attach in CDM A+F (especially with fribronectin-coated plates). So, if on the day after passaging the colonies look like EBs, you have to wait an additional day without changing the medium. Usually, the colonies spread during the second night after passaging.

8. Cells in CDM only grow at 5% CO_2 (so, do not put them in a 10% CO_2 incubator).

9. Human ESCs reach confluency when colonies are almost touching each other and most importantly when the size of the colonies is sufficient to avoid complete dissociation upon collagenase or dispase treatment (Fig. 2).

10. In addition to hESCs, the CDM A+F can be used to grow mouse epiblast stem cells, human induced pluripotent stem cells (12), and *Callithrix jacchus* ESCs (N. zur Nieden, personal communication).

11. Pluripotency of hESCs grown in CDM A+F can be checked regularly by analyzing pluripotency markers, such Oct-4, Nanog, and Sox-2, using immunostaining (Fig. 1). Homogeneity of cell population can be validated using flow cytometry to determine the fraction of cells expressing the cell surface markers Tra-1-60, SSAEA-3, and SSEA-4. Finally, karyotype analyses should be performed every 30 passages to check for genomic instability.

References

1. Thomson, J.A., Itskovitz-Eldor, J., Shapiro, S.S., Waknitz, M.A., Swiergiel, J.J., Marshall, V.S., et al. (1998) Embryonic stem cell lines derived from human blastocysts. *Science* **282**, 1145–1147.

2. Amit, M., Carpenter, M.K., Inokuma, M.S., Chiu, C.P., Harris, C.P., Waknitz, M.A., et al. (2000) Clonally derived human embryonic stem cell lines maintain pluripotency and proliferative potential for prolonged periods of culture. *Dev. Biol.* **227**, 271–278.

3. Xu, R.H., Peck, R.M., Li, D.S., Feng, X., Ludwig, T., and Thomson, J.A. (2005) Basic FGF and suppression of BMP signaling sustain undifferentiated proliferation of human ES cells. *Nat. Methods* **2**, 185–190.

4. Xu, C., Inokuma, M.S., Denham, J., Golds, K., Kundu, P., Gold, J.D., et al. (2001) Feeder-free growth of undifferentiated human embryonic stem cells. *Nat. Biotechnol.* **19**, 971–974.

5. Rosler, E.S., Fisk, G.J., Ares, X., Irving, J., Miura, T., Rao, M.S., et al. (2004) Long-term culture of human embryonic stem cells in feeder-free conditions. *Dev. Dyn.* **229**, 259–274.

6. Vallier, L., Reynolds, D., and Pedersen, R.A. (2004) Nodal inhibits differentiation of human embryonic stem cells along the neuroectodermal default pathway. *Dev. Biol.* **275**, 403–421.

7. Sato, N., Meijer, L., Skaltsounis, L., Greengard, P., and Brivanlou, A.H. (2004) Maintenance of pluripotency in human and mouse embryonic stem cells through activation of Wnt signaling by a pharmacological GSK-3-specific inhibitor. *Nat. Med.* **10**, 55–63.

8. Vallier, L., Alexander, M., and Pedersen, R.A. (2005) Activin/Nodal and FGF pathways cooperate to maintain pluripotency of human embryonic stem cells. *J. Cell. Sci.* **118**, 4495–4509.

9. James, D., Levine, A.J., Besser, D., and Hemmati-Brivanlou, A. (2005) TGFbeta/activin/nodal signaling is necessary for the maintenance of pluripotency in human embryonic stem cells. *Development* **132**, 1273–1282.

10. Levenstein, M.E., Ludwig, T.E., Xu, R.H., Llanas, R.A., VanDenHeuvel-Kramer, K., Manning, D., et al. (2006) Basic fibroblast growth factor support of human embryonic stem cell self-renewal. *Stem Cells* **24**, 568–574.

11. Ludwig, T.E., Bergendahl, V., Levenstein, M.E., Yu, J., Probasco, M.D., and Thomson, J.A. (2006) Feeder-independent culture of human embryonic stem cells. *Nat. Methods* **3**, 637–646.

12. Brons, I.G., Smithers, L.E., Trotter, M.W., Rugg-Gunn, P., Sun, B., Chuva de Sousa Lopes, S.M., et al. (2007) Derivation of pluripotent epiblast stem cells from mammalian embryos. *Nature* **448**, 191–195.

13. Ludwig, T.E., Levenstein, M.E., Jones, J.M., Berggren, W.T., Mitchen, E.R., Frane, J.L., et al. (2006) Derivation of human embryonic stem cells in defined conditions. *Nat. Biotechnol.* **24**, 185–187.

14. Yao, S., Chen, S., Clark, J., Hao, E., Beattie, G.M., Hayek, A., et al. (2006) Long-term self-renewal and directed differentiation of human embryonic stem cells in chemically defined conditions. *Proc. Natl. Acad. Sci. U.S.A.* **103**, 6907–6912.

15. Lu, J., Hou, R., Booth, C.J., Yang, S.H., and Snyder, M. (2006) Defined culture conditions of human embryonic stem cells. *Proc. Natl. Acad. Sci. U.S.A.* **103**, 5688–5693.

16. Furue, M.K., Na, J., Jackson, J.P., Okamoto, T., Jones, M., Baker, D., et al. (2008) Heparin promotes the growth of human embryonic stem cells in a defined serum-free medium. *Proc. Natl. Acad. Sci. USA* **105**, 13409–13414.

17. Dravid, G., Ye, Z., Hammond, H., Chen, G., Pyle, A., Donovan, P., et al. (2005) Defining the role of Wnt/beta-catenin signaling in the survival, proliferation, and self-renewal of human embryonic stem cells. *Stem Cells* **23**, 1489–1501.

18. Melchior, K., Weiss, J., Zaehres, H., Kim, Y.M., Lutzko, C., Roosta, N., et al. (2008) The WNT receptor FZD7 contributes to self-renewal signaling of human embryonic stem cells. *Biol. Chem.* **389**, 897–903.

19. Bendall, S.C., Stewart, M.H., Menendez, P., George, D., Vijayaragavan, K., Werbowetski-Ogilvie, T., et al. (2007) IGF and FGF cooperatively establish the regulatory stem cell niche of pluripotent human cells in vitro. *Nature* **448**, 1015–1021.

20. Osafune, K., Caron, L., Borowiak, M., Martinez, R.J., Fitz-Gerald, C.S., Sato, Y., et al. (2008) Marked differences in differentiation propensity among human embryonic stem cell lines. *Nat. Biotechnol.* **26**, 313–315.

21. Rajala, K., Hakala, H., Panula, S., Aivio, S., Pihlajamaki, H., Suuronen, R., et al. (2007) Testing of nine different xeno-free culture media for human embryonic stem cell cultures. *Hum. Reprod.* **22**, 1231–1238.

Chapter 4

Functional Assays for Human Embryonic Stem Cell Pluripotency

Michael D. O'Connor, Melanie D. Kardel, and Connie J. Eaves

Abstract

Realizing the potential that human embryonic stem cells (hESCs) hold, both for the advancement of biomedical science and the development of new treatments for many human disorders, will be greatly facilitated by the introduction of standardized methods for assessing and altering the biological properties of these cells. The 7-day in vitro alkaline phosphatase colony-forming cell (AP$^+$-CFC) assay currently offers the most sensitive and specific method to quantify the frequency of undifferentiated cells present in a culture. In this regard, it is superior to any phenotypic assessment protocol. The AP$^+$-CFC assay, thus, provides a valuable tool for monitoring the quality of hESC cultures, and also for evaluating quantitative changes in pluripotent cell numbers following manipulations that may affect the self-renewal and differentiation properties of the treated cells. Two other methods routinely used to evaluate hESC pluripotency involve either culturing the cells under conditions that promote the formation of nonadherent differentiating cell aggregates (termed embryoid bodies), or transplanting the cells into immunodeficient mice to obtain teratomas containing differentiated cells representative of endoderm, mesoderm, and ectoderm lineages.

Key words: Human embryonic stem cell, Induced pluripotent stem cell, Functional assay, Colony-forming cell, Embryoid body, Teratoma

1. Introduction

When kept under specific supportive conditions, human embryonic stem cells (hESCs) can be expanded indefinitely in culture while retaining the potential to give rise to all differentiated cell types found in the adult (1–4). Recently, it has been reported that adult human cells can be reprogrammed to a similar pluripotent state through the use of specific sets of exogenously delivered genes (5–7). Both sources of pluripotent human cells are of immense interest to biomedicine as they offer tractable systems for elucidating

Nicole I. zur Nieden (ed.), *Embryonic Stem Cell Therapy for Osteo-Degenerative Diseases*, Methods in Molecular Biology, vol. 690,
DOI 10.1007/978-1-60761-962-8_4, © Springer Science+Business Media, LLC 2011

molecular events involved in normal and pathological developmental processes, while also providing a resource for generating useful cell types for transplantation or novel therapeutic screening and toxicological studies (8–11).

Traditionally, hESCs have been maintained on mouse embryonic fibroblasts (MEFs) or in the presence of MEF-conditioned medium (1, 2, 12). However, the quality of these cultures was often highly unpredictable due to variations in the hESC-supportive ability of different batches of critical reagents, particularly the MEFs and the serum or serum substitute used to supplement the growth medium. Commercialization of recently reported, defined, feeder-independent hESC media is helping circumvent the unpredictability of MEF-based hESC culture protocols, enabling consistent maintenance of karyotypically normal, undifferentiated hESCs over long periods of time (3, 4), as is discussed in more detail in Chapters 2 and 3. A concurrent and equally important development has been the standardization of methods to monitor the undifferentiated cell content of hESC cultures (13), to facilitate data comparisons within and between experiments and laboratories, and in particular during culture scale-up for clinical applications.

Assessment of hESC cultures historically relied on direct visual assessment of their morphology by experienced observers. Detection of pluripotency-associated antigens and genes via flow cytometry or reverse transcriptase (RT)-PCR has become increasingly popular due to the greater objectivity and quantifiability of these methods relative to morphological analysis (14). Markers routinely assessed through mRNA profiling are described in Chapter 5. These methods also have the advantage of being relatively rapid and uncomplicated to perform. However, even in the murine (m) system, flow cytometry and RT-PCR are less sensitive at detecting changes in ESC properties than are functional assays of their pluripotency. For mESCs, the most stringent functional assay involves determining the frequency of cells present that can contribute to germ-line chimerism after their injection into blastocysts – a property that was found to be lost more rapidly than the expression of well-established pluripotency-associated antigens (i.e. SSEA-1) or genes (e.g. *Oct-4*) following exposure of the cells to differentiation-inducing conditions (15). In addition to chimera formation, three other properties have been used to assess retention of pluripotency by mESCs: (1) the formation of colonies of alkaline phosphatase-positive (AP$^+$) cells (15), (2) the formation of embryoid bodies (EBs) from single cells plated in semisolid medium (15, 16), and (3) the formation of teratomas containing differentiated cells of multiple types following injection of the cells directly into adult hosts (17, 18). For hESCs, obvious ethical constraints preclude the use of tests for chimera-forming ability, but methods to assess the other three properties have been developed and optimized (1, 2, 13, 19). Interestingly, as previously shown for mESCs, loss of AP[+] clonogenic activity after exposure of hESCs to differentiation-inducing conditions preceded loss of expression of standard pluripotency-related antigens (13). Together, these findings point to the greater specificity of functional assays, as compared to changes in currently used phenotypic markers, to detect the earliest changes in the undifferentiated status of variously manipulated ESCs.

In this chapter, we present detailed protocols for the detection of three functionally defined properties of undifferentiated hESCs. The most powerful and sensitive of these detects the ability of hESCs to form AP[+] colonies and allows the composition of test cell populations to be quantified at the single cell level. The second protocol is used to determine whether the test cells can generate multilineage EB structures, under conditions and in a time frame optimized for unmanipulated hESCs. The third protocol is used to determine whether the cells have retained or lost an ability to generate multilineage teratomas. Each protocol includes details for appropriate cell harvesting, followed by all of the subsequent steps required to complete the procedure. For the AP[+]-colony-forming cell (AP[+]-CFC) assay, this involves culturing appropriate numbers of single hESCs under defined conditions, then fixing the resultant colonies and staining them for alkaline phosphatase activity. For the EB assay, we describe culture conditions optimized for EB formation from undifferentiated hESCs and indicate how loss of pluripotent cells and gain of differentiated cells in EBs can be detected. For the teratoma assay, we describe transplantation and harvesting procedures using immunocompromised mice as hosts, as well as a method for determining whether cell types representative of all three germ layers are present. To help guide investigators in the best use of these protocols for different types of investigations, we include a summary of advantages and limitations of each (Table 1).

2. Materials

2.1. Alkaline Phosphatase Colony-Forming Cell Assay

1. TrypLE Express (Invitrogen).

2. Y-27632 (Calbiochem): Powder is dissolved in sterile water to make up a 1 mM stock solution (100×). Sterilize by passing the solution through a 0.2-μm filter and store at –20°C. This solution is light-sensitive.

3. mTeSR1™ hESC medium (StemCell Technologies).

4. Matrigel™ hESC-qualified Matrix (BD Biosciences): Thaw stock solution at 4°C and dilute aliquots as needed using cold Dulbecco's Modified Eagle's Medium (DMEM) to a final concentration of ~83 μg/mL.

Table 1
Comparison of methods for assessing undifferentiated hESC activity in test cell suspensions

Method	Detection sensitivity	Time of assay	Advantages	Limitations
AP⁺-CFC assay	$\sim 10^2 - 10^4$ cells	7 days	Rapid Sensitive Enumerates individual cells	Does not detect pluripotency or self-renewal directly but infers this from AP⁺ staining and historical evidence that these AP⁺-CFCs are pluripotent and self-renewing
EB formation	$\sim 10^4 - 10^5$ cells	~15 days	Relatively rapid Relatively sensitive	Not quantitative (conditions for clonal generation of EBs from hESCs have not yet been determined) Evidence for pluripotency usually indirect (e.g. RT-PCR) rather than by demonstration of generation of differentiated cells Efficiency of EB formation is highly sensitive to medium composition
Teratoma formation	$\sim 10^5 - 10^6$ cells	8–12 weeks	Can demonstrate a broad range of hESC pluripotentiality via the generation of morphologically recognizable differentiated cell types	Not quantitative Low detection sensitivity Expensive Time-consuming

5. Phosphate-buffered saline (PBS, StemCell Technologies).

6. Conical tubes, 15 mL, Falcon (BD Biosciences) or similar.

7. Hemocytometer (e.g. VWR).

8. 40-μm cell strainer.

9. Alkaline Phosphatase detection kit (Sigma-Aldrich): fixative and diazonium salt staining solution are prepared as recommended by the manufacturer (see Note 1).

2.2. Embryoid Body Formation

1. PBS (StemCell Technologies).

2. 0.05% trypsin/Ethylenediaminetetraacetic acid (EDTA) (Invitrogen).

3. Calcium chloride ($CaCl_2$): Powder is dissolved in sterile water to make up a 100 mM stock solution. Sterilize by passing solution through a 0.2-µm filter. Make up a calcium-supplemented trypsin solution as follows: 1 mL of 0.05% trypsin/EDTA, 1 mL of PBS, 10 µL of 100 mM $CaCl_2$ for each 6-cm dish of test cells to be assessed (scale all volumes accordingly for other culture volumes).

4. 60-mm petri dish.

5. HES medium: DMEM/F12 with 20% Knockout Serum Replacement (Invitrogen), 1 mM L-glutamine, 0.1 mM non-essential amino acids, 0.1 mM 2-mercaptoethanol.

6. Plastic serological pipettes (e.g. Falcon 2 mL).

7. RNA purification kit, such as Absolutely RNA microprep kit (Stratagene).

8. Reverse transcriptase for cDNA synthesis, i.e. SuperScript II Reverse Transcriptase (Invitrogen).

9. SYBR Green reagent for quantitative PCR (qPCR), such as Power SYBR Green PCR Mix (Applied Biosystems).

10. Applied Biosystems 7500 Real Time PCR System or similar.

2.3. Teratoma Assay

1. PBS (StemCell Technologies).

2. Dispase, 1 mg/mL in DMEM.

3. Cell scrapers.

4. TrypLE Express (Invitrogen).

5. 60-mm culture dish (i.e. BD Biosciences).

6. Matrigel™ (BD Biosciences): Thaw at 4°C and dilute to 1–2 mg/mL using cold DMEM. Keep on ice to avoid gel formation prior to use.

7. 1-mL syringes and 21-gauge needles.

8. 6- to 12-week-old NOD/SCID mice or variant of NOD/ SCID mice (Jackson Labs).

9. Anesthetic (e.g. IsoFlo; isoflurane).

10. 10% buffered formalin: dilute 100% formalin [also known as ~40% (m/v) formaldehyde solution] tenfold in PBS.

11. Access to histological services for preparation of paraffin blocks and Hematoxylin- & Eosin-stained sections (see Note 2).

3. Methods

The methods presented here assume a basic ability to maintain undifferentiated hESCs. For protocols detailing the methods of undifferentiated hESC culture, please refer to other chapters in this collection. The methods are supplied in increasing order of the time required to perform the assay, and also decreasing order of assay detection sensitivity (see Table 1).

3.1. Matrigel™-Coating of 6-Well Plates

1. Add diluted Matrigel™ to 6-well plates at a volume of 1 mL per well.

2. Incubate for 1–2 h at room temperature or 4°C for up to 1 week.

3. Excess Matrigel™ solution is removed immediately prior to use, and the wells should not be rinsed.

3.2. Alkaline Phosphatase Colony-Forming Cell Assay

This assay exploits the selective ability of single undifferentiated hESCs, which are themselves AP$^+$, to form colonies of AP$^+$ progeny, to measure the frequency (and hence the total number) of undifferentiated hESCs in a given culture (13). Previous studies have validated the association of this ability with retained pluripotency and self-renewal activity. This assay is useful for detecting the first or subtle changes in the undifferentiated hESC content of variously manipulated cultures, since effects on AP$^+$ colony-forming activity appear before changes in the expression of cell surface antigens become apparent (13) (see also Fig. 1). Here, we describe a feeder-free AP$^+$-CFC assay protocol using mTeSR1™, since this is more readily maintained as a reproducible method and also easier to perform than when using MEF-based conditions (see Note 3).

1. Remove the medium from the hESCs and replace with fresh mTeSR1™ containing 10 μM Y-27632 (1.5 mL for a culture in a 60-mm dish; scale all volumes accordingly for other

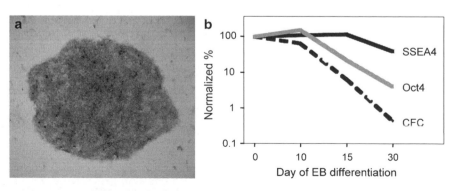

Fig. 1. Representative data from hESC CFC assays. (a) An alkaline phosphatase-stained H9 colony fixed and stained 7 days after plating on MEFs. (b) Following exposure to differentiation conditions, loss of CFCs occurs more rapidly and more precipitously than loss of cells expressing pluripotency-related antigens (CA1 hESCs).

culture dishes) and incubate at 37°C in 5% CO_2 for 1 h (see Note 4).

2. Remove the Y-27632-containing medium and wash the cells with 3 mL of PBS.

3. Remove the PBS and add 2 mL of TrypLE per dish. Incubate at 37°C in 5% CO_2 for 7–10 min.

4. Wash the single cells off the culture dish and transfer to a 15-mL conical tube.

5. Wash the culture surface with 3 mL of PBS to remove residual cells and add this to the 15-mL tube. Gently triturate the sample to break up remaining aggregates.

6. Centrifuge at $300 \times g$ for 5 min and then remove the supernatant.

7. Resuspend the cells in 1 mL of mTeSR1™ containing 10 µM Y-27632, triturate gently and then filter cells through a 40-µm cell strainer to remove any residual cell aggregates.

8. Remove a small aliquot for cell counts in a hemocytometer and determine the cell concentration (see Note 5).

9. Add between 10^2–10^4 cells to each Matrigel™ coated well (see Subheading 2.1, item 4) containing 1 mL of mTeSR1™ and 10 µM Y-27632 (see Note 6) and culture at 37°C in 5% CO_2.

10. Replace the culture medium after 24 h with mTeSR1™ that does not contain Y-27632 and culture for an additional 6 days (a total of 7 days).

11. Fix and stain colonies for alkaline phosphatase activity at the end of the culture period using the AP detection kit.

12. Remove mTeSR1™ and wash once with 2 mL of PBS. Add 1 mL fixative per well for 30 s at room temperature.

13. Remove fixative and wash once with 2 mL of PBS. Add 1 mL of diazonium salt staining solution and incubate at room temperature in the dark for 15 min (see Note 7).

14. Remove the staining solution and wash once with 2 mL of PBS, then store in 2 mL of PBS.

15. Count the number of colonies containing >30 AP+ cells. Determine the frequency of CFCs present in the initial test cell suspension (i.e. # AP+ colonies/# cells plated; see Note 8).

3.3. Embryoid Body Formation

This method of EB formation supports hESC differentiation into cell types normally derived from all the three germ layers (ectoderm, mesoderm, and endoderm). This method provides a relatively rapid and simple way of assessing whether a test culture likely contains cells with pluripotent differentiation capacity, although this is not stringently evaluated at a clonal level. Unlike the murine EB assay where EBs are formed from single cells (15), hESC-derived EBs are generated from aggregates of hESCs.

1. Remove the medium from the test culture and wash with 3 mL of PBS.

2. Remove the PBS and add the calcium-supplemented trypsin solution. Incubate for 10 min at 37°C in 5% CO_2, then carefully remove and discard the trypsin solution without disturbing the hESCs.

3. Add 2 mL of HES medium and gently scrape the plate with a 2-mL serological pipette to remove the cells.

4. Transfer the hESC aggregates to a 15-mL conical tube. Rinse the plate with 3 mL of PBS and add this to the 15-mL tube. Centrifuge at $300 \times g$ for 5 min.

5. Remove the supernatant and resuspend the aggregates in 1 mL of HES medium. Triturate gently, only enough to just break up large aggregates (~3 times).

6. Remove 100 µL (10%) of the aggregates and transfer to a 1.5-mL microfuge tube.

7. Centrifuge at $300 \times g$ for 5 min, remove the supernatant, and resuspend the cells in RNA lysis buffer. Store at –80°C for use as the undifferentiated control in the RNA analyses to be performed later.

8. Transfer the remainder of the aggregate suspension to a 60-mm petri dish containing 4 mL of HES medium (see Note 9) and culture at 37°C under 5% CO_2. This is day 0 of EB culture.

9. Change the culture medium every 2–3 days by transferring the EBs to a 15-mL conical tube and allowing the EBs to settle to the bottom (~2–3 min). Remove most of the supernatant, leaving ~0.5 mL so as not to disturb the EBs, and then add 5 mL of fresh medium. Transfer the suspension back to the petri dish and return it to the incubator (see Note 10).

10. On day 15 of EB culture, harvest the resulting EBs (see Fig. 2a) for RNA analysis (see Note 11). Transfer the EBs to a 15-mL conical tube, centrifuge at $300 \times g$ for 5 min, remove

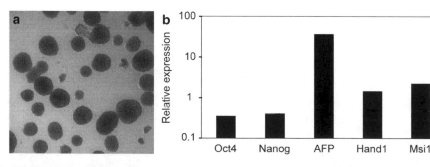

Fig. 2. EB differentiation data from H9 hESCs. (a) EBs after 9 days of nonadherent culture in HES medium. (b) Gene expression in 15-day EBs (relative to undifferentiated hESCs) shows evidence of pluripotent differentiation, i.e. decreased pluripotency gene expression and increased expression of differentiation-related genes (see Table 2).

Table 2
Primer sequences for EB analysis with quantitative PCR

Gene	Putative lineage	Forward primer sequence	Reverse primer sequence
GAPDH	Reference	CCCATCACCATCTTCCAGGAG	CTTCTCCATGGTGGTGAAGACG
OCT-4	hESCs	GTGGAGGAAGCTGACAACAA	CTCCAGGTTGCCTCTCACTC
NANOG	hESCs	AACTGGCCGAAGAATAGCAA	CATCCCTGGTGGTAGGAAGA
AFP	Endoderm	GTAGCGCTGCAAACAATGAA	TCTGCAATGACAGCCTCAAG
MSX1	Mesoderm	CGAGAAGCCCGAGAGGAC	GGCTTACGGTTCGTCTTGTG
HAND1	Mesoderm	AACTCAAGAAGGCGGATGG	CGGTGCGTCCTTTAATCCT
MSI1	Ectoderm	CTTTGATTGCCACAGCCTTC	ACTCGTGGTCCTCAGTCAGC

the supernatant, and resuspend in RNA lysis buffer. Store the sample at –80°C.

11. Purify RNA from both the EB sample and the undifferentiated control sample (step 7) according to the manufacturer's instructions. Synthesize cDNA and perform qRT-PCR for selected pluripotency-related genes and sufficient differentiation-related genes to assess representation of all the three germ layers (see Note 12). Examples of primer sequences to detect evidence of pluripotent differentiation are shown in Table 2 (14). Perform qRT-PCR using the following program on an Applied Biosystems 7500 Real Time PCR System: 50°C for 2 min, 95°C for 10 min, followed by 40–50 two-step cycles of 95°C for 15 s and 60°C for 1 min (14).

12. Normalize the expression level of each gene to an appropriate housekeeping gene and then determine the relative normalized expression of each gene between the undifferentiated control and the EB samples. The levels of pluripotency-related gene transcripts (e.g. for Oct-4 and Nanog) should decrease in the EB samples, and the level of differentiation-related gene transcripts (e.g. for AFP, MSX1, and MSI1) should increase if differentiation has occurred in the EB samples (see Fig. 2b).

3.4. Teratoma Assay   This assay is typically performed to establish whether a new hESC line or hESCs grown under new conditions possess the ability to generate differentiated cells from all the three germ layers (i.e. endoderm, mesoderm, and ectoderm) (1–3, 12) (see Fig. 3). While multiple engraftment sites are possible, such as intramuscular, kidney capsule, and subcutaneous, we find that subcutaneous

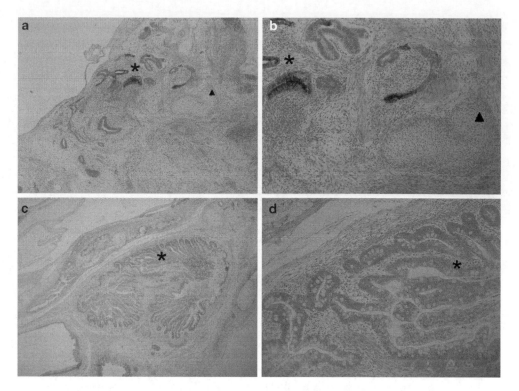

Fig. 3. Representative images of cell types found within teratomas after subcutaneous injection of hESCs in Matrigel™ into NOD/SCID mice. (**a**, **b**) CA1-derived teratoma (1 × 10⁶ cells/graft, 7.5 weeks postengraftment): cartilage (*triangle*, mesoderm); retinal pigment epithelium (*asterisk*, ectoderm). (**b**) is a higher magnification of the regions indicated in (**a**). (**c**, **d**) CA2-derived teratoma (8 × 10⁵ cells/graft, 12 weeks postengraftment): gut-like structures (*asterisk*, endoderm). (**d**) is a higher magnification of the region indicated in (**c**).

injection of hESCs in Matrigel™ enables ~100% teratoma formation with simple teratoma extraction (20). Since 10^5–10^6 hESCs are required per graft, it is necessary to expand the hESCs being analyzed to provide 5–10 million cells on the day that the teratoma assay is to be performed. This number corresponds to approximately two to four 6-cm culture dishes using mTeSR1™-based culture. Before attempting this assay, ensure that all steps comply with the animal protocols permitted by your institution.

1. To harvest hESCs as aggregates, remove the culture medium from a 60-mm dish, rinse with 5 mL of PBS and then discard the PBS.

2. Add 3.5 mL of 1 mg/mL dispase and incubate at 37°C/5% CO_2 for ~4 min, or until the edges of the hESC colonies begin to detach from the culture dish (see Note 13).

3. Remove the dispase and rinse the plate three times, each time with 5 mL of PBS.

4. After discarding the final PBS wash, add 3 mL of mTeSR1™ to each dish and gently scrape the cells off the bottom of each dish using a 16-cm cell scraper.

5. Transfer the detached hESC aggregates into a 15-mL tube. Rinse the dishes with 5 mL of mTeSR1™ and transfer the cells to the 15-mL tube.

6. Centrifuge the sample at $300 \times g$ for 5 min, then remove the supernatant and gently resuspend the hESC aggregates in 2 mL of mTeSR1™ using a 2-mL plastic pipette, taking care not to triturate to a single cell suspension.

7. As it is often useful to know the average hESC number per aggregate, take 5 µL of the hESC aggregates and place in one well of a flat-bottom 96-well plate. Add 20 µL of PBS and count the number of aggregates with length × width dimensions >70 × 50 µm (see Note 14). Determine the concentration of aggregates (# of aggregates/5 µL).

8. Take ~10% of the hESC aggregate solution (equal to "x" aggregates), centrifuge at $300 \times g$ for 5 min, and then remove the supernatant. Add 1 mL of TrypLE and incubate at 37°C/5% CO_2 for 10 min. Triturate the cell sample to generate a single cell suspension and perform a cell count using a hemocytometer to determine the total cell number obtained ("c"). Use this to determine the average cell number per aggregate (i.e. c/x; see Note 15).

9. Aliquot the hESC aggregates into sterile 1.5-mL microfuge tubes to give 10^5–10^6 hESCs per tube.

10. Centrifuge the microfuge tubes at $300 \times g$ for 5 min and place on ice without removing the supernatant.

11. Anesthetize mice (e.g. inhalation of isoflurane).

12. With a P200 pipette and sterile tip, take one microfuge tube and carefully remove and discard the supernatant covering the hESC aggregates.

13. Using a 1-mL syringe and 21-gauge needle, aspirate 100 µL to 200 µL of diluted Matrigel™ and gently resuspend the hESC aggregates, then draw the solution into the syringe ensuring that all air-bubbles are removed.

14. Use the thumb and forefinger of one hand to create a skin fold in the back flank of an anesthetized mouse, and with the other hand, carefully place onto this skin fold the needle/syringe containing the Matrigel™/hESC solution.

15. With EXTREME care, insert the needle into the skin without piercing the muscle below so that the needle moves freely under the skin. USE UTMOST CAUTION TO ENSURE THAT THE NEEDLE DOES NOT EXIT THE SKIN AND PIERCE ANY PART OF YOUR HANDS.

16. Slowly inject the solution beneath the skin and carefully remove the needle so that none of the solution escapes. A bolus can be felt under the skin once the injection is complete.

17. A maximum of one to four sites can be injected per animal, as permitted by your institutional animal care protocols.

18. Return mice to their cages and monitor until the anesthetic has worn off.

19. Assess the mice each week for 8–12 weeks to monitor for appearance of teratomas; once an assay end point has been reached (e.g. growth to a diameter of 2 cm), euthanize the mice, remove the teratomas, and place them in fixative (10% buffered formalin).

20. Embed teratomas in paraffin wax, then cut 5–6 μm sections and stain with hematoxylin and eosin.

21. Analyze sections using a microscope to determine the range of cell types present (endodermal, mesodermal, and ectodermal), photographing representative cell types (see Note 16). See also Fig. 3.

4. Notes

1. Fixative can be stored at 4°C for up to 1 month but should be warmed to room temperature before use. The diazonium salt staining solution should always be made fresh immediately before use.

2. If you do not have access to such facilities, follow the methods described for sectioning paraffin-embedded specimen and H&E staining in Chapter 21.

3. The AP$^+$-CFC assay can be adapted to feeder-dependent conditions with the following modifications: substitute plates containing mitotically inactivated MEFs used for standard hESC cultures (i.e. CF1 or CD1) for Matrigel™-coated dishes; substitute HES media (see Subheading 2.2, item 5) containing 4 ng/mL bFGF (StemCell Technologies) for mTeSR1™.

4. Use of the Y-27632 inhibitor allows increased survival and detection of AP$^+$-CFCs in this assay (13, 21). Some cell lines (e.g. trypsin-adapted lines) that survive well as single cells may not require the use of this inhibitor.

5. See Chapter 6 for a description on how to take hemocytometer cell counts with trypan blue exclusion.

6. The optimal number of cells for plating will depend on the AP$^+$-CFC content of a particular test cell suspension. Therefore, plating at multiple cell concentrations may be required to ensure that dishes with a suitable colony density for counting are obtained.

7. Colonies should be stained immediately after fixation, as increasing the time interval between fixation and staining decreases the strength of the alkaline phosphatase staining. If colonies cannot be counted immediately, plates should be stored in PBS at 4°C.

8. From highly undifferentiated mTeSR1™-based H9 hESC maintenance cultures, AP+-CFC frequencies range typically from 5 to 15% when using Y-27632. Since the AP+-CFC frequency is variable between different cell lines and is sensitive to changes in culture conditions (e.g. time between passaging, amount of spontaneous hESC differentiation), the optimal number of cells to be plated may vary accordingly and will have to be empirically determined.

9. Many different media formulations have been reported for EB formation and culture. We recommend HES medium (see Subheading 2.2, item 5) for the purpose of this assay as we have found that this medium supports EB formation from aggregates at relatively high efficiency and also allows the differentiation of cells within the EBs into cell lineages derived from all the three germ layers. Use of alternate media may lead to decreased EB formation and/or survival, altered differentiation kinetics, or skewing of differentiation towards a particular cell type. For example, when serum-containing medium is used instead of HES medium, we have found that differentiation-related gene expression changes are seen faster, but this is offset by a marked decrease in EB formation/survival.

10. Very dense EB cultures may need more frequent media changes or may require splitting into multiple petri dishes so that the culture medium is depleted less rapidly of essential components.

11. Changes in gene expression indicative of pluripotent differentiation are reliably detected after 15 days of EB culture under the conditions detailed here. However, it is possible that some hESC lines may show different differentiation kinetics under the same conditions.

12. While there are many genes that can be chosen to demonstrate pluripotent differentiation via qRT-PCR, we have found that this list of genes routinely shows changes consistent with pluripotent differentiation under the EB conditions described (14).

13. It is possible to substitute 1 mg/mL collagenase for 1 mg/mL dispase, in which case one PBS wash after incubation with collagenase is sufficient.

14. We use an eyepiece graticule calibrated to our objective lenses to calculate the length and width dimensions of hESC aggregates.

15. The hESCs remaining from the single cell suspension can be collected in RNA lysis buffer and stored at −80°C for subsequent analyses.

16. Confirmation of cell type identifications should be made by a trained pathologist.

References

1. Thomson, J. A., Itskovitz-Eldor, J., Shapiro, S. S., Waknitz, M. A., Swiergiel, J. J., Marshall, V. S., et al. (1998) Embryonic stem cell lines derived from human blastocysts. *Science* **282**, 1145–1147.

2. Reubinoff, B. E., Pera, M. F., Fong, C. Y., Trounson, A., and Bongso, A. (2000) Embryonic stem cell lines from human blastocysts: somatic differentiation in vitro. *Nat. Biotechnol.* **18**, 399–404.

3. Ludwig, T. E., Levenstein, M. E., Jones, J. M., Berggren, W. T., Mitchen, E. R., Frane, J. L., et al. (2006) Derivation of human embryonic stem cells in defined conditions. *Nat. Biotechnol.* **24**, 185–187.

4. Ludwig, T. E., Bergendahl, V., Levenstein, M. E., Yu, J., Probasco, M. D., and Thomson, J. A. (2006) Feeder-independent culture of human embryonic stem cells. *Nat. Methods* **3**, 637–646.

5. Takahashi, K., Tanabe, K., Ohnuki, M., Narita, M., Ichisaka, T., Tomoda, K., et al. (2007) Induction of pluripotent stem cells from adult human fibroblasts by defined factors. *Cell* **131**, 861–872.

6. Yu, J., Vodyanik, M. A., Smuga-Otto, K., Antosiewicz-Bourget, J., Frane, J. L., Tian, S., et al. (2007) Induced pluripotent stem cell lines derived from human somatic cells. *Science* **318**, 1917–1920.

7. Park, I. H., Zhao, R., West, J. A., Yabuuchi, A., Huo, H., Ince, T. A., et al. (2008) Reprogramming of human somatic cells to pluripotency with defined factors. *Nature* **451**, 141–146.

8. Lerou, P. H., and Daley, G. Q. (2005) Therapeutic potential of embryonic stem cells. *Blood Rev.* **19**, 321–331.

9. Lensch, M. W., and Daley, G. Q. (2006) Scientific and clinical opportunities for modeling blood disorders with embryonic stem cells. *Blood* **107**, 2605–2612.

10. Daley, G. Q. (2007) Towards the generation of patient-specific pluripotent stem cells for combined gene and cell therapy of hematologic disorders. *Hematology Am. Soc. Hematol. Educ. Program* **2007**, 17–22.

11. Murry, C. E., and Keller, G. (2008) Differentiation of embryonic stem cells to clinically relevant populations: lessons from embryonic development. *Cell* **132**, 661–680.

12. Amit, M., Shariki, C., Margulets, V., and Itskovitz-Eldor, J. (2004) Feeder layer- and serumfree culture of human embryonic stem cells. *Biol. Reprod.* **70**, 837–845.

13. O'Connor, M. D., Kardel, M. D., Iosfina, I., Youssef, D., Lu, M., Li, M. M., et al. (2008) Alkaline phosphatase-positive colony formation is a sensitive, specific and quantitative indicator of undifferentiated human embryonic stem cells. *Stem Cells* **26**, 1109–1116.

14. Ungrin, M., O'Connor, M. D., Eaves, C. J., and Zandstra, P. W. (2007) Phenotypic analysis of human embryonic stem cells. *Curr. Protoc. Stem Cell Biol.* **2**, 3.1–3.25.

15. Palmqvist, L., Glover, C. H., Hsu, L., Lu, M., Bossen, B., Piret, J. M., et al. (2005) Correlation of murine embryonic stem cell gene expression profiles with functional measures of pluripotency. *Stem Cells* **23**, 663–680.

16. Wiles, M. V. (1993) Embryonic stem cell differentiation in vitro. *Methods Enzymol.* **225**, 900–918.

17. Martin, G. R. (1981) Isolation of a pluripotent cell line from early mouse embryos cultured in medium conditioned by teratocarcinoma stem cells. *Proc. Natl. Acad. Sci. U.S.A.* **78**, 7634–7638.

18. Takahashi, K., and Yamanaka, S. (2006) Induction of pluripotent stem cells from mouse embryonic and adult fibroblast cultures by defined factors. *Cell* **126**, 663–676.

19. Itskovitz-Eldor, J., Schuldiner, M., Karsenti, D., Eden, A., Yanuka, O., Amit, M., et al. (2000) Differentiation of human embryonic stem cells into embryoid bodies compromising the three embryonic germ layers. *Mol. Med.* **6**, 88–95.

20. Lawrenz, B., Schiller, H., Willbold, E., Ruediger, M., Muhs, A. and Esser, S. (2004) Highly sensitive biosafety model for stem-cell-derived grafts. *Cytotherapy* **6**, 212–222.

21. Watanabe, K., Ueno, M., Kamiya, D., Nishiyama, A., Matsumura, M., Wataya, T., et al. (2007) A ROCK inhibitor permits survival of dissociated human embryonic stem cells. *Nat. Biotechnol.* **25**, 681–686.

Chapter 5

Using Cadherin Expression to Assess Spontaneous Differentiation of Embryonic Stem Cells

Helen Spencer, Maria Keramari, and Christopher M. Ward

Abstract

Embryonic stem cells (ESCs) are pluripotent cells derived from preimplantation embryos and can be maintained in an undifferentiated state over prolonged periods in vitro. In addition, ESCs can be induced to differentiate into cells representative of the three primary germ layers. As such, ESCs are a useful system for studying early developmental events in vitro and have the potential to provide a ubiquitous supply of somatic cells for use in regenerative medicine. However, significant differences in the expression pattern of various cell surface markers between murine and human ESCs, e.g. the SSEA series, necessitate the use of separate markers for determining the undifferentiated state of these cells. We have recently shown that an E- to N-cadherin switch occurs during spontaneous differentiation of both murine and human ESCs. Here we describe the use of E-cadherin and N-cadherin proteins and transcript expression for assessing the proportion of undifferentiated and spontaneously differentiated cells within ESC populations. In summary, loss of cell surface E-cadherin and/or gain of N-cadherin protein expression provides a useful nondestructive assay for the determination of the proportion of spontaneously differentiated cells within an ESC population. In addition, presence of N-cadherin transcripts in an ESC population is indicative of spontaneous differentiation of a proportion of the cells.

Key words: Embryonic stem cells, E-cadherin, N-cadherin, Pluripotency, Nanog, Oct-4, Differentiation

1. Introduction

1.1. Embryonic Stem Cells

Embryonic stem cells (ESCs) are pluripotent cells derived from preimplantation embryos that have the capacity for unlimited self-renewal (1). Along with their unique ability for undifferentiated proliferation in vitro, they are able to form all adult cell types. The use of murine ESCs (mESCs) in the generation of knockout and transgenic mice has contributed significantly to our understanding of gene function (2). In addition, mESCs have provided

Nicole I. zur Nieden (ed.), *Embryonic Stem Cell Therapy for Osteo-Degenerative Diseases*, Methods in Molecular Biology, vol. 690, DOI 10.1007/978-1-60761-962-8_5, © Springer Science+Business Media, LLC 2011

a unique in vitro system with which to investigate pathways involved in pluripotency, self-renewal, and differentiation. Culture conditions for the maintenance of undifferentiated ESCs vary considerably, an example being the maintenance conditions for human ESCs (hESCs) just described in the last three chapters, with several proprietary media formulations available. In our laboratory, we culture mESCs in the presence of fetal bovine serum (FBS) and leukemia inhibitory factor (LIF) in the absence of a fibroblast feeder layer.

ESC derivation from non-human primates such as rhesus monkey and marmoset (3, 4) and parallel advances in human in vitro fertilization embryo culture conditions (5) led to the derivation of hESCs in 1998 (6). Human ESCs can be maintained in an undifferentiated and karyotypically stable state for prolonged periods and differentiate into cells representative of the three primary germ layers. These cells represent a valuable in vitro system to study aspects of early embryogenesis; they also provide the opportunity to treat a wide range of human disorders/diseases. Though LIF supports the undifferentiated growth of mESCs, it does not support growth in expanding cultures of hESCs (6, 7). Therefore, hESCs are routinely cultured in the presence of basic fibroblast growth factor (bFGF) and Knockout™ serum replacement on fibroblast feeder layers (6).

Of critical importance in ESC studies is the ability to rapidly assess the undifferentiated phenotype of the cells to ensure a relatively homogenous pluripotent population for subsequent studies. While the pluripotent markers Oct-4 and Nanog are routinely used to assess the undifferentiated state of both murine and human ESCs, methods for the analysis of these markers are destructive to the cells. Furthermore, transcription of each of the pluripotency transcription factor triad Oct-4, Nanog, and Sox-2 may be enhanced as the result of the onset of differentiation (8–10), before finally subsiding. Consequently, considerable research has been performed to assess cell surface markers suitable for the non-destructive analysis of ESC pluripotency (11).

1.2. Species-Specific Variation in Cell Surface Markers of ESC Pluripotency and Differentiation

Stage-specific embryonic antigen-1 (SSEA-1) is a cell surface lactoseries oligosaccharide antigen present in the mouse inner cell mass in trophectoderm and in ESCs (11, 12). Unlike the assessment of mESC pluripotency using transcript expression, which utilizes destructive techniques, SSEA-1 can be determined on live cell populations. Following induction of mESC differentiation, SSEA-1 is downregulated and is commonly used as a pluripotency marker (11). However, Cui et al. observed that mESCs exhibit heterogeneous cell surface SSEA-1 expression and that a proportion of undifferentiated ESCs can be SSEA-1 negative (13). Using fluorescence-activated cell sorting (FACS) to separate SSEA-1 positive and negative populations, they observed that both populations exhibited a mixed SSEA-1 phenotype (13). Our lab has also shown that the expression of SSEA-1 does not necessarily reflect the pluripotent state of mESCs (14). Therefore, although SSEA-1 expression can provide some indication of the undifferentiated state of a live mESC population, it should be used in conjunction with other markers of pluripotency.

In mESCs, downregulation of SSEA-1 during differentiation is associated with increased expression of other SSEA family members such as SSEA-3 and SSEA-4. In contrast, hESCs exhibit high expression levels of SSEA-3 and -4 in the undifferentiated state, and these are decreased following induction of differentiation, while SSEA-1 expression is increased (11). This suggests that common cell surface antigen markers might not exist between murine and human ESCs. However, we have assessed expression of various members of the cadherin family and have demonstrated that E- and N-cadherin can be used as cross-species markers for determining the undifferentiated status of ESCs (15, 16).

1.3. E-Cadherin as a Marker of ESC Pluripotency

E-cadherin (E-cad) is a 120 kDa cell surface glycoprotein responsible for calcium-dependent cell–cell adhesion in most epithelial tissues and is expressed at high levels on ESCs (17). Abrogation of E-cad in wild-type murine and human ESCs results in significantly decreased cell–cell contact, although the cells maintain an undifferentiated phenotype (15, 16). Loss of E-cad is a defining characteristic of epithelial–mesenchymal transition (EMT), a process essential for the ingression of epiblast cells within the primitive streak during early embryogenesis and associated with tumor cell metastasis (18). We have shown that spontaneous differentiation of ESCs results in rapid loss of cell surface E-cad protein, suggesting that differentiation induction is associated with an EMT event (15, 16). As we show in Fig. 1a, E-cad is expressed on the majority of the population of undifferentiated ESCs, whereas expression is rapidly lost from a proportion of the population following differentiation (Fig. 1).

Loss of cell surface E-cad in early differentiating ESCs is not associated with significant downregulation of SSEA-1. Similar to what is observed in mESCs, hESCs also exhibit rapid loss of cell surface E-cad protein upon spontaneous differentiation and this is associated with the absence of Oct-4 protein expression (15). Therefore, cell surface E-cad expression is a very useful marker for determining the proportion of spontaneously differentiated cells within an ESC population using a nondestructive cell assay.

1.4. N-Cadherin as a Marker of ESC Differentiation

N-cadherin (N-cad) is a member of the cadherin glycoprotein superfamily and is involved in calcium-dependent adhesion. During tissue reorganization in embryonic development, cells can undergo an EMT event where E-cad is downregulated at the cell surface. For example, epiblast ingression within the primitive

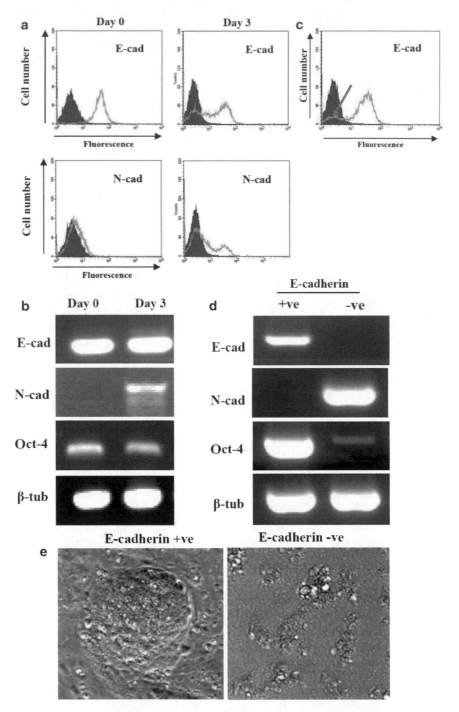

Fig. 1. Analysis of E-cadherin and N-cadherin proteins and transcripts in murine ESCs. (**a**) Mouse ESC line MESC20 was cultured in ESC medium (day 0) or for 3 days in differentiation inducing medium. Cells were assessed for cell surface expression of E-cadherin (E-cad) and N-cadherin (N-cad) using flow cytometry. The profiles for E- and N-cadherin at day 3 demonstrate differentiation of a proportion of the cells within the population. (**b**) RT-PCR expression analysis for *E-cad*, *N-cad*, *Oct-4*, and β-*tubulin* (β-*tub*) transcripts in cells treated as described in (**a**). Note that while E-cad and Oct-4 transcripts remain unaltered, the level of N-cad transcripts is upregulated in the differentiation-induced population. Therefore, detection of N-cad transcripts in an ESC population is a good indicator of the presence of spontaneously differentiated cells. (**c**) An example of E-cad expression in an ESC population cultured in ESC medium. The *arrow*

streak is not only associated with loss of E-cad, but also increased expression of cell surface N-cad (18). Our studies have shown that spontaneous ESC differentiation is associated with an E- to N-cad switch (15, 16). In contrast to E-cad, N-cad expression is very low in undifferentiated ESCs and upregulated in a proportion of the population following differentiation, with N-cad expressing cells exhibiting significantly decreased levels of Oct-4 protein and transcripts (Fig. 1).

These results demonstrate that determining the loss of cell surface E-cad and/or gain of N-cad protein provides a rapid and nondestructive technique for assessing the proportion of spontaneously differentiated cells within an ESC population. In addition, presence of *N-cad* transcripts in an ESC population is indicative of spontaneous differentiation of a proportion of the cells.

2. Materials

All reagents should be stored at room temperature unless otherwise stated.

2.1. Mouse ESC Culture Reagents and Media

1. 1× Phosphate-buffered saline (PBS) at pH 7.4 (Invitrogen).

2. Gelatin from porcine skin, type A cell culture tested. Make a 0.1% (w/v) solution in ddH_2O, autoclave to dissolve, and store at 4°C (see Note 1). Use within 1 month.

3. Fetal bovine serum (FBS). FBS for use with ESCs must be screened prior to use to ensure low toxicity and ability to maintain the undifferentiated state of the cells. Alternatively, serum can be purchased prescreened from serum suppliers. We have used Hyclone (Fisher Scientific) and Invitrogen ES screened serum to successfully maintain germline competence in the E14TG2a ESC line. FBS should be aliquoted into sterile containers under aseptic conditions. Store at –20°C for future use.

4. ESC medium: mESCs are maintained in Knockout™ Dulbecco's Modified Eagle's Medium (K-DMEM, Invitrogen) supplemented with heat-inactivated FBS, nonessential amino acids (NEAA) solution (100×, 10 mM), and L-glutamine.

Fig. 1. (continued) indicates spontaneously differentiated cells within the population. (**d**) MESC20 murine ESCs were cultured for 3 days in differentiation-inducing medium and cells were subsequently sorted by fluorescent-activated cell sorting (FACS) based on the expression of the cell surface E-cad antigen. E-cad⁺ and E-cad⁻ cells were then assessed for the expression of *E-cad*, *N-cad*, *Oct-4*, and β-*tub* transcripts by RT-PCR. (**e**) E-cad⁺ and E-cad⁻ cells isolated by FACS were plated into ESC media and assessed for formation of characteristic ESC colonies by phase contrast microscopy. Note that E-cad⁺ cells grew into characteristic ESC colonies, whereas E-cad⁻ cells exhibited significant cell death and differentiation. Images in these figures have been previously published in full or in part by ASCB (http://www.molbiolcell.org/cgi/content/full/18/8/2838#F1).

For example, to each 500 mL of K-DMEM add 50 mL heat-inactivated FBS, 6 mL NEAA (1% v/v), and 6 mL L-glutamine to a final concentration of 1% (v/v). Add (v/v) 2-mercaptoethanol to a final concentration of 50 μM and 1,000 U/mL leukemia inhibitory factor (LIF) (see Note 2).

5. Knockout Serum Replacement™ (KSR, 500 mL, Invitrogen). KSR should be aliquoted into sterile containers under aseptic conditions. Store at –20°C for future use.

6. ESC differentiation medium: To provide a positive control for spontaneously differentiated cells, mESCs should be cultured for 1–3 days in K-DMEM (Invitrogen) supplemented with 10% KSR, 1% (v/v) NEAA, 1% (v/v) L-glutamine, and 50 μM 2-mercaptoethanol (see Notes 3 and 4).

7. 1× trypsin/ethylenediaminetetraacetic acid (EDTA): 0.05% trypsin, 0.02% EDTA·4Na. Store at 4°C.

8. Freezing medium: for 10 mL, mix 1 mL dimethyl sulfoxide (DMSO) with 9 mL ESC medium.

9. Disposable hemocytometer (i.e. Kova Ltd).

10. Disposable 10- and 25-mL pipettes, 10-, 20-, 200-, and 1,000-μL filter tips, designated for tissue culture use only, and 1.8 mL cryovials.

11. Inverted phase contrast microscope with 10× and 25× objectives (i.e. Nikon Eclipse TS100).

12. Centrifuge suitable for 20 mL universal containers, for example Star Labs.

2.2. Human ESC Culture Reagents and Media

1. Irradiated mouse embryonic fibroblast feeder cells (VH Bio). Store in liquid nitrogen.

2. Cell dissociation buffer (Invitrogen). Store at 4°C.

3. H1 (WiCell Research Institute, WI) hESCs are maintained in DMEM/F12 supplemented with KSR (see above), 1% (v/v) NEAA, 1% (v/v) L-glutamine, 50 μM 2-mercaptoethanol, and 4 ng/mL (Invitrogen) on irradiated mouse embryonic fibroblast feeder cells (see Notes 2 and 4).

4. Freezing medium: For 10 mL of freezing medium, use 1 mL of DMSO and 9 mL FBS.

5. Disposable hemocytometer (i.e. Kova Ltd).

6. Plastic ware as described above.

2.3. Fluorescent Flow Cytometry

1. Fluorescent-activated cell sorting (FACS) buffer: For 500 mL FACS buffer, add 1 g bovine serum albumin (BSA) (0.2% w/v) and 0.5 g sodium azide (0.1% w/v) to 500 mL 1× PBS at pH 7.4 (Invitrogen). Store at 4°C. This is used as the antibody diluent and for all washes.

2. Cell fixative solution: 1% (v/v) solution of formaldehyde in 1× PBS.

3. 1.5 mL microcentrifuge tubes, such as Eppendorf.

4. Flow cytometer, i.e. FACSCalibur (Becton Dickinson) or similar.

5. Primary antibodies: mouse anti-human E-cadherin clone SHE78.7 (Invitrogen), rat anti-mouse E-cadherin clone DECMA-1 (Sigma), and mouse anti-N-cadherin clone GC-4 (Sigma).

6. Secondary antibodies: Goat anti-mouse conjugated to either FITC or PE (i.e. Santa Cruz) and goat anti-rat conjugated to either FITC or PE (i.e. Santa Cruz).

7. FACS tubes (i.e. Becton Dickinson).

2.4. Transcript Analysis in ESCs

2.4.1. RNA Extraction and Purification

1. Trizol® reagent (Invitrogen). Store at 4°C.

2. DNaseI and buffer (Invitrogen). Store at –20°C.

3. RNasin (Promega). Store at –20°C.

4. DNase/RNase free water (Invitrogen).

5. Termination mix, supplied with DNaseI enzyme (Promega). Store at –20°C.

6. Chloroform and phenol solutions.

7. 3 M sodium acetate (pH 4.8).

8. Ethanol, 70 and 95%.

9. Glycogen. Store at –20°C.

2.4.2. Reverse Transcription of Extracted mRNA to cDNA

1. Oligo dT primers. Store at –20°C.

2. AMV reverse transcriptase and RT buffer. Store at –20°C.

3. RNasin. Store at –20°C.

4. DNase/RNase-free water.

5. dNTP mix. Store at –20°C.

2.4.3. Polymerase Chain Reaction

1. ReddyMix (Abgene). Store at –20°C.

2. Thin-walled 0.2 mL polymerase chain reaction (PCR) tubes, i.e. GeneFlow Ltd.

3. PCR thermal cycler (Jencons Plc or similar).

4. Agarose.

5. Ethidium bromide solution (10 mg/mL).

6. Tris–borate–EDTA buffer (TBE).

7. Molecular weight standard, for instance PCR 100 bp Low Ladder (Sigma).

8. 5× Loading buffer (i.e. BioLine).

9. Horizontal electrophoresis unit (Jencons Plc).

10. Gel documentation system, BioDoc-It, M-26X (Jencons Plc) or similar.

3. Methods

3.1. Preparation of Gelatin-Treated Tissue Culture Plates

1. In a laminar flow cabinet, add enough of the 0.1% gelatin solution to cover the bottom of the plate or flask (for a ⌀ 100-mm dish, this is typically about 6–7 mL) (see Note 5).

2. Store dishes containing this solution overnight at 4 or 37°C for 1 h (see Note 6).

3. Remove the gelatin solution from the plates and air dry in a laminar flow cabinet until all solution has evaporated (see Note 7).

4. Store the plates at 4°C for a maximum of 1 month.

3.2. Murine ESC Culture

3.2.1. Thawing Cells

1. Remove cryovial from liquid nitrogen and leave on ice for 1 min.

2. Thaw vial at 37°C in a water bath. Only leave in for as short as possible, but until completely thawed (see Note 8).

3. Transfer the contents of the cryotube to 10 mL of medium in a 20 mL screw-capped bottle and then pellet the cells by centrifugation ($1,000 \times g$, 3 min).

4. Aspirate freezing media and gently resuspend the cell pellet in the required volume of medium (9 mL for a 100-mm diameter dish or three wells of a 6-well plate) and transfer to a gelatin-treated tissue culture plate.

3.2.2. Maintenance of mESCs

1. To subculture ESCs, aspirate the medium and gently wash the cells twice with 1× PBS.

2. Add 4 mL of trypsin/EDTA to a 100-mm dish, remove after 10 s, and place in a tissue culture incubator (37°C and 5% CO_2).

3. After 60 s, remove the plate and gently tap to detach the cells.

4. Resuspend cells in 3 mL medium and transfer 0.5 mL of this to a fresh gelatin-treated tissue culture dish (approximately 3×10^6 cells) containing 8 mL of medium; this results in a 1:6 dilution of the original culture.

5. Gently move the plate from side to side to ensure an even distribution of the cells and place in a tissue culture incubator at 37°C and 5% CO_2.

6. Observe ESCs daily and passage at approximately 70–80% confluency, typically every 2 days (see Note 9).

3.2.3. Freezing mESCs

1. Trypsinize cells from an approximately 70% confluent 100-mm dish (see Subheading 3.2.2).

2. Pellet cells by centrifugation at $1,000 \times g$ for 3 min.

3. Resuspend cell pellet in 1 mL of freezing medium and aliquot 250 µL into individual cryotubes.

4. To freeze cells slowly, wrap vials in several layers of paper towels, taking care to keep the tubes upright (see Notes 10 and 11).

5. The wrapped vials should be stored upright overnight at –70°C and transferred to liquid nitrogen the following day for long-term storage.

3.2.4. Induction of Spontaneous Differentiation in mESCs

1. Aspirate the medium and gently wash the cells twice with 1× PBS.

2. Add 4 mL of trypsin/EDTA to a 100-mm dish, remove after 10 s, and place in a tissue culture incubator (37°C and 5% CO_2).

3. After 60 s, remove the plate and gently tap to detach the cells.

4. Resuspend cells in 3 mL of medium and transfer 0.5 mL of this to a fresh gelatin-coated tissue culture dish (approximately 3×10^6 cells) containing 8 mL of the same medium. For alternative ways to induce differentiation in ESCs, see Chapters 9, 10, and 14.

5. Assess cell surface expression of E- and N-cadherin at days 1, 2, and 3 (see Subheading 3.5).

3.3. Human ESC Culture

3.3.1. Thawing Cells

The procedure is the same as for mESCs except gently resuspend the cell pellet in the required volume of medium (9 mL for three wells of a 6-well plate). Transfer to a gelatin-treated tissue culture plate (see Subheading 3.1) containing a fresh fibroblast feeder layer (see Note 12).

3.3.2. Maintenance of hESCs

1. To subculture hESCs, aspirate the medium and gently wash the cells twice with 1× PBS.

2. Add fresh medium and, under an inverted phase contrast microscope, gently scrape the undifferentiated colonies to produce small clumps of cells.

3. Transfer 200 µL of disaggregated cells to a fresh dish coated with gelatin and mitotically inactivated fibroblast feeder cells.

4. Gently place the culture dish in a tissue culture incubator at 37°C and 5% CO_2.

5. Observe hESCs daily and passage approximately every 5–7 days. Add fresh medium every 2 days.

3.3.3. Freezing hESCs

The procedure is the same as for mESCs except that cells are resuspended at 1.75×10^6 cells/mL freezing medium. Aliquot 1 mL into individual cryotubes (see Note 10).

3.3.4. Induction of Spontaneous Differentiation of hESCs

Cells should be cultured as described in Subheading 3.3.2 except that the hESCs are allowed to proliferate for between 8 and 12 days in the absence of subculturing.

3.4. Assaying Cell Surface Marker Expression Using Fluorescent Flow Cytometry

FACS analysis is a quick and reliable method to detect cell surface antigens expressed on individual cells. For the detection of cell surface antigen expression in ESCs:

1. Wash cells with 1× PBS twice and remove cells from plates by covering with dissociation buffer and incubating at 37°C for 2 min (see Note 13).

2. Once cells are detached, add 9 mL of medium and pellet by centrifugation (see Note 14).

3. Wash cells twice in 1× PBS and resuspend cells in FACS buffer to give a final concentration of 1×10^7 cells/mL.

4. Aliquot 100 µL of the cell suspension into a 1.5 mL microcentrifuge tube.

5. Pellet cells by centrifugation at $1,000 \times g$ for 2 min and decant the supernatant.

6. Resuspend the pellet in 100 µL appropriate antibody diluted at 1:100 in FACS buffer. Incubate on ice for 30 min (see Note 15).

7. Repeat step 5 and wash the cells once in 1 mL of FACS buffer.

8. Incubate the cells with 100 µL of fluorophore-conjugated secondary antibody (diluted in FACS buffer) for 30 min on ice and in the dark.

9. Pellet cells by centrifugation at $1,000 \times g$ for 2 min. Decant the supernatant.

10. Wash the cells once with 1 mL of FACS buffer.

11. Resuspend the cell pellet in 500 µL of cell fixative solution and transfer cells to a 5 mL FACS tube. Cells can be stored at 4°C in the dark for up to 1 week prior to analysis, although rapid analysis is recommended.

12. Assess cell surface fluorescence using a Becton Dickinson FACSCalibur (see Note 16).

3.5. RT-PCR Analysis

3.5.1. Extraction, Purification, and DNase Treatment of RNA

1. Remove media from ESCs and follow manufacturer's instructions for extraction of RNA using Trizol.

2. Resuspend RNA pellet in 50 µL of DNase-free water.

3. Treat RNA with DNaseI using the following method: 50 μg of total RNA, 10 μL of 10× DNaseI buffer, 5 μL of DNaseI, 2 μL of RNasin and bring to a total volume of 50 μL using RNase-free water. Incubate for at least 1 h at 37°C (see Note 17).

4. Add 10 μL of 10× termination mix to stop the reaction (see Note 18).

5. Add 50 μL of phenol and 30 μL of chloroform to the solution and mix by vortexing for 5 s.

6. Centrifuge at $16,300 \times g$ for 10 min at 4°C.

7. Transfer upper layer to a fresh microcentrifuge tube, add 55 μL of chloroform, vortex for 5 s, and centrifuge at $16,300 \times g$ for 10 min at 4°C.

8. Transfer upper layer to a fresh microcentrifuge tube and add 10% (v/v) of 3 M sodium acetate (pH 4.8), 2.5 volumes of 95% ethanol, and 20 μg of glycogen (see Note 19).

9. Vortex the solution and leave on ice for 1 h.

10. Centrifuge the sample at $16,300 \times g$ for 15 min at 4°C.

11. Remove the supernatant, add 50 μL of 95% ethanol, and centrifuge at $16,300 \times g$ for 15 min at 4°C.

12. Remove the supernatant, add 50 μL of 70% ethanol, and centrifuge at $16,300 \times g$ for 15 min at 4°C.

13. Remove the supernatant, air dry the pellet, and resuspend in 50 μL of RNase-free water.

14. Use 1 μL of the RNA solution to carry out a PCR reaction (see Subheading 3.5.3) using β-*tubulin* forward and reverse primers. If a PCR band is evident, it demonstrates contamination by genomic DNA and the DNase treatment should be repeated from step 3 onwards.

3.5.2. Formation of cDNA from RNA

1. Add 0.5 μg of total RNA to 5 μL of Oligo dT and bring volume to 40 μL using RNase-free water.

2. Mix well and incubate at 65°C for 10 min.

3. Immediately place tubes on ice for 5 min.

4. Add 20 μL of 5× RT buffer, 1 μL of RNasin, 10 μL of (2.5 mM) dNTP, 2 μL of AMV transcriptase enzyme, and 17 μL of RNase-free water to the solution.

5. Incubate the solution at 42°C for 1 h.

6. Inactivate AMV transcriptase by heating to 98°C for 5 min.

7. Centrifuge the solution at $16,300 \times g$ for 1 min and transfer supernatant to a fresh tube.

8. Store the cDNA solution at −20°C until required.

3.5.3. Polymerase Chain Reaction and Gel Electrophoresis

1. To 5 µL of ReddyMix in a 0.2 mL thin-walled PCR tube, add 0.5 µL of forward and reverse primers (50 pmol/µL stock solution), 0.5 µL of cDNA, and 3.5 µL of ddH$_2$O to provide a final reaction volume of 10 µL (primer sequences are shown in Tables 1 and 2).

2. Program the PCR thermal cycler as follows: (1) 94°C for 3 min, (2) 60°C for 1 min, (3) 72°C for 1 min, (4) 94°C for 30 s, (5) 60°C for 30 s, (6) 72°C for 30 s, (7) repeat steps 4–6 another 34 times, (8) 72°C for 10 min, and (9) hold at 4°C.

3. Separate PCR products by agarose gel electrophoresis: 2% (w/v) agarose gel containing 0.16 µg/mL ethidium bromide dissolved in TBE buffer (see Note 20). Leave gel to set at room temperature. To each PCR reaction (10 µL), add 1 µL loading buffer and then add this mix to a single well of the gel. Separate products along with a DNA standard at 100 V for 1 h in TBE buffer using a horizontal gel electrophoresis unit.

4. View PCR products using a BioDoc-It, M-26X gel documentation system.

Table 1
Primer sequences for mouse undifferentiated and differentiated markers

Gene	Forward primer sequence (5′ to 3′)	Reverse primer sequence (5′ to 3′)	Size (bp)
β-*Tubulin*	GGAACATAGCCGTAAACTGC	CACTGTGCCTGAACTTACC	317
Oct-4	AGAAGGAGCTAGAACAGTTTGC	CGGTTACAGAACCATACTCG	415
E-cadherin	CGAGAGAGTTACCCTACATA	GTGTTGGGGGCATCATCATC	214
N-cadherin	CCCAAGTCCAACATTTCCATCC	AAAGCCTCCAGCAAGCACG	781

Table 2
Primer sequences for human undifferentiated and differentiated markers

Gene	Forward primer sequence (5′ to 3′)	Reverse primer sequence (5′ to 3′)	Size (bp)
β-*Tubulin*	GGAACATAGCCGTAAACTGC	TCACTGTGCCTGAACTTACC	317
Oct-4	AGAAGGAGCTAGAACAGTTTGC	CGGTTACAGAACCATACTCG	415
E-cadherin	TCGACACCCGATTCAAAGTGG	TTCCAGAAACGGAGGCCTGAT	194
N-cadherin	CCGACGAATGGATGAAAGACC	TTGCAGCCTATGCCAAAGC	438

4. Notes

1. The gelatin will not dissolve until the solution is autoclaved.

2. All media should be warmed to 37°C prior to use, with the exception of the trypsin/EDTA solution, which should be at room temperature.

3. This medium has been optimized in our lab to induce a rapid E- to N-cadherin switch in mESCs.

4. Do not heat-inactivate KSR.

5. All work should be carried out in a laminar flow class II cabinet using disposable sterile plastic ware.

6. We find it better to incubate dishes overnight at 4°C.

7. Gelatin solution can be left on plates for up to 7 days at 4°C and removed prior to culture of ESCs.

8. Cell thawing must be performed quickly to minimize cell death. Care must be taken when removing vials from liquid nitrogen as these can explode when exposed to temperature changes. Always wear appropriate safety wear (including goggles) when handling cryovials.

9. It is possible to split ESCs grown in medium containing 10% ESC screened FBS at less than 1:6. However, this is cell line and serum dependent, so care must be taken when determining the optimum dilution for subculture of the user's own cell line. Undifferentiated ESCs form three-dimensional colonies with few spreading cells, although the exact morphology is cell line and serum dependent.

10. To maintain cell integrity, cells should be frozen slowly in medium containing a cryopreservant.

11. Alternatively, a Mr. Frosty cryocontainer (ThermoFisher) may be used.

12. For more details on fibroblast feeder layers, see Chapters 2 and 16.

13. Dissociation buffer must be used for hESCs as trypsin will remove the epitope recognized by Ab SHE78.7.

14. Addition of PBS directly to dissociated or trypsinized cells can cause cell aggregation. Therefore, always resuspend cells in medium.

15. If the protein of interest is not on the cell surface, e.g. Oct-4, then it is necessary to include a permeabilization step (10 min in 0.1% Triton X-100 diluted in FACS buffer) prior to incubation with primary antibody.

16. Not all fluorophores can be detected with the FACSCalibur.

17. We usually leave the incubation for between 1.5 and 2 h.

18. Alternatively, stop the reaction by incubation of the solution at 95°C for 5 min.

19. Addition of glycogen enhances the precipitation of low amounts of RNA. In addition, it provides a visible pellet that aids removal of the supernatant.

20. Ethidium bromide is likely to be carcinogenic. Addition of ethidium bromide to the molten gel should be performed in a fume hood to prevent accidental inhalation.

References

1. Smith, A.G. (2001) Embryo-derived stem cells: of mice and men. *Annu. Rev. Cell Dev. Biol.* **17**, 435–462.

2. Thomas, K.R., and Capecchi, M.R. (1987) Site-directed mutagenesis by gene targeting in mouse embryo-derived stem cells. *Cell* **51**, 503–512.

3. Thomson, J.A. (1995) Isolation of a primate embryonic stem cell line. *Proc. Natl. Acad. Sci. U.S.A.* **92**, 7844–7848.

4. Thomson, J.A. (1996) Pluripotent cell lines derived from common marmoset (Callithrix jacchus) blastocysts. *Biol. Reprod.* **55**, 254–259.

5. Gardner, D.K. (1998) Changes in requirements and utilization of nutrients during mammalian preimplantation embryo development and their significance in embryo culture. *Theriogenology* **49**, 83–102.

6. Thomson, J.A., Itskovitz-Eldor, J., Shapiro, S.S., Waknitz, M.A., Swiergiel, J.J., Marshall, V.S., et al. (1998). Embryonic stem cell lines derived from human blastocysts. *Science* **282**, 1145–1147.

7. Humphrey, R.K., Beattie, G.M., Lopez, A.D., Bucay, N., King, C.C., Firpo, M.T., et al. (2004). Maintenance of pluripotency in human embryonic stem cells is STAT3 independent. *Stem Cells* **22**, 522–530.

8. Niwa, H., Miyazaki, J., and Smith, A.G. (2000) Quantitative expression of Oct-3/4 defines differentiation, dedifferentiation or self-renewal of ES cells. *Nat. Genet.* **24**, 372–376.

9. Kopp, J.L., Ormsbee, B.D., Desler, M., and Rizzino, A. (2008) Small increases in the level of Sox2 trigger the differentiation of mouse embryonic stem cells. *Stem Cells* **26**, 903–911.

10. Darr, H., Mayshar, Y., and Benvenisty, N. (2006) Overexpression of NANOG in human ES cells enables feeder-free growth while inducing primitive ectoderm features. *Development* **133**, 1193–1201.

11. Henderson, J.K., Draper, J.S., Baillie, H.S., Fishel, S., Thomson, J.A., Moore, H., et al. (2002). Preimplantation human embryos and embryonic stem cells show comparable expression of stage-specific embryonic antigens. *Stem Cells* **20**, 329–337.

12. Solter, D. and Knowles, B.B. (1978) Monoclonal antibody defining a stage-specific mouse embryonic antigen (SSEA-1). *Proc. Natl. Acad. Sci. U.S.A.* **75**, 5565–5569.

13. Cui, L., Johkura, K., Yue, F., Ogiwara, N., Okouchi, Y., Asanuma, K., et al. (2004) Spatial distribution and initial changes of SSEA-1 and other cell adhesion-related molecules on mouse embryonic stem cells before and during differentiation. *J. Histochem. Cytochem.* **52**, 1447–1457.

14. Ward, C.M., Barrow, K., Woods, A.M., and Stern, P.L. (2003) The 5T4 oncofoetal antigen is an early differentiation marker of mouse ESC cells and its absence is a useful means to assess pluripotency. *J. Cell Sci.* **116**, 4533–4542.

15. Eastham, A.M., Spencer, H., Soncin, F., Ritson, S., Merry, C.L., Stern, P.L., et al. (2007) Epithelial–mesenchymal transition events during human embryonic stem cell differentiation. *Cancer Res.* **67**, 11254–11262.

16. Spencer, H.L., Eastham, A.M., Merry, C.L., Southgate, T.D., Perez-Campo, F., Soncin, F., et al. (2007) E-cadherin inhibits cell surface localization of the pro-migratory 5T4 oncofetal antigen in mouse embryonic stem cells. *Mol. Biol. Cell* **18**, 2838–2851.

17. Larue, L., Antos, C., Butz, S., Huber, O., Delmas, V., Dominis, M., et al. (1996) A role for cadherins in tissue formation. *Development* **122**, 3185–3194.

18. Cavallaro, U. and Christofori, G. (2004) Cell adhesion and signalling by cadherins and Ig-CAMs in cancer. *Nat. Rev. Cancer* **4**, 118–132.

Chapter 6

Generation of Human Embryonic Stem Cells Carrying Lineage Specific Reporters

Parinya Noisa, Alai Urrutikoetxea-Uriguen, and Wei Cui

Abstract

The distinctive properties of human embryonic stem cells (hESCs) enable them to provide unique models to study the network of signaling pathways that regulate organogenesis, generate disease models, produce cells and tissues for therapies, and identify new drugs for treatment. Genetic modification of hESCs is a powerful tool to assist the above studies. Generation of lineage-specific fluorescent protein reporter hESC lines will greatly benefit investigators to monitor specific cell lineages in a live, easy, and timely manner. This technique will facilitate high throughput screening to identify molecules important in regulating specific cell fate commitment. In addition, such reporter cell lines enable researchers to enrich certain cell populations by fluorescent activated cell sorting (FACS) for either downstream biological analysis or in vivo applications. We have shown that hESCs can be stably transfected with a plasmid in which expression of the green fluorescent protein (GFP) is under the control of the Oct-4 promoter using chemical transfection. The expression pattern of transgenic Oct-4-GFP reflects that of endogenous Oct-4.

Key words: Human embryonic stem cell, Reporter cell line, Stable transfection, Cell culture, Differentiation

1. Introduction

Human embryonic stem cells (hESCs) are able to grow in culture indefinitely while maintaining their pluripotency, and are able to differentiate into most, if not all, cell types in the body (1, 2). Therefore, hESCs have the potential to provide an unlimited source of all human cell types. Given those unique properties of hESCs, a considerable amount of studies have been done, and range from basic developmental biology to potential cell therapeutic applications. However, since culture and differentiation of hESCs are regulated by both intrinsic and extrinsic factors in a complicated paracrine and autocrine signaling network (3, 4), the

Nicole I. zur Nieden (ed.), *Embryonic Stem Cell Therapy for Osteo-Degenerative Diseases*, Methods in Molecular Biology, vol. 690, DOI 10.1007/978-1-60761-962-8_6, © Springer Science+Business Media, LLC 2011

optimal conditions for culture and specific differentiation of hESCs remain ambiguous, although progress is increasing rapidly ((5), see also Chapter 2). The generation of hESC reporter lines that contain lineage-specific markers can provide invaluable tools for investigating the factors that control self-renewal and lineage specific differentiation. What is explained using the Oct-4 promoter as an example in this chapter, finds applications in osteoblast and chondrocyte differentiation through the use of osteoblast and chondrocyte-specific promoters, i.e. for osteocalcin, Cbfa1, and aggrecan. In particular, introduction of fluorescent reporter genes under the control of cell-type specific promoters offers several advantages over other reporters as they enable investigators to observe kinetic changes of marker gene expression without terminating the culture. In addition, specialized cell types can be enriched by fluorescence activated cell sorting (FACS) for further analysis or applications.

Currently, three technical strategies have been applied for genetic manipulation of embryonic stem cells: transgenic approach, knock-in approach, and artificial chromosome approach, each of which has advantages and limits (5). Here we will focus our discussion on the transgenic approach, the most common method used for hESCs. The transgenic approach involves the random genomic integration of a construct containing a cell or tissue-specific promoter-driven fluorescent reporter gene, as well as a selection marker gene under the control of a constitutive expression promoter, such as a viral promoter. The DNA delivery could be achieved by plasmid transfection with electroporation (6), nucleofection (7) or chemical-based transfection reagents (8), or by viral transduction (9). Since the reporter gene construct is randomly integrated into the host genome, the site of integration could affect transgene expression (10). In addition, rearrangement of the transgene could also change the transgene expression pattern. Therefore, it is important to characterize the transfected clones to confirm that reporter gene expression is correspondent to the expression of the endogenous gene. We have previously generated stably transfected Oct4-GFP hESC reporter lines with chemical based transfection (8) and will use it as an example to further discuss in detail the generation of hESC reporter lines.

2. Materials

2.1. Preparation of MEFs and MEF-CM

1. Dissecting scissors, sterile.
2. γ-Irradiator, for instance IBL 637.
3. Phosphate-buffered saline (PBS) without Ca^{2+} and Mg^{2+}.

4. 2× P/S solution: Prepare PBS containing penicillin (50 U/mL)/streptomycin (50 μg/mL) (stock is 100×) to a final concentration of 2× in 50 mL Falcon tubes, 30–40 mL per tube. Store at 4°C.

5. Plastic ware: 50 mL Falcon tubes, Bijou tubes and tissue culture treated plastic, such as six-well plates and 100-mm dishes.

6. 0.05% trypsin/0.02% Ethylenediaminetetraacetic acid (EDTA, Sigma).

7. Mouse Embryonic Fibroblast (MEF) culture medium: Dulbecco's Modified Eagle's Medium (DMEM), high glucose without L-glutamine containing 10% fetal bovine serum (FBS, Sigma) and 2 mM L-glutamine.

8. Freezing medium: MEF medium containing 10% DMSO.

9. T-flasks, T75, T150, T225-cm^2 (Corning, BD Biosciences, or Nunc).

10. Gelatin solution, 2% in water, tissue culture grade (Sigma). Store at 4°C upon arrival. Prepare 0.5% gelatin by diluting 2% gelatin with embryo transfer water.

11. Human basic fibroblast growth factor recombinant (hbFGF) (PeproTech). Prepare an hbFGF stock by dissolving it in PBS containing 0.2% BSA (Fraction V) at 10 μg/mL. Store at –20°C or at –80°C in 0.5–1 mL aliquots.

12. hESC culture medium (also called SR medium) contains Knockout Dulbecco's Modified Eagle's Medium (K-DMEM, Invitrogen) supplemented with 20% Knockout Serum Replacement (KSR, Invitrogen), 0.1 mM nonessential amino acids (NEAA, stock is 100×), 1 mM L-glutamine, and 0.1 mM 2-mercaptoethanol to final concentrations as indicated.

13. L-broth: 1% tryptone, 0.5% yeast extract, and 0.5% NaCl.

2.2. Culture of hESCs

1. Matrigel™-growth factor reduced, a product of Becton Dickinson, distributed by VWR. Matrigel™ stock solution: Prepare by slowly thawing it overnight at 4°C to avoid gel formation. Add equal volume of ice-cold K-DMEM and mix well, then aliquot 1 mL into a prechilled 15-mL tube and store at –20°C (see Note 2). Make a Matrigel working solution by slowly thawing an aliquot of Matrigel stock (1 mL) on ice when needed. Dilute Matrigel stock 1:15 by adding 14 mL cold K-DMEM into the tube to make a final dilution of 1:30. Mix the solution thoroughly.

2. Collagenase IV solution: dissolve collagenase in K-DMEM at 200 U/mL, then filter through a 0.22-μm filter. Store the collagenase solution at –20°C in 5–10 mL aliquots until use (see Note 1).

3. 0.02% (0.5 mM) EDTA solution.

4. Differentiation medium: similar to SR medium except replacing 20% KSR with 20% FBS (Sigma).

5. Antibiotic agent as needed, i.e. G418 sulfate.

2.3. Transfection Reagents

1. Lipofectamine 2000 (Invitrogen). Kit includes Optimem.

2. Qiagen plasmid maxi kit (Qiagen).

3. QiaEx II gel extraction kit (Qiagen).

3. Methods

Similar to all the other cell types, transgene expression in hESCs could be achieved by a number of methods, such as electroporation, nucleofection, chemical transfection, and viral transduction. The first two approaches normally require single-cell suspension, which could affect the reattachment of hESCs after transfection (11). Viral transduction is a commonly used approach to obtain populational cells carrying transgenes as it is an efficient method for introducing foreign genes into a host genome. However, a good reporter cell line not only requires expression of the reporter gene, but also requires a specific expression pattern which reflects the endogenous promoter/enhancer gene expression. Surrounding DNA, structures of the transgene integration site have a significant effect on transgene expression and transgene integration, with most viral transduction being random. Therefore, individual clones of transduced cells may not have the same expression pattern. In our laboratory, chemical transfection is the commonly used method, although it may not have very high efficiency.

In addition to various methods of introducing transgenes into cells, several other factors could also affect efficiency of introducing transgenes into hESCs, such as culture of hESCs, propagation approaches, and DNA/transfectant ratio. For example, we routinely culture hESCs in feeder-free conditions with a mouse embryonic fibroblasts-conditioned medium (MEF-CM) and propagate the cells with collagenase IV treatment (12). However, hESCs are treated with EDTA split if transfection is going to be carried out. We will discuss these methods in more detail in the following sections.

3.1. Preparation of MEF-CM for hESCs Culture

3.1.1. Preparation of MEF

1. Terminate day 13–14 pregnant CD-1 mice according to animal regulations. Dissect the uterus and put into a 100-mm dish, then dissect and terminate each embryo and place them into the 50-mL tubes containing P/S solution, one per tube (see Note 3).

2. Wash the embryos three times in P/S solution.

3. Prepare Bijou tubes by adding 2 mL of trypsin/EDTA to each tube.

4. Remove viscera from the embryo, then cut carcass into small pieces and transfer into the Bijou tubes containing trypsin/EDTA (one to two embryos per tube depending on their size).

5. Take the tubes to the cell culture room and incubate them in a 37°C incubator for 10 min, then vortex to break the clumps.

6. Incubate the tubes again at 37°C for another 5 min.

7. Add 3 mL of MEF medium to each Bijou tube to reach a volume of 5 mL and mix.

8. Leave the cell suspension to allow the bigger tissue lumps to settle down at room temperature.

9. Transfer about ~4 mL of supernatant (single cell suspension) to a T75-cm² flask containing 10–15 mL of MEF medium.

10. Check all flasks in the next 1–2 days and make sure they have not been contaminated. Trypsinize and combine all the cells, then count the number in a hemocytometer. See Chapters 5 and 8 for details on hemocytometer counts.

11. Freeze cells in freezing medium, 1×10^7 cells/vial and label them as P0 (see Note 4).

3.1.2. Preparation of MEF-CM

1. Resuscitate one vial of frozen MEF into a T75-cm² (or T150-cm²) flask (P1) and replate into two T225-cm² flasks (P2) when they are confluent.

2. Once confluent, split again 1:4–6 depending on the status of the cells up to passage four.

3. Remove medium when most of the flasks are >90% confluent, then wash with approximately 10 mL of PBS.

4. Add 3 mL of trypsin/EDTA to each flask and incubate the cells at 37°C for about 5 min. When the cells have rounded up, then tap the flasks to loosen the cells. Top off with MEF medium to 7–10 mL once cells are detached. Collect all trypsinized cells together into 50-mL tubes and count the cell number.

5. Calculate the number of flasks required for seeding the cells after irradiation (about 19×10^6 cells/T225-cm²).

6. Prepare flasks before irradiating the cells. Add 12 mL of gelatin to each T225-cm² flask. Place them in the 37°C incubator for 5 min or leave them in the culture hood for 10–15 min.

7. Meanwhile, take the 50-mL tube containing MEF cells to the irradiator and irradiate the cells at 40 grays, then centrifuge the cells in 50-mL Falcon tubes at $200 \times g$ for 5 min in a universal centrifuge.

8. Resuspend the cells in an appropriate volume of MEF medium to a cell concentration of about 4×10^6 cells/mL.

9. Remove the gelatin from the T225-cm² flasks and immediately add 45 mL of MEF medium, then transfer ~19×10^6 cells to each T225-cm² flask.

10. Label each flask with the number and date, and then place them into a 37°C, 5% CO_2 incubator.

11. The next day, replace the MEF medium with 150 mL SR medium supplemented with 4 ng/mL of hbFGF (equals 0.4 µl/mL of hbFGF stock solution).

12. Collect medium from the flasks the following day into 150-mL collection bottles, which have been labeled with the collection date and flask number, then add the same volume of fresh SR medium containing 4 ng/mL of hbFGF to the flask (see Note 5).

13. Freeze the MEF-CM immediately in a –80°C freezer until use (see Note 6).

14. Repeat steps 12–13 for 7–8 days and check for contamination in each flask one to two times during the collection period (see below). If contamination is observed in a flask, discard the flask and corresponding collected MEF-CM.

15. Prepare two sets of 30 mL universal tubes and label each set with the same numbers as shown on the flasks. Add 3 mL of L-broth solution to each tube and transfer 2 mL of collected medium to the tube with the same number, 1 mL/tube. Incubate one set of tubes in a 37°C shaker overnight for bacterial contamination and incubate the other set of tubes at 25–30°C overnight for yeast contamination (see Note 7).

3.2. Feeder-Free Culture of hESCs

3.2.1. Preparation of Matrigel-Coated Plates

1. Place 1 mL of Matrigel working solution into each well of a six-well plate.

2. Incubate the plate at least overnight at 4°C or 1–2 h at room temperature if in an emergency. The plate containing the Matrigel solution can be kept at 4°C for up to 1 week.

3. Remove the Matrigel working solution immediately before use.

3.2.2. Routine Culture and Propagation of hESCs with Collagenase

1. Thaw the MEF-CM. Aliquot the exact amount to be used into a sterile tube. Add 4–8 ng/mL of fresh hbFGF (0.4–0.8 µL/mL) into the MEF-CM (see Note 8) and warm it to 37°C. The MEF-CM containing freshly added hbFGF will be called CM thereafter.

2. Take Matrigel-coated plates from the fridge. Remove Matrigel solution from the plates and immediately add 1–2 mL of CM.

3. Aspirate the medium from the hESCs and add 1 mL/well of collagenase IV stock (200 U/mL) to the cells in six-well plates.

4. Incubate for 5–10 min in a 37°C incubator until the edge of the hESC colonies become thick (see Note 9).

5. Remove the collagenase and gently wash the cells once with 2 mL of PBS/well.

6. Add 1 mL of CM into the well.

7. Gently scrape cells with a cell scraper or a 5 mL pipette, then add another 2 mL of CM. Dissociate the cells into small clusters (50–500 cells) by gently pipetting up and down.

8. Transfer the cells into the newly prepared Matrigel-coated plates. The final volume of the medium should be 3 mL per well (see Note 10).

9. Return the plate to the incubator. Be sure to gently shake the plates from left to right and back and forth to obtain an even distribution of cells.

10. Feed the cells daily with CM.

3.3. Transfection of hESCs

3.3.1. Preparation of hESCs with EDTA Split for Transfection

1. Prepare hESCs 24 h before transfection with EDTA split to obtain better transfection efficiency (see Note 11, Fig. 1).

2. Prepare Matrigel-coated plates as described before.

3. Remove media from hESCs (90–100% confluent in six-well plates), then wash with PBS.

4. Add 1 mL per well of 0.5 mM EDTA and incubate at 37°C for 6–10 min.

5. Aspirate the EDTA and add 1 mL of CM, then pipette (using a P-1000 pipette) a few times (about three to five times) to get cluster cells containing five to ten cells per cluster (see Note 12).

6. Add another 2 mL of CM and transfer 1 mL to each well of the new Matrigel-coated plate (1:3 split ratio).

7. Transfer the plate (3 mL total volume/well) back to the incubator, which will then be ready for transfection next day.

3.3.2. Determine the Working Concentration for the Selection Drug

1. Prepare hESCs as described above for transfection except reduced in proportion to 12-well plates.

2. Next day, prepare fresh CM containing a range of concentrations of the selection drug (typically five concentrations).

3. Replace the culture medium with above selection medium, including controls. One concentration is normally tested in duplicate wells.

4. Change selection medium daily and record cell death. Continue culturing cells for 2 weeks. The optimal concentration is the lowest one that kills all cells within 10–14 days (see Note 13).

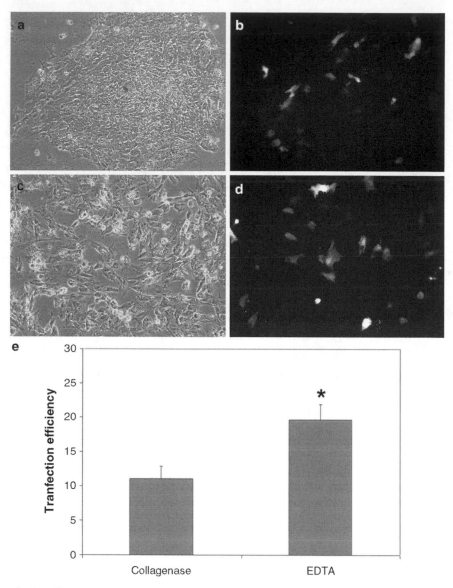

Fig. 1. Transfection efficiency in collagenase and EDTA prepared hESCs. Collagenase split hESCs (**a**, **b**) showed more compact morphology than EDTA split hESCs (**c**, **d**), which affected transfection efficiency. EDTA split cells showed higher transfection efficiency than collagenase split ones (**e**). *Asterisk* indicates $P < 0.001$.

3.3.3. Preparation of Plasmid DNA for Transfection

1. Prepare plasmid DNA by maxi prep with a Qiagen plasmid maxi kit, following the manufacturer's instructions.

2. Linearize the plasmid with the appropriate restriction enzyme and clean the DNA with a QiaEx II gel extraction kit following the manufacturer's instructions.

3. OD to obtain DNA concentration.

3.3.4. Transfection of hESCs with Lipofectamine 2000

1. For each well of hESCs, mix 2 µg of linearized DNA and 125 µL of Optimem in a clean tube; mix 6 µL of Lipofectamine 2000 with 125 µL of Optimem in another clean tube.

2. Add the DNA mix to the Lipofectamine mix to make a total volume of 250 µL of the transfection mix (see Note 14), then incubate the tube at room temperature for 10–20 min.

3. Replace the existing medium with 1 mL of fresh CM.

4. Add 250 µL of transfection mix drop-wise to the cells and incubate them at 37°C, 5% CO_2 for 6 h.

5. Replace the transfection mix with 3 mL of fresh CM and return to the incubator.

6. Feed the cells daily with fresh CM. Apply the selection drug 24–72 h later (see Note 15).

3.3.5. Picking and Expanding Transfected Colonies

1. Drug resistant colonies are visible 2–3 weeks after transfection. Count the number of colonies.

2. Prepare a 24-well culture plate and coat the wells with Matrigel overnight at 4°C as described above.

3. Replace the Matrigel with 0.5 mL of CM.

4. Remove the culture medium from the transfected hESCs and add 1 mL of collagenase, then incubate for 8–10 min similar to the collagenase propagation.

5. Aspirate the collagenase and wash cells with PBS.

6. Add 0.5 mL of CM, then using a P-1000 (set for 0.2–0.3 mL) pick up a colony and transfer to the above 24-well plate, one colony per well.

7. Change the pipette tip for each colony and add more CM if required, until all colonies have been transferred.

8. Expand colony as routine hESC culture.

3.4. Characterization of Stably Transfected Colonies

1. Checking for reporter gene expression: since transgene expression may be affected by the integration site, not all drug-resistant clones will express the reporter gene as expected. Select colonies in which reporter gene expression exhibits the same pattern as the endogenous gene whose promoter/enhancer is used to drive reporter genes. This is normally performed by differentiating the transfected hESCs through embryoid body formation (see Chapter 4 for details).

2. Stably transfected hESCs should also maintain the same characteristics of parental cells. Check for expression of cell surface specific antigens: SSEA-4 and Tra-1-60/81 by immunostaining and Fig. 2 for an example.

3. Check for karyotype of transfected colonies as described in Chapter 3.

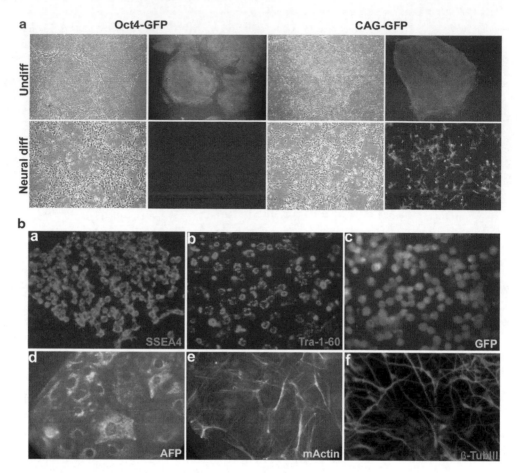

Fig. 2. Characterization of transfected reporter hESC lines. (**a**) Expression of green fluorescent protein (GFP) reporter gene under the control of the Oct-4 promoter is restricted to undifferentiated hESCs, but is dramatically downregulated after neural differentiation. In contrast, expression of GFP under the control of the constitutive CAG promoter exhibited consistent expression independent of differentiation status. (**b**) Stably transfected hESC clones maintained characteristics of parental hESCs, showing specific surface staining for SSEA-4 (a) and Tra-1–60 (b). (c) is the GFP staining of the same cells as (b). (c) Furthermore, the transfected hESCs are able to differentiate into cells of the three germ layers after embryoid body formation which are represented by positive staining of α-fetoprotein (AFP, (d)), muscle actin (mActin, (e)) and b-tublin III (β-TubIII, (f)).

4. Notes

1. Defrost and warm the collagenase solution to 37°C before use. If the solution is not finished, the remaining solution can be stored at 4°C for up to 1 week.

2. Geltrex™ Reduced Growth Factor Basement Membrane Matrix (Invitrogen) is an alternative product for Matrigel. It is important to keep everything (tube, solution) cold during the preparation of the Matrigel stock.

3. We noticed that it is difficult to obtain sufficient fibroblasts if less than day-13 gestation fetuses are used.

4. In our hands, the frozen MEF can be stored in liquid nitrogen for a year, after which however they exhibit a significant decline of proliferation in culture.

5. Make sure each collection bottle has been labeled with the corresponding flask number and collection date (or day). For example, the bottle which contains medium collected from flask 1 on the first day of collection will be labeled as flask 1/day 1 (or actual date of collection). This will enable you to eliminate a particular bottle of collection if any of the flasks have problems after certain days of collection (e.g. contamination).

6. MEF-CM can be stored at –80°C for up to a year in our laboratory with no effect on culturing hESCs.

7. In our laboratory, MEF-CM is normally collected continuously for 7–8 days and the contamination test is usually carried out on day 4 and day 8 of collection. If any of the flasks are found to be contaminated, the corresponding collection will be discarded. For example, if the tests on day 4 of the collection are all negative but flask 6 is found contaminated on day 8 test, then the day 5–8 MEF-CM collected from flask 6 will be removed and discarded from the freezer.

8. If MEF-CM has been frozen for more than 1–2 weeks, it is topped off with L-glutamine to a final concentration of 2 mM.

9. Each cell line may have a different sensitivity to collagenase treatment and also collagenase may vary slightly from batch to batch. Therefore, please check the cells to determine incubation time. The ideal time is when hESC colonies still adhere to the plate with a thickened edge but spontaneous differentiated cells have rounded up and can be easily removed following PBS washing.

10. When confluent, the optimal split ratio is usually 1:3. However, if hESC colonies are too compact, the hESCs tend to differentiate. It is necessary to separate the colonies before cells reach maximum confluence. Therefore, split ratio should be less than 1:3.

11. Routine culture of hESCs with collagenase usually produces compact colonies in which each cell has a small surface area. This will restrict the contact of foreign DNA with the cells and affect transfection efficiency (Fig. 1a, b). Splitting cells with 0.5 mM EDTA results in smaller and looser colonies that permit better DNA penetration and more efficient transfection (Fig. 1c, e).

12. Similar to the collagenase split, each cell line may respond to EDTA differently. Therefore, please check and determine the incubation time. Stop the incubation before cells round up and detached from the plates. Unlike splitting with collagenase, the cells treated with EDTA may readhere to the plate

after addition of CM. Therefore, do not treat more than two wells at one time.

13. As hESCs grow into colonies, it is difficult to plate the exact same number into each well. This is a simple way to determine optimal selection drug concentration. In our laboratory, the optimal concentrations have been determined using this method, 200–400 ng/mL for G418 and 0.75–1 μg/mL for puromycin.

14. If more than one well of cells are to be transfected with the same DNA, please multiply all reagents with the number of wells.

15. Timing of selection depends on the drugs to be used. The principle is to try to maintain cells in small clusters rather than single cells because single hESCs tend to differentiate. G418 can be applied 24 h after transfection as it only kills proliferating cells and the selection process takes longer. In contrast, puromycin kills both dividing and nondividing cells. Its selection is much quicker. Therefore, we normally apply puromycin 72 h after transfection.

References

1. Reubinoff, B. E., Pera, M. F., Fong, C. Y., Trounson, A., and Bongso, A. (2000) Embryonic stem cell lines from human blastocysts: somatic differentiation in vitro. *Nat. Biotechnol.* **18**, 399–404.

2. Thomson, J. A., Itskovitz-Eldor, J., Shapiro, S. S., Waknitz, M. A., Swiergiel, J. J., Marshall, V. S., and Jones, J. M. (1998) Embryonic stem cell lines derived from human blastocysts. *Science.* **282**, 1145–1147.

3. Niwa, H. (2007) How is pluripotency determined and maintained? *Development.* **134**, 635–646.

4. Walker, E., Ohishi, M., Davey, R. E., Zhang, W., Cassar, P. A., Tanaka, T. S., et al. (2007) Prediction and testing of novel transcriptional networks regulating embryonic stem cell self-renewal and commitment. *Cell Stem Cell.* **1**, 71–86.

5. Giudice, A. and Trounson, A. (2008) Genetic modification of human embryonic stem cells for derivation of target cells. *Cell Stem Cell.* **2**, 422–433.

6. Tomishima, M. J., Hadjantonakis, A. K., Gong, S., and Studer, L. (2007) Production of green fluorescent protein transgenic embryonic stem cells using the GENSAT bacterial artificial chromosome library. *Stem Cells.* **25**, 39–45.

7. Hohenstein, K. A., Pyle, A. D., Chern, J. Y., Lock, L. F., and Donovan, P. J. (2008) Nucleofection mediates high-efficiency stable gene knockdown and transgene expression in human embryonic stem cells. *Stem Cells.* **26**, 1436–1443.

8. Gerrard, L., Zhao, D., Clark, A. J., and Cui, W. (2005) Stably transfected human embryonic stem cell clones express OCT4-specific green fluorescent protein and maintain self-renewal and pluripotency. *Stem Cells.* **23**, 124–133.

9. Gallo, P., Grimaldi, S., Latronico, M. V., Bonci, D., Pagliuca, A., Gallo, P., et al. (2008) A lentiviral vector with a short troponin-I promoter for tracking cardiomyocyte differentiation of human embryonic stem cells. *Gene Ther.* **15**, 161–170.

10. Alami, R., Greally, J. M., Tanimoto, K., Hwang, S., Feng, Y. Q., Engel, J. D., et al. (2000) Beta-globin YAC transgenes exhibit uniform expression levels but position effect variegation in mice. *Hum. Mol. Genet.* **9**, 631–636.

11. Watanabe, K., Ueno, M., Kamiya, D., Nishiyama, A., Matsumura, M., Wataya, T., et al. (2007) A ROCK inhibitor permits survival of dissociated human embryonic stem cells. *Nat. Biotechnol.* **25**, 681–686.

12. Xu, C., Inokuma, M. S., Denham, J., Golds, K., Kundu, P., Gold, J. D., and Carpenter, M. K. (2001) Feeder-free growth of undifferentiated human embryonic stem cells. *Nat. Biotechnol.* **19**, 971–974.

Chapter 7

Manipulations of MicroRNA in Human Pluripotent Stem Cells and Their Derivatives

Stephanie N. Rushing, Anthony W. Herren, Deborah K. Lieu, and Ronald A. Li

Abstract

Human embryonic stem cells (hESCs) and induced pluripotent stem cells (iPSCs) reprogrammed from somatic cells can self-renew while maintaining their pluripotency to differentiate into virtually all cell types. In addition to their potential for regenerative medicine, hESCs and iPSCs can also serve as excellent *in vitro* models for the study of human organogenesis and disease models, as well as drug toxicity screening. MicroRNAs (miRNAs) are nonencoding RNAs of ~22 nucleotides that function as negative transcriptional regulators via degradation or inhibition by RNA interference (RNAi). MiRNAs play essential roles in developmental pathways. This chapter provides a description of how miRNAs can be introduced into hESCs/iPSCs or their derivatives for experiments via lentivirus-mediated gene transfer.

Key words: RNA interference, MicroRNA, Human embryonic stem cells, Directed differentiation

1. Introduction

Human embryonic stem cells (hESCs), isolated from the inner cell mass of blastocysts, can self-renew while maintaining their pluripotency to differentiate into virtually all cell types (1). More recently, direct reprogramming of adult somatic cells to become pluripotent hES-like cells (a.k.a. induced pluripotent stem cells or iPSCs) has been achieved (2, 3). Maintenance of pluripotency and induction of differentiation of hESC/iPSCs involve the complex interplay of a range of extrinsic and intrinsic factors, such as microenvironmental cues, gene expression and regulation, *trans*-acting molecules, such as transcription factors, microRNAs (miRNAs), and other small molecules. While the topics of pluripotency and differentiation have been extensively covered elsewhere, this chapter focuses

Nicole I. zur Nieden (ed.), *Embryonic Stem Cell Therapy for Osteo-Degenerative Diseases*, Methods in Molecular Biology, vol. 690, DOI 10.1007/978-1-60761-962-8_7, © Springer Science+Business Media, LLC 2011

on the use of miRNAs and RNA interference (RNAi) for studying hESC/iPSCs and their derivatives. MiRNAs are noncoding RNAs of ~22 nucleotides that mediate gene expression by functioning as negative transcriptional regulators via degradation or inhibition by RNAi (4, 5). While RNAi was initially discovered in *Caenorhabditis elegans* in the form of small-interfering (si) RNA-mediated silencing of viruses and other foreign DNA sources, many aspects of the pathway are conserved in miRNA-mediated gene silencing. Primary miRNA (pri-miRNA) sequences located within the genome resemble those of genes and are thought to undergo similar regulation, transcription, and posttranscriptional modification (6, 7). The transcribed precursor miRNA (pre-miRNA) sequence forms a hairpin-loop structure, which is recognized and cleaved by Dicer-1, a member of the highly conserved family of RNase III enzymes, in conjunction with Loquacious to produce mature miRNAs that are typically 21–23 nucleotides in length; miRNAs are then incorporated into the RNA-induced silencing complex (RISC), where the single-stranded mature miRNA binds the target transcript (or multiple transcripts) with sequence complementarity for degradation, deadenylation, or repression, to subsequently result in posttranscriptional silencing (8, 9).

To date, there are 706 known human miRNAs registered in the Sanger database and their corresponding pri-miRNA sequences are located at discreet loci along the genome: 53% (375/706) of miRNAs overlap genes, while 47% (331/706) are located within intergenic regions. Of the genic miRNAs, 81% (305/375) are exclusively found within introns, 11% (41/375) are exclusively found within exons, and the remaining 8% (29/375) are found within both exons or introns, primarily due to alternatively spliced transcripts (10). MiRNAs are thought to regulate ~30% of human genes (11). For instance, the miR-302 cluster (miR-302a, -302a*, -302b, -302b*, -302c, -302c*, -302d, and -302d*) located within intron 8 of LARP7 and the 371/372/373 cluster (miR-371-3p, -371-5p, -372, and -373) located within an intergenic region of chromosome 19 have been shown to associate with pluripotency (12–14). In mouse models, several miRNAs have been implicated in heart (e.g. miR-1, -18b, -20b, -21, -106a, -126, -133, -138, and -208), liver (miR-30, -122), CNS (miR-101, -124, -127, -128, -131, and -132, -134), and hematopoietic (miR-150, -155, -181) developments (15–24).

Experimentally, recombinant viruses such as lentiviral and adenoviral vectors can be conveniently used as delivery vehicles for introducing specific miRNAs into undifferentiated hESCs/iPSCs or their derivatives for investigating the corresponding functional consequences in such processes as pluripotency maintenance, differentiation, and maturation of specific lineages to obtain scientific insights. It is to be noted that adenovirus-mediated gene transfer is only transient in nature. In contrast, lentiviral gene transfer leads to transgene integration into the host genome and, therefore, to persistent modification.

2. Materials

2.1. Human ES/iPS Cell Culture

1. Human ESCs (some potential sources include the H1, H7, H9 lines, etc. from University of Wisconsin, Madison, WI; HES-1 to -6 from ESI and BG01-3 from Novocell) and iPSCs.

2. Gelatin diluted 0.1% in sterile water (Sigma-Aldrich).

3. Irradiated CF-1 mouse embryonic feeder (mEF) cells obtained from 13.5-day embryos (feeder-free systems are also available, please refer to other chapters or reviews for further details) (25, 26).

4. Dulbecco's Modified Eagle's Medium (DMEM) (Invitrogen) modified for cell culture: 20% fetal bovine serum (FBS; HyClone), 2 mM L-glutamine (Invitrogen), 0.1 mM 2-mercaptoethanol (Invitrogen), and 1% nonessential amino acids (Invitrogen). Store in the dark at 4°C.

2.2. Detecting miRNAs in hESCs/iPSCs and Their Derivatives

1. Collagenase IV (Invitrogen) diluted 1 mg/mL in DMEM.

2. Phosphate-buffered saline (PBS; Invitrogen). Store at 4°C.

3. Trizol (Qiagen). Store at 4°C. Use under chemical hood.

4. Chloroform.

5. Phase Lock Gel Heavy 2.0 mL (PLG) tubes (5 PRIME/Thermo Fisher Scientific).

6. miRNeasy Mini Kit (Qiagen).

7. Distilled water (diH$_2$O).

8. Paraflow miRNA microarray (Atactic Technologies).

9. Bioanalyzer (Agilent).

10. TaqMan stem-loop primers and qRT-PCR kit (Applied Biosystems).

2.3. Lentiviral Gene Transfer

1. Lentiviral vectors are available from academic labs (e.g. the Trono lab, Geneva, Switzerland, (27)) or commercial sources (e.g. pLenti4/V5-DEST and pLenti6/V6-DEST from Invitrogen).

2. Restriction enzymes (New England Biolabs, Inc.).

3. Primers.

4. Agarose.

5. T4 ligase.

6. Stbl3 *Escherichia coli* cells (Invitrogen). Store at −80°C.

7. Where appropriate, antibiotic, as specified in the lentiviral vector for the selection of positively transformed Stbl3 cells. Store at 4°C.

8. LB medium (MP Biomedicals LLC).

9. MiniPrep Kit (Qiagen).

10. MaxiPrep Kit (Qiagen).

2.4. Production of Recombinant Lentiviral Particles

1. Lentiviral vector suspended in water. Store at −20°C.

2. Lentivirus packaging plasmids: pMD2.G and psPAX2 (Addgene).

3. HEK293T/17 (293T) cell line (American Type Culture Collection).

4. 293T medium: DMEM with 10% FBS (HyClone), 2 mM L-glutamine (Invitrogen), 0.1 mM nonessential amino acids (Invitrogen), and 1% penicillin/streptomycin (Invitrogen). Store at 4°C.

5. 0.45-μm pore disposable sterile filter units (Nalgene/Thermo Fisher Scientific).

6. Trypsin, 0.05% (Invitrogen).

7. Poly-d-lysine diluted to 5 μg/mL in sterile water (BD Biosciences).

8. PBS (Invitrogen). Store at 4°C.

9. Lipofectamine 2000 reagent (Invitrogen). Store at 4°C.

10. OptiMEM medium (Invitrogen). Store at 4°C.

11. Ethanol, 75%.

12. Dulbecco's phosphate-buffered saline (DPBS) with Ca^{2+}, and Mg^{2+} containing 0.1% BSA.

2.5. Transducing hES/iPS Cells with Lentiviral Vector

1. Lentiviral particles suspended in PBS.

2. Polybrene diluted to 6 mg/mL in distilled water (Millipore).

2.6. Generation and Differentiation of a Pure Stably Transduced ES-Derived Cell Line

1. DMEM modified for suspension culture: 15% FBS (HyClone), 1% nonessential amino acids (Invitrogen), 1 mM L-glutamine (Invitrogen), 0.5 U/mL penicillin (Invitrogen), and 0.5 mg/mL streptomycin (Invitrogen). Store at 4°C.

3. Methods

3.1. Human ES/iPS Cell Culture

1. Human ESCs or iPSCs are grown to 70–80% confluence on a layer of mitotically inactivated mouse embryonic feeder cells (mEF) or a feeder-free system (see Subheading 2.1, step 3) on Matrigel-coated six-well plates. Each well of cells is immersed in 2 mL of culture medium.

2. Undifferentiated pluripotent stem cells can be maintained by changing the media daily. If maintained properly, hESCs and

iPSCs will maintain a normal karyotype, high levels of pluripotency markers, and characteristic colony morphology (1).

3. If the desired miRNAs to be introduced into hESCs/iPSCs are already known, proceed to Subheading 3.3 below.

3.2. Detecting miRNAs in hESCs/iPSCs and Their Derivatives

1. Comparing miRNA expression levels in undifferentiated and terminally differentiated cells via microarray is a simple way to determine the known miRNAs that are expressed in a specific cell type and their expression profile changes during differentiation, specification, and development (see Note 1).

2. hESCs/iPSCs in culture are lifted from the plate surface by applying 1 mL of collagenase IV to each well. A typical yield of 0.5–0.8 μg RNA/μL is expected for two 9.6 cm^2 wells on a six-well plate, or approximately 1–2 billion hESCs (see Note 2). The plate is allowed to incubate for approximately 30 min at 37°C. The cells are pelleted by centrifuging at $1,000 \times g$ for 1 min and washed twice with cold PBS. The cell pellet is then resuspended in 1 mL of Trizol per 1×10^7 cells and pipetted up and down 15–20 times with a P200 micropipettor. At this point, the Trizol suspension can be stored long term at −80°C.

3. To precipitate the RNA, one part chloroform is added to four parts Trizol suspension (see Note 3). Mix by inverting gently. Let sit at room temperature for 5 min. While waiting, centrifuge a PLG heavy tube at maximum speed (or approximately $15,000 \times g$ for 2 min). Invert the chloroform/Trizol suspension to create a transiently homogenous suspension before transferring it to the PLG tube. Centrifuge for 10 min at maximum speed to separate the organic (below gel) and aqueous (above gel) phases (see Note 4). Carefully transfer the aqueous phase, containing all RNAs, to a new microfuge tube.

4. Qiagen provides a simple way to extract whole RNA from the aqueous phase, including miRNAs, in the miRNeasy Mini Kit. The concentration of RNA yielded will vary depending on the quantity, quality, and conditions of the cells extracted. When eluting RNA, elute once with 30 μL diH$_2$O and a second time with fresh diH$_2$O to maximize yield.

5. In brief, 5 μg of total RNA is enriched for miRNAs by isolating the small RNA fraction. The enriched small RNA fraction must be analyzed for quality using a Bioanalyzer (Agilent) or similar method. The enriched small RNAs are then labeled with Cy3 dye for a single channel experiment. Dual channel experiments, comparing two cell types or treated and non-treated cells, require the labeling of one RNA sample with Cy3 and the other sample with Cy5. Labeled RNA is hybridized to the microarray chip and scanned for Cy3 or Cy5 signal

intensity, corresponding to the expression of specific microRNAs. See Zhu et al. for additional details regarding the preparation of RNA for microarray (28). A broad variety of techniques for analyzing miRNA expression by microarray has been reviewed by Li and Ruan (29).

6. Differential miRNA expression can be determined by comparing miRNA expression to endogenous controls included on the microarray chip, using ANOVA or a similar statistical analysis. Significantly up- or downregulated miRNAs can be selected for further investigation. Previous literature may provide some insight in narrowing down potential candidates. The Sanger Institute hosts miRBase, a Web-based repository of information on all known miRNAs (30). Putative mRNA targets may also be taken into account. There are several algorithms (PICTAR, miRANDA) available for calculating predicted miRNA:mRNA pairs based upon sequence similarity, free energy, and other parameters.

7. Candidate miRNAs should be validated by qRT-PCR. Applied Biosystems has developed a technique for miRNA qRT-PCR, which utilizes stem-loop primers capable of accurately detecting levels of miRNAs in a qRT-PCR. Techniques for assaying miRNA expression using this system have been described elsewhere (31, 32). qRT-PCR results are normalized to negative controls and endogenous controls, such as small nucleolar RNAs (snoRNAs) or stably expressed miRNAs (33). Endogenous controls will vary depending on the cell types used and experimental design (see Note 5).

3.3. Lentiviral
Gene Transfer

1. Beginning with a lentiviral backbone, a pre-miRNA can be inserted into the vector by molecular subcloning. Once a candidate pre-miRNA is discerned, primers are designed in Primer3 or a similar program to amplify the region containing this sequence from human genomic DNA in a standard PCR.

2. An aliquot of the PCR is used for a restriction enzyme digestion. Choose restriction enzymes that are unique to the desired insertion region of the lentiviral backbone. These restriction enzyme sequences should also flank the pre-miRNA. If the appropriate restriction enzyme sites do not exist on the respective sides of the pre-miRNA, they can be generated during the PCR by introducing the base pairs recognized by the restriction enzyme. After restriction enzyme digestion, the PCR products are run on a 0.7% agarose gel at 110 V for approximately 40 min. The correct bands are excised and extracted from the gel and ligated to the lentiviral backbone using T4 ligase.

3. The ligation reaction is transformed into thawed 50 μL aliquots of Stbl3 *E. coli* competent cells, as per manufacturer

instructions (see Note 6). The solution is mixed gently by hand and allowed to incubate on ice for 30 min. The cells are then heat shocked by immersing in a 42°C hot bath for 30–45 s and immediately quenching on ice for 3 min. 200 μL of SOC medium is added to the cells. The cells are placed in a shaker set at 37°C and 200 rpm for 1 h. The cell suspension is transferred and spread evenly onto an LB agarose plate with the antibiotic matching the selectable marker in the lentiviral backbone that confers antibiotic resistance. The plates are allowed to incubate overnight at 37 or 30°C, depending on the size of the lentiviral vector (see Note 7).

4. The following day, approximately eight colonies are chosen for amplification. Using a small pipette tip, each colony is transferred to a different test tube containing 4 mL of LB media with the appropriate antibiotic for selecting positively transformed cells. These tubes are allowed to incubate overnight in the shaker set at 37°C and 200 rpm.

5. On the next day, plasmid is purified from 2 to 3 mL of each bacterial clone using the Qiagen MiniPrep Kit (see Note 8). The clones are screened by restriction enzyme digestion or PCR, where one restriction site or primer is specific to the lentiviral backbone and another restriction site or primer is specific to the pre-miRNA insert. A small aliquot of each MiniPrep reaction is cut with the same restriction enzymes used to introduce the pre-miRNA into the lentiviral backbone.

6. The product of this digestion is run on an agarose gel to determine the band sizes. If the correct vector has been generated, the same colony can be amplified as described in step 4, but using 200 mL of LB with the appropriate antibiotic. The Qiagen MaxiPrep Kit is used to isolate plasmid from this larger volume of cells. The vector is further validated by sequencing.

7. For cell type-specific expression of the pre-miRNA of interest, an appropriate promoter can be introduced directly upstream of the pre-miRNA as in steps 1–6 above. For example, a promoter such as myosin light chain (34) can be used to confer cardiac specific expression to pre-miRNAs, since the activity of this promoter is restricted to cardiac cells. Other tissue-specific promoters can be employed as needed. If constitutive expression is desired, a constitutively active promoter (such as CMV, EF1α, or CAG) can be used instead (35). Similarly, an inducible promoter can be employed to express or repress the miRNA of interest in hESCs/iPSCs and/or their derivatives at different time points and at various expression levels (36). An additional fluorescent or antibiotic resistance reporter is also introduced into the vector, downstream

of the pre-miRNA sequence, to aid in identifying cells over-expressing the pre-miRNA (see Note 9).

3.4. Production of Recombinant Lentiviral Particles

1. HEK293T cells can be maintained in T175-cm^2 flasks immersed in 20 mL of 293T media. Cells are kept at 37°C and split once a week (see Note 10). The 293T cell line is split by washing the plate with 10 mL of sterile PBS twice. Then, 5 mL of trypsin is added to the plate and incubated at 37°C. After 5 min, 10 mL of medium is added to stop the trypsin reaction. The cells are centrifuged for 5 min at 1,000×g to pellet the cells. The pellet is resuspended in the appropriate amount of 293T medium for the desired split ratio to end with a final volume of 20 mL per flask. Typically, a 1:3 split or ~2×10^7 cells should result in 85–90% confluence on the next day.

2. It is recommended to start seeding from one confluent T175-cm^2 flask of 293T cells grown in 293T medium and expand these cells to nine T175-cm^2 flasks for the day of transfection. The final nine flasks for transfection are prepared by coating with 7 mL of 5 µg/mL poly-d-lysine for 2 h at room temperature. The plates are washed three times with 10 mL of sterile ddH$_2$O or PBS for 1 min and dried by placing them in the hood.

3. In one conical tube, 15 µg of transfer vector, 30 µg of packaging plasmid psPAX2, 7.5 µg of envelope plasmid pMD2.G, and 4.3 mL OptiMEM media are combined for each flask of 293T cells (for nine flasks, multiply all volumes by 9, see Note 11). In a second conical tube, 150 µL lipofectamine and 4.3 mL OptiMEM media are combined for each flask of 293T cells. After 5 min at room temperature, the contents of both tubes are mixed together and allowed to incubate at room temperature for an additional 20 min.

4. The media from each flask of 293T cells is aspirated and replaced by 15 mL of fresh OptiMEM media. The vector mix is then distributed evenly to the contents of the flask. After 5–6 h of incubation at 37°C, the media is aspirated and replaced with 20 mL of 293T medium. The flask is allowed to incubate at 37°C and 5% CO$_2$ overnight.

5. The following day, the flask can be assayed for fluorescent reporter expression by fluorescence microscopy (see Note 12).

6. Approximately 48 h after plating, >90% of the 293T cells should be transfected. At the 48-h mark, the supernatant from the 293T cell flasks is collected in 50-mL conical tubes and stored at 4°C for up to 3 days or –80°C for long-term storage. The supernatant is replaced with another 20 mL of fresh 293T medium per flask and collected again 24 h later at the 72-h posttransfection time point. The supernatants from

+48 and +72 h are centrifuged at 200-400 g units for 5 min to remove large particles of cell debris. At this point, unpurified supernatants are of low titer (~10^6 viral particles/mL) and can be applied directly to cells or frozen at –80°C.

7. If a higher titer (~10^9 viral particles) is desired, viral particles can be concentrated by ultracentrifugation. In this case, the nine flasks of 293T with 20 mL of viral supernatant collected at +48 and +72 h time points yield 360 mL of total volume. Three centrifuge tubes containing 60 mL of supernatant each are centrifuged twice consecutively. To concentrate viral particles, the ultracentrifuge is prepared by setting to 4°C and applying the vacuum for 1–1.5 h prior to centrifugation. Ultracentrifuge tubes are sterilized by soaking in 75% ethanol for 15 min. The tubes must be washed with PBS three times to remove all traces of ethanol. In the culture hood, the 0.45-μm virus filter membrane is prewetted with 20 mL of 293T medium to reduce unspecific binding of virus to the filter. Vacuum suction is applied to remove filtrate (see Note 13). The virus suspension is immediately applied to the filter. The virus filtrate is pipetted into the prepared ultracentrifuge tube. If balances are used, ensure the difference between weights is less than 0.1 g. The tubes are placed into the ultracentrifuge and centrifuged at 4°C for 2 h at 80,000 g units (see Note 14).

8. The viral supernatant is disinfected by pouring it into pure bleach. In the culture hood, the tubes are inverted over a Kimwipe to remove excess supernatant. The pellet should be translucent and barely visible. A large pellet may indicate excessive cellular debris due to a high rate of posttransfection cell death. When the tubes are free of all supernatant, a total volume of 100–300 μL of 4°C cold DPBS/0.1% BSA is added to the viral particles, split evenly among the centrifuge tubes. The tubes are sealed with parafilm and allowed to shake overnight on a platform shaker in the cold room at 4°C to dissolve the pellet.

9. On the following day, the contents of the tubes are pooled, pipetted up and down 5–10 times (see Note 15), aliquoted, and immediately stored at –80°C (see Note 16).

10. Successful generation of virus is assayed by virus titration. Twenty-four hours prior to titration, cells (see Note 17) are resuspended in 293T medium at 0.5×10^6 cells/mL in a total volume of approximately 10 mL. 100 μL of the hESC suspension is added to each well in a 96-well plate. Ten microfuge tubes are prepared by adding 90 μL of 293T medium to each tube. The virus is then series diluted by adding 10 μL to the first tube and mixed by pipetting. With a clean pipette tip, 10 μL of virus suspension from the first tube is transferred to

the second tube. The serial dilution is repeated for all remaining tubes for a total of ten dilutions (dilution factor 10^1–10^{10}). Then, 45 µL of each virus dilution is added to each aspirated well in the 96-well plate in duplicate. Twenty-four hours later, an additional 150 µL of fresh 293T medium is added. The reporter for positive viral transduction, usually fluorescence, is assayed after 2–3 days.

11. When assaying, the well containing transduced cells at the greatest dilution is used to determine the positively transduced cell count. Viral titer is calculated by dividing the number of positively transduced cells by the volume of the viral dilution added (45 µL) multiplied by the dilution factor, 10^n, where n is the value of the dilution. For example, 20 colonies of GFP$^+$ cells in the sixth dilution are equal to 2×10^6 infection units per 5 µL or 4×10^5 infection units per 1 µL. Take the average of duplicate dilutions. A reasonable titer after concentration is ~10^9 transducing unit/µL.

3.5. Transducing hES/iPS Cells with Lentiviral Vector

1. To transduce the hESCs, colonies are manually detached and resuspended in 1 mL culture medium supplemented with a final concentration of 8 µg/mL polybrene.

2. Lentiviral particles are then added to the cell suspension. The volume of virus depends on the viral titer and must be optimized (see Note 18).

3. The cells are incubated with the lentivirus for 5–6 h at 37°C.

4. After incubation, hESCs inoculated with the lentivirus are plated on mEF or Matrigel depending on the culture system (see other chapters) with an additional 1 mL of mEF-conditioned medium.

3.6. Generation and Differentiation of a Pure Stably Transduced ES-Derived Cell Line

1. Positively transduced colonies are typically identifiable by the reporter of choice (e.g. GFP). Using a glass needle, colonies are manually excised from the bottom of the plate. The media containing excised colonies is transferred via pipette into a 5-cm culture dish containing 10 mL of suspension media. The suspension culture is allowed to incubate at 37°C.

2. The media in the suspension culture is changed approximately every 48 h. To change the media, the contents of the culture dish are pipetted into a conical tube and allowed to pellet naturally or by centrifuging at $1,000 \times g$ for 1 min. The majority of the supernatant is aspirated off and replaced with fresh suspension media. The contents of the conical tube are then transferred back to the culture dish and returned to the incubator.

3. After 7 days in suspension culture, the colonies are replated onto a gelatin-coated 6-well plate. Cells maintained in suspension

medium will begin to spontaneously differentiate into cells from all three germ layers. Different components (growth factors, small molecules, miRNAs, etc.) can be introduced to the medium to promote differentiation into specific lineages.

4. Spontaneously contracting embryoid bodies (EB) are typically visible via microscopy approximately 4 days following replating, as contractile cardiomyocytes are a common cell type to arise during EB differentiation.

4. Notes

1. It is possible to purchase whole RNA, including miRNAs, from a variety of adult tissues (Ambion) or RNA can be extracted from postmortem tissue samples.

2. Human ESCs are rich in mRNA content but adult cells typically have reduced amounts of RNA. The yield will vary greatly depending on the cell type and methods of isolation. Biopsies must be well preserved by quickly transferring them to ice after excision. These technical problems can be remedied by purchasing RNA directly from a supplier.

3. The ratio of chloroform:Trizol can be adjusted to optimize RNA yield. Increasing the amount of chloroform may improve final yield.

4. The PLG tube is not absolutely necessary for RNA precipitation. It is possible to allow the layers to separate by centrifugation and pipette away the upper aqueous layer. The PLG tube maximizes purity of the aqueous phase without any carryover of the organic phase, due to the layers being physically separated by the phase lock gel.

5. While miR-26b is a recommended endogenous control for miRNA microarray, due to its stable expression across many adult tissues, it is not stable across development. We have had success using miR-188-5p, miR-296-5p, RNU38B, and RNU48 as small RNA endogenous controls with stable expression across embryonic, fetal, and adult cardiomyocytes.

6. Stbl3 are recommended over other competent cell types for the amplification of lentiviral plasmids to reduce the occurrence of random recombination events common with lentiviral vectors.

7. For a large vector over 11 kb, it is recommended to grow bacteria slower at 30°C to reduce the likelihood of errors due to recombination events. If growing bacterial agars at 30°C, then continue to amplify bacteria at this temperature in all

subsequent steps. Otherwise, bacteria can be grown normally at 37°C.

8. Save at least 1 mL of each bacterial clone at 4°C to amplify any correct clones following screening.

9. Care should be taken not to duplicate reporters in the vector. Most commercially available lentiviral vectors contain constitutively active promoters driving reporters for the selection of positively transduced cells. A different promoter and reporter should be selected for the expression and detection of the pre-miRNA, respectively.

10. Ideally, low passage number cells should be used. Passage numbers above 20 or slow growing cells are not recommended.

11. Amounts given are per flask. Multiply these amounts by the number of flasks used. For example, if using nine flasks of 293T, then 135 μg of transfer vector, 270 μg of packaging plasmid psPAX2, 67.5 μg of envelope plasmid pMD2G, and 38.7 mL of OptiMEM media are combined in a conical tube.

12. In transfected 293T cells, the lentiviral long-terminal repeats (LTRs) will act as the promoter to drive expression of the viral genome, including one's transgenes, for packaging. Therefore, any fluorescence observed will be from the protein with the first start codon translated after the 5′ LTR. In other words, the viral LTR will override any transgenic promoters until the promoter activity of the viral 5′ LTR is destroyed during the reverse transcription step of viral transduction. Hence, the self-inactivating feature of modern lentivirus systems to enhance biosafety and reduce viral promoter interference of transgenes.

13. Save an aliquot of the 293T medium filtrate in case it is needed for use later when balancing the centrifuge tubes.

14. With a Sharpie marker, draw a circle on the centrifuge tube where one would expect the pellet to be located. This will aid in identifying the pellet later in case it is difficult to see.

15. Avoid frothing and introducing bubbles when pipetting, as this will decrease the titer.

16. Avoid repeated freeze/thaws of the virus. This will substantially decrease the titer.

17. HEK293T cells can be used in place of hESCs for a rough approximation of viral titer. The titer will vary from cell type to cell type due to differences in transduction efficiency.

18. It is best to perform a series of dilutions to find an optimal virus titer. The smallest amount of virus that still produces visible fluorescence will result in fewer potentially negative insertion events. A rough estimate is 50 TDU/cell.

References

1. Thomson, J.A., Itskovitz-Eldor, J., Shapiro, S.S., Waknitz, M.A., Swiergiel, J.J., Marshall, V.S., et al. (1998) Embryonic stem cell lines derived from human blastocysts. *Science* **282**, 1145–1147.

2. Yu, J., Vodyanik, M.A., Smuga-Otto, K., Antosiewicz-Bourget, J., Frane, J.L., Tian, S., et al. (2007) Induced pluripotent stem cell lines derived from human somatic cells. *Science* **318**, 1917–1920.

3. Takahashi, K., Tanabe, K., Ohnuki, M., Narita, M., Ichisaka, T., Tomoda, K., et al. (2007) Induction of pluripotent stem cells from adult human fibroblasts by defined factors. *Cell* **131**, 861–872.

4. Bartel, D.P. (2004) MicroRNAs: genomics, biogenesis, mechanism, and function. *Cell* **116**, 281–297.

5. Lee, R.C. and Ambros, V. (2001) An extensive class of small RNAs in *Caenorhabditis elegans*. *Science* **294**, 862–864.

6. Lee, Y., Kim, M., Han, J., Yeom, K.H., Lee, S., Baek, S.H., et al. (2004) MicroRNA genes are transcribed by RNA polymerase II. *EMBO* **23**, 4051–4060.

7. Cai, X., Hagedorn, C.H., and Cullen, B.R. (2004) Human microRNAs are processed from capped, polyadenylated transcripts that can also function as mRNAs. *RNA* **10**, 1957–1966.

8. Hannon, G.J. (2002) RNA interference. *Nature* **418**, 244–251.

9. Meister, G and Tuschl, T. (2004) Mechanisms of gene silencing by double-stranded RNA. *Nature* **431**, 343–349.

10. Griffiths-Jones, S. (2004) The microRNA Registry. *Nucleic Acid Res.* **32**(Database Issue), D109–D11. http://microrna.sanger. ac.uk/sequences/

11. Lewis, B.P., Burge, C.B., and Bartel, D.P. (2005) Conserved seed pairing, often flanked by adenosines, indicates that thousands of human genes are microRNA targets. *Cell* **120**, 15–20.

12. Barroso-del Jesus, A., Lucena-Aguilar, G., and Menendez, P. (2009) The miR-302-367 cluster as a potential stemness regulator in ESCs. *Cell Cycle* **8**, 394–398.

13. Card, D.A., Hebbar, P.B., Li, L., Trotter, K.W., Komatsu, Y., Mishina, Y., et al. (2008) Oct4/Sox2-regulated miR-302 targets cyclin D1 in human embryonic stem cells. *Mol. Cell Biol.* **28**, 6426–6438.

14. Suh, M.R., Lee, Y., Kim, J.Y., Kim, S.K., Moon, S.H., Lee, J.Y., et al. (2004) Human embryonic stem cells express a unique set of microRNAs. *Dev. Biol.* **270**, 488–498.

15. Zhao, Y., Samal, E., and Srivastava, D. (2005) Serum response factor regulates a muscle-specific microRNA that targets Hand2 during cardiogenesis. *Nature* **436**, 214–220.

16. van Rooij, E., Sutherland, L.B., Qi, X., Richardson, J.A., Hill, J., and Olson, E.N. (2007) Control of stress-dependent cardiac growth and gene expression by a microRNA. *Science* **316**, 575–579.

17. Zhao, Y., Ransom, J.F., Li, A., Vedantham, V., von Drehle, M., Muth, A.N., et al. (2007) Dysregulation of cardiogenesis, cardiac conduction, and cell cycle in mice lacking miRNA-1-2. *Cell* **129**, 303–317.

18. Sayed, D., Rane, S., Lypowy, J., He, M., Chen, I.Y., Vashistha, H., et al. (2008) MicroRNA-21 targets Sprouty2 and promotes cellular outgrowths. *Mol. Biol. Cell* **19**, 3272–3282.

19. Fish, J.E., Santoro, M.M., Morton, S.U., Yu, S., Yeh, R.F., Wythe, J.D., et al. (2008) miR-126 regulates angiogenic signaling and vascular integrity. *Dev. Cell* **15**, 272–284.

20. Morton, S.U., Scherz, P.J., Cordes, K.R., Ivey, K.N., Stainier D.Y., and Srivastava, D. (2008) microRNA-138 modulates cardiac patterning during embryonic development. *Proc. Natl. Acad. Sci. USA* **105**, 17830–17835.

21. Chen, C.Z., Li, L., Lodish, H.F., and Bartel, D.P. (2004) MicroRNAs modulate hematopoietic lineage differentiation. *Science* **303**, 83–86.

22. Schratt, G.M., Tuebing, F., Nigh, E.A., Kane, C.G., Sabatini, M.E., Kiebler, M., et al. (2006) A brain-specific microRNA regulates dendritic spine development. *Nature* **439**, 283–289.

23. Hand, N.J., Master, Z.R., EauClaire, S.F., Weinblatt, D.E., Matthews, R.P., and Friedman, J.R. (2009) The microRNA-30 family is required for vertebrate hepatobiliary development. *Gastroenterology* **136**, 1081–1090.

24. Lagos-Quintana, M., Rauhut, R., Yalcin, A., Meyer, J., Lendeckel, W., and Tuschl, T. (2002) Identification of tissue-specific microRNAs from mouse. *Curr. Biol.* **12**, 735–739.

25. Amit, M. and Itskovitz-Eldor, J. (2006) Feeder-free culture of human embryonic stem cells. *Methods Enzymol.* **420**, 37–49.

26. Ludwig, T.E., Bergendahl, V., Levenstein, M.E., Yu, J., Probasco, M.D., and Thomson, J.A.

(2006) Feeder-independent culture of human embryonic stem cells. *Nat. Methods* **3**, 637–646.

27. Zufferey, R., Dull, T., Mandel, R.J., Bukovsky, A., Quiroz, D., Naldini, L., and Trono, D. (1998) Self-inactivating lentivirus vector for safe and efficient in vivo gene delivery. *J. Virol.* **72**, 9873–9880.

28. Zhu, Q., Hong, A., Sheng, N., Zhang, X., Matejko, A., Jun, K.Y., et al. (2007) MicroParaflo biochip for nucleic acid and protein analysis. *Methods Mol. Biol.* **382**, 287–312.

29. Li, W. and Ruan, K. (2009) MicroRNA detection by microarray. *Anal. Bioanal. Chem.* **394**, 1117–1124.

30. Griffiths-Jones, S., Saini, H.K., van Dongen, S., and Enright, A.J. (2008) miRBase: tools for microRNA genomics. *NAR* **36**(Database Issue), D154–D158. http://microrna.sanger.ac.uk/sequences/

31. Wayman, G.A., Davare, M., Ando, H., Fortin, D., Varlamova, O., Cheng, H.Y.M., et al. (2008) An activity-regulated microRNA controls dendritic plasticity by down-regulating p250GAP. *Proc. Natl. Acad. Sci. USA* **105**, 9093–9098.

32. Neumueller, R.A., Betschinger, J., Fischer, A., Bushati, N., Poernbacher, I., Mechtler, K., et al. (2008) Mei-P26 regulates microRNAs and cell growth in the Drosophila ovarian stem cell lineage. *Nature* **454**, 241–245.

33. Livak, K.J. and Schmittgen, T.D. (2001) Analysis of relative gene expression data using real-time quantitative PCR and the $2\text{-}\Delta\Delta C_T$ method. *Methods* **25**, 402–408.

34. Fu, J.D., Jiang, P., Rushing, S., Liu, J., Chiamvimonvat, N., and Li, R.A. (2010) Na^+/Ca^{2+} exchanger is a determinant of excitation-contraction coupling in human embryonic stem cell-derived ventricular cardiomyocytes. *Stem Cells Dev.* **19**, 773–782.

35. Xue, T., Cho, H.C., Akar, F.G., Tsang, S.Y., Jones, S.P., Marban, E., Tomaselli, G.F., Li, R.A. (2005) Functional Integration of electrically active cardiac derivatives from genetically-engineered human embryonic stem cells with quiescent recipient ventricular cardiomyocytes insights into the development of cell based pacemakes. *Circulation* **111**, 11–20.

36. Fu, J.D., Jung. Y., Chan, C.W., Li, R.A. (2008) An inducible transgene exexpression system for regulated phenotypic modification of human embryonic stem cells. *Stem Cells Dev.* **17**, 315–24.

Chapter 8

Large-Scale Expansion of Mouse Embryonic Stem Cells on Microcarriers

Ana Fernandes-Platzgummer, Maria Margarida Diogo, Cláudia Lobato da Silva, and Joaquim M.S. Cabral

Abstract

A large-scale stirred culture system for the expansion of mouse embryonic stem cells (mESCs) in spinner flasks under serum-free conditions was established using macroporous microcarriers for cell attachment and growth. This type of microcarrier was chosen as it potentially offers more protection to cells against shear stress in the absence of serum compared to microporous ones. In addition, methods to characterize ESCs after large-scale expansion were established. The pluripotency of expanded mESCs was evaluated based on both flow cytometry and alkaline phosphatase staining. Envisaging the application of ESCs as a potential source of neural progenitors, the neural commitment potential of cells after expansion in the spinner flask was also determined by culturing cells in serum-free adherent monolayer conditions.

Key words: Embryonic stem cells, Expansion, Large-scale, Microcarriers, Serum-free medium, Neural commitment, Spinner flask, Pluripotency

1. Introduction

Embryonic stem cells (ESCs) are pluripotent cells, which have the capacity for self-renewal and can give rise to differentiated cells of the three embryonic germ layers: ectoderm, endoderm, and mesoderm (1). ESC derivatives are potentially very attractive for many applications in cellular therapies (2), tissue engineering (3), and drug screening (4, 5) and can be potentially used as a reliable alternative to animal models. In particular, the generation of pure populations of neural progenitors from ESCs and their further differentiation into neurons, astrocytes, and oligodendrocytes (6) allows the potential use of these cells for the cure of neurodegenerative

Nicole I. zur Nieden (ed.), *Embryonic Stem Cell Therapy for Osteo-Degenerative Diseases*, Methods in Molecular Biology, vol. 690, DOI 10.1007/978-1-60761-962-8_8, © Springer Science+Business Media, LLC 2011

diseases and for neural drug testing. Similarly, expanding adequate numbers of ESCs for the differentiation of osteo- and chondro-progenitors is equally important for the treatment of osteodegen-erative diseases. All these types of applications require a wide available source of both undifferentiated ESCs and their differen-tiated derivatives, which constitutes an enormous challenge in terms of the large-scale in vitro expansion and controlled differ-entiation of ESCs.

Standard procedures for the expansion of ESCs rely on the use of static culture systems, such as T-flasks and tissue culture petri dishes. However, these systems have serious limitations con-cerning their nonhomogeneous nature, resulting in concentra-tion gradients in the culture medium (7), and are also limited in their productivity by the number of cells that can be supported by a given surface area (8). In order to circumvent these limitations, the scale-up of murine ESC (mESC) expansion has been per-formed using simple laboratory scale stirred bioreactors, the so-called spinner flasks.

mESCs have been expanded in these types of bioreactors, both in the form of aggregates (9) or with the use of microporous microcarriers (10, 11) for cell attachment. In general, these pro-cedures rely on the use of serum-containing medium supple-mented with leukemia inhibitory factor (LIF). However, this medium can cause cells to acquire karyotypic changes due to the presence of fetal bovine serum (FBS), which is poorly defined and potentially exposes ESCs to animal pathogens (12). Several attempts have been made in order to develop serum-free formula-tions that are capable of maintaining ESC properties during expansion. A specific serum-free medium was developed that maintains the undifferentiated state of mESCs during prolonged expansion, including single cells (12). In addition to LIF, this medium is supplemented with bone morphogenetic protein 4 (BMP4), whose molecular signals are necessary for suppression of neural differentiation in the absence of serum (12). mESCs can also be expanded under serum-free conditions using a proprietary serum replacement designed to directly replace FBS. Nevertheless, serum-free conditions potentially exacerbate the harmful effects of shear stress on cultured cells (13). To circumvent these harm-ful effects, the use of macroporous microcarriers can be advanta-geous in offering a more protective environment to cells (14), favoring cell expansion. This strategy was recently followed with success for the scale-up of mESC expansion under serum-free conditions (15) and will be described in this chapter. As a model cell line, 46C mESCs were used. In this cell line, the open reading frame of the *Sox1* gene, an early marker of the neuroectoderm in the mouse embryo, is replaced with the coding sequence for the Green Fluorescent Protein (GFP), which allows the monitoring of neural commitment (Sox1-GFP expression) by fluorescence microscopy and flow cytometry (16, 17).

2. Materials

2.1. Thawing and Expansion of mESCs Under Static Conditions Prior to Spinner Flask Inoculation

1. Phosphate buffered saline (PBS) solution. Make solution by dissolving PBS powder in 1 L of water. Filter the solution and store at room temperature.

2. ESC medium: Combine Knockout Dulbecco's Modified Eagle's Medium (K-DMEM) with 15% Knockout Serum Replacement (KSR, Invitrogen), 1% l-glutamine [stock solution is 200 mM (100×)], 1% penicillin (50 U/mL)/streptomycin (50 µg/mL), 1% nonessential amino acids (NEAA, stock is 100×), and 0.1% 2-mercaptoethanol. Store this medium for up to 15 days at 4°C. Exposure of KSR or serum-free complete media to light and to 37°C should be minimized.

3. KSR expansion medium: ESC medium as described above, but supplemented with human LIF. Human LIF was produced in 293-HEK cells (see Note 1). Store at 4°C for up to 1 week only and make fresh when necessary. Protect from light and heat.

4. Serum-free ESGRO complete clonal-grade expansion medium (Millipore). Store medium in aliquots at −20°C and protect from light.

5. 2.5% trypsin solution, diluted to 0.025% in PBS. A total volume of 50 mL is prepared by using 500 µL of 2.5% trypsin, 650 µL of 0.1 M Ethylenediaminetetraacetic acid (EDTA), and 50 µL of heat-inactivated chicken serum in PBS. Sterilize the solution by filtration and store in aliquots at −20°C. After thawing, the trypsin aliquots can be maintained at 4°C.

6. Accutase® solution (see Note 2). Store at 4°C after thawing.

7. Gelatin, 2% in water, tissue culture grade, sterile, Type B, cell culture tested. Store at 4°C. Make a 0.1% gelatin working solution by diluting 12.5 mL of 2% gelatin in 237.5 mL of PBS and store at 4°C.

8. 0.4% trypan blue dye solution. Store at room temperature. Prepare a 0.1% trypan blue solution from the 0.4% trypan blue dye by diluting in PBS (i.e., 1 mL trypan blue:3 mL PBS). Store at room temperature.

9. Tissue culture treated plastic ware (i.e., Corning) and Falcon tubes.

10. Hemocytometer.

11. Optical microscope, i.e., Olympus CK40.

2.2. Cell Culture Monitoring in the Spinner Flask

2.2.1. Stirred Culture

1. Spinner Flask *Stem Span* (StemCell Technologies) of 50 mL working volume equipped with an impeller with 90° normal paddles and a magnetic stir bar.

2. Microcarriers *Cultispher®* S (Sigma). Store at room temperature.

3. Stirring plate (Variomag Biosystem Direct).

4. Serum-containing media: High-glucose DMEM + 10% FBS (ES cell qualified, Invitrogen).

2.2.2. Cell Counts and Viability

1. 1× PBS solution (see above).

2. 0.1% trypan blue dye solution (see above). Store at room temperature.

3. 2.5% trypsin solution: Prepare 37.5 mL of 1% trypsin in PBS with 15 mL of 2.5% trypsin solution, 1.5 mL of heat-inactivated chicken serum, and 1.5 mL of 0.1 M EDTA in PBS. Filter-sterilize the solution and store in 10 mL aliquots at –20°C. After thawing, the trypsin aliquots can be maintained at 4°C.

4. Hemocytometer.

5. Optical microscope, i.e. Olympus CK40.

2.2.3. Determination of the Expression of Pluripotency Markers

1. Microcentrifuge tubes, 1.5 or 2 mL and microcentrifuge, i.e. Hermle Z 300 K.

2. FACSCalibur flow cytometer (Becton Dickinson Biosciences).

3. FACSFlow™ sheath fluid (Becton Dickinson Biosciences).

4. Primary antibodies: Mouse anti-Oct-3/4 (C-10) (Santa Cruz Biotechnology), rabbit anti-Nanog (Chemicon), and mouse antihuman CD15 FITC (BD).

5. Secondary antibodies: Alexa Fluor® 488 goat antimouse IgG (Invitrogen) and Alexa Fluor® 488 goat antirabbit IgG secondary antibody (Invitrogen).

6. 10× PBS solution. Make solution by dissolving PBS powder in 100 mL of water. Filter the solution and store at room temperature.

7. 2% paraformaldehyde (PFA) solution. Dissolve 2 g of PFA in 100 mL of PBS (see Note 3). Filter before use and maintain at 4°C.

8. 30% BSA solution. Store at 4°C upon arrival. Make a 5% BSA solution in PBS. To do this, dilute 1 mL of the 30% BSA solution in 5 mL of PBS. Filter before use and store at 4°C.

9. 1% saponin solution in PBS: Dissolve 1 g of saponin in 100 mL of PBS. Filter before use and store at 4°C.

10. Normal goat serum (NGS). Make a 3% NGS solution in PBS by diluting 3 mL of NGS in 97 mL of PBS. In addition, make a 1% NGS solution in PBS by diluting 10 mL of 3% NGS in 20 mL of PBS. Filter both NGS solutions and store at 4°C. Finally, prepare blocking solution for the antibody staining procedure: 150 μL of 3% NGS and 150 μL of 1% saponin for intracellular staining (Oct-4, nanog) and 300 μL of 3% NGS for surface staining.

2.2.4. Alkaline Phosphatase Staining

1. Solution of 10% cold neutral-buffered formalin. Store at room temperature.

2. Fast Violet B Salt. Store at 4°C. Dissolve one capsule in 48 mL of Milli-Q water. Aliquots are stable at −20°C.

3. 0.25% Naphthol AS-MX Phosphate Alkaline Solution. Store at 4°C.

4. Reagent X: Add 4% (v/v) Naphthol AS-MX Phosphate Alkaline Solution 0.25% to a prethawed aliquot of Fast Violet Solution. Protect from light and use immediately.

2.2.5. Neural Commitment of mESCs and Flow Cytometric Quantification of Neural Conversion

1. Neural differentiation medium RHB-A (Stem Cell Sciences). Store at 4°C or in aliquots at −20°C. Avoid freeze–thaw cycles and protect from light.

2. FBS (Invitrogen). Store at −20°C. Make a 4% solution of FBS in PBS. In order to do so, dilute 4 mL of FBS in 96 mL of PBS. Filter solution before use and store at 4°C.

3. Microcentrifuge tubes, 1.5 or 2 mL and microcentrifuge, i.e. Hermle Z 300 K.

4. FACSCalibur flow cytometer (Becton Dickinson Biosciences).

5. FACSFlow™ sheath fluid (Becton Dickinson Biosciences).

3. Methods

3.1. Thawing and Expansion of mESCs Under Static Conditions

1. Remove a cryogenic vial of frozen 46C mESCs (approximately 1 mL) from the liquid nitrogen tank and quickly thaw in a 37°C water bath.

2. Resuspend the contents of the cryogenic vial in 4 mL of prewarmed (37°C) ESC medium (or ESGRO complete medium).

3. Centrifuge the cell suspension at $800 \times g$ for 2 min, discard the supernatant, and resuspend the pellet in 5 mL of prewarmed (37°C) expansion medium (or ESGRO complete medium).

4. Seed the cell suspension into a 60-mm tissue culture dish previously coated with 0.1% gelatin. See Chapters 2 and 5 for details on how to gelatin-coat plastic ware.

5. Let the cells grow at 37°C in a 5% CO_2 fully humidified atmosphere.

6. After 48 h of expansion, wash cells twice with PBS and incubate with Accutase® or with 0.025% trypsin at 37°C for 2 min (see Note 2).

7. After the cells have completely detached from the plastic, dilute the cell suspension in ESC medium (see Note 4), transfer to a Falcon tube, and centrifuge for 2 min at $800 \times g$.

8. Resuspend the cells in expansion medium and replate at a density of approximately 2×10^4 cells/cm^2.

9. After 48 h of expansion, repeat the passaging procedure. In each passage determine the viable and dead cells by counting in a hemocytometer under an optical microscope using the trypan blue dye exclusion test. In order to do so, the cell suspension is mixed (1:1) with the trypan blue staining solution and the viable (unstained cells) and dead cells (blue-stained cells) are identified and counted. Double the number of counted cells to get a cell count per milliliter taking the predilution with trypan blue into account.

3.2. Expansion of mESCs Under Stirred Conditions

1. Before cell expansion, hydrate the *Cultispher*® S microcarriers (see Note 5) overnight, sterilize by autoclaving for 20 min at 120°C, decant the water, and equilibrate the microcarriers in KSR expansion medium or serum-containing medium for at least 12 h (see Note 6).

2. Mix 46C mESCs (5×10^4 cells/mL), previously expanded for at least two passages (see Note 7) under static conditions in a 60-mm culture plate, with 1 mg/mL of *Cultispher*® S microcarriers in KSR expansion medium or serum-free ESGRO complete expansion medium.

3. Incubate cells and microcarriers at 37°C in one sixth of the final medium volume (5 mL) for 30 min, and gently agitate every 10 min.

4. Gently add fresh prewarmed (37°C) medium until half of the final volume (15 mL) then transfer the cell suspension to the spinner flask.

5. After a 24 h seeding period with intermittent stirring (15 min of stirring at 30–40 rpm followed by 60 min of no stirring), add medium up to the final volume of 30 mL and adjust the speed to 40 rpm.

6. Feed the cells every day by replacing 50% of the medium with fresh prewarmed medium (see Note 8). Figure 1 shows a scanning electron microscopy (18) of 46C mESCs adherent to *Cultispher*® S microcarriers after expansion on the spinner flask.

3.3. Monitoring of Cell Culture in the Spinner Flask

3.3.1. Cell Counts and Viability

1. Collect 0.5 mL of duplicate samples of an evenly mixed culture from the spinner flask every day.

2. After the microcarriers settle down, remove 0.3 mL of the supernatant. Wash the microcarriers with 2 mL of 37°C prewarmed PBS. Afterwards, incubate the microcarriers until complete dissolution in a 37°C water bath after adding 0.8 mL of 1% trypsin (see Note 9).

3. After dissolution of the microcarriers, add 1 mL of 0.1% trypan blue solution and determine the number of viable and

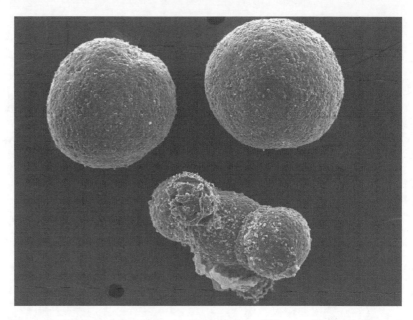

Fig. 1. Scanning electron microscopy (SEM) of 46C mESCs growing on *Cultispher* S microcarriers in the spinner flask.

dead cells with a hemocytometer. Figure 2 shows the expansion of 46C mESCs on the Cultispher® S microcarriers in the spinner flask system for 8 days, both in terms of viable cells per milliliter and total cell fold increase. Two different serum-free media were compared in terms of cell expansion, the K-DMEM supplemented with KSR and the ESGRO complete clonal-grade medium.

3.3.2. Determination of the Expression of Pluripotency Markers

After expansion in the spinner flask for 8 days under serum-free conditions, the cell suspension is analyzed for the presence of antibody markers specific to mESCs by flow cytometry, after intracellular (Oct-4 and Nanog) or surface (SSEA-1) staining:

1. Take a 2 mL sample of the spinner flask culture.

2. Detach cells from the microcarriers using 1% trypsin for 15–20 min, inactivate the trypsin with twice the volume of serum-containing medium, and centrifuge cell suspension for 5 min at $200 \times g$. Cells are then resuspended in 5 mL of ESC medium.

3. Perform a cell count with trypan blue exclusion as explained above.

4. Aliquot 10^6 cells per antibody/negative control, centrifuge, and resuspend in 10 mL of PBS.

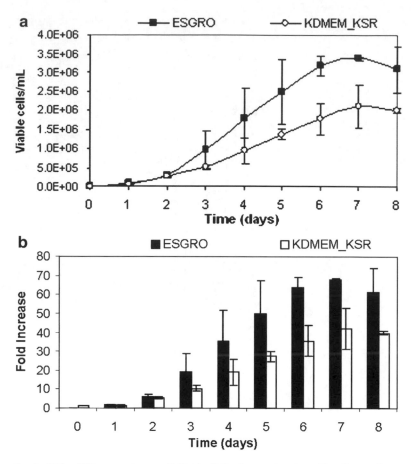

Fig. 2. 46C mESC expansion on *Cultispher®* S microcarriers in a *StemSpan* spinner flask under serum-free conditions. (**a**) Growth curve in terms of viable cells per milliliter. (**b**) Cell expansion in terms of fold increase in total cell numbers. Cells were inoculated at 5×10^4 cells/mL on 1 mg *Cultispher®* S in ESGRO complete medium and K-DMEM supplemented with KSR. *Error bars* represent standard deviation of two independent experiments.

5. Centrifuge for another 7 min at $200 \times g$, resuspend the pellet in 10 mL of 2% PFA solution, and incubate at room temperature for 20 min (see Note 9).

6. Centrifuge PFA-fixed cells for 7 min at $200 \times g$, wash them with 5 mL of 1% NGS, and centrifuge again for 7 min at $200 \times g$. Perform this procedure twice.

7. Resuspend cells in 3% NGS (0.5 mL for each antibody and negative control tested).

8. Transfer cell suspension to 1.5 mL tubes previously coated with 1 mL of 5% BSA solution for at least 15 min, centrifuge the tubes at $1,000 \times g$ for 3 min, and aspirate the supernatant from each tube carefully.

9. Resuspend the cell pellet with the respective blocking solution and/or detergent depending on if you are doing intra- or extracellular staining.

10. Incubate at room temperature for 15 min.

11. Centrifuge all 1.5 mL tubes at $1,000 \times g$ for 3 min at room temperature, remove the supernatant, resuspend in 300 µL of 3% NGS, and incubate at room temperature for 15 min.

12. Centrifuge all tubes at $1,000 \times g$ for 3 min at room temperature, carefully aspirate the supernatant, then add the primary antibody diluted in 3% NGS in a final volume of 300 µL (in the case of the negative controls, cells are maintained in 3% NGS).

13. Incubate the tubes for 1 h and 30 min at room temperature in the dark.

14. Centrifuge all tubes at $1,000 \times g$ for 3 min at room temperature and resuspend the pellet with 500 µL of 1% NGS. Perform this procedure twice. This step does not apply to the negative control tube.

15. Add the secondary antibody diluted in 3% NGS to all tubes in a final volume of 300 µL and incubate the tubes for 45 min at room temperature in the dark.

16. Centrifuge all tubes at $1,000 \times g$ for 3 min at room temperature and resuspend the pellet with 500 µL of 1% NGS. Perform this procedure two more times.

17. Add 0.5 mL of PBS, transfer the resuspended pellets to flow cytometry tubes, and perform flow cytometric analysis (see Note 10). Figure 3a shows examples of the expression profiles obtained by flow cytometry for 46C ESCs expanded for 8 days in the spinner flask. The ESCs were obtained after intracellular staining with the anti-Oct-4 and anti-Nanog antibodies and after surface staining with the anti-SSEA-1 antibody. In this example, more than 95% of the 46C ESCs analyzed expressed the three different pluripotency markers.

3.3.3. Alkaline Phosphatase Staining

In addition to flow cytometry analysis, the pluripotency of mESCs after expansion in the spinner flask can also be determined based on the activity of alkaline phosphatase:

1. Wash the samples of microcarriers containing mESCs with PBS and fix in 10% cold neutral-buffered formalin for 15 min.

2. After fixing, wash cells and keep in distilled water for another 15 min.

3. Following the washing step, incubate the cells with Reagent X for 1 h in a dark environment and wash three times with distilled water.

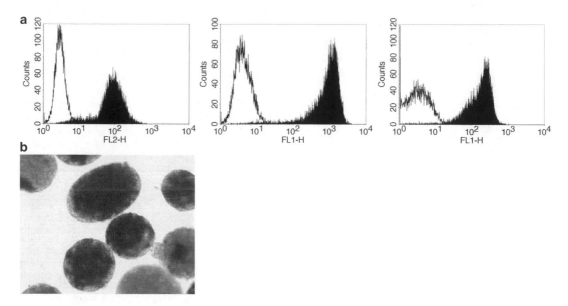

Fig. 3. Evaluation of pluripotency of 46C mESCs cultured on *Cultispher*® S microcarriers under stirred conditions in serum-free medium for 8 days. (**a**) Pluripotency was evaluated by flow cytometry based on the expression of markers specific to mESCs [Oct-4 (A1), SSEA-1 (A2), and Nanog (A3)]. (**b**) Alkaline phosphatase staining (optical microscope photograph at ×200 amplification).

4. Keep cells in distilled water and observe under an optical microscope. Figure 3b shows 46C ESCs adherent to the *Cultispher*® S microcarriers following alkaline phosphatase staining after 8 days in culture (see Note 11).

3.3.4. Neural Commitment of 46C mESCs

The neural commitment potential of 46C ESCs cells is determined by plating the cells obtained after expansion in the spinner flask under serum-free adherent monolayer conditions (16, 17):

1. Take the samples of microcarrier-containing cells from the spinner flask and dissolve them with 1% trypsin.

2. After the complete dissolution of the microcarriers, expand the cells at a relatively high density (10^5 cells/cm^2) for 24 h in ESGRO complete medium (17).

3. After expansion at high density, detach the cells from the plate with 0.025% trypsin or Accutase® (see Note 2), resuspend in serum-free RHB-A medium, and replate in two wells of a 12-well plate (1 mL/well) precoated with 0.1% gelatin (see Note 12), at a density of 10^4 cells/cm^2.

4. Change the medium every 2 days.

5. The use of 46C mESCs allows the quantification of neural conversion by flow cytometry based on GFP expression (16). After 6 days in culture with neural differentiation medium, detach cells from the plate and resuspend in PBS containing 4% FBS.

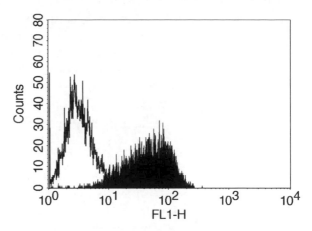

Fig. 4. Evaluation of neural commitment potential of 46C ESCs cultured on *Cultispher®* S microcarriers under stirred conditions in serum-free medium. The percentage of neural progenitors was determined by flow cytometry using undifferentiated 46C mESCs as the negative control (*white*), and after culturing 46C mESCs for 6 days using the RHB-A medium (*black*).

6. Analyze the cells by flow cytometry using appropriate analysis software (e.g. CellQuest, BD Biosciences). Set gates at ten units of fluorescence, which excludes more than 99% of undifferentiated cells. Exclude cell debris and dead cells from the analysis based on electronic gates using the forward scatter (size) and side scatter (cell complexity) criteria. Undifferentiated 46C ESCs are used as a negative control. Figure 4 shows a typical flow cytometric profile of 46C ESCs after expansion in the spinner flask using serum-free ESGRO complete medium, followed by neural commitment for 6 days under adherent monolayer conditions in RHB-A medium. In this example, more than 90% of the 46C ESCs that expanded in the spinner flask underwent neural commitment, assessed by GFP expression along the FL1 axis.

4. Notes

1. Human LIF was produced in HEK-293 cells. The optimal dilution of LIF for ESC expansion was previously determined by testing the effect of different dilutions of LIF on standard ESC cultures. The optimal dilution was determined based either on ESC fold increase, expression of pluripotency markers (Oct-4, SSEA-1, and Nanog), or direct microscope observation of ESC morphology. LIF formulations (ESGRO, Millipore) can also be used to supplement ESC expansion medium (1,000 U/mL).

2. When mESCs are cultured in the absence of serum, Accutase® or a low concentrated trypsin solution (0.025%) should be used for passaging in order to preserve cell attachment properties. This step is especially critical when cells are plated for neural differentiation.

3. PFA powder should be initially dissolved in a low volume of water at a high temperature (lower than 70°C) in order to facilitate dissolution. The pH should be set to 7.3 and the final volume completed with 10× PBS.

4. When using trypsin for cell detachment, its action must be stopped using serum-containing medium, which typically contains trypsin inhibitors. When using Accutase® for cell detachment, cell suspension can be diluted in serum-free medium and centrifuged immediately.

5. *Cultispher®* S microcarriers are spherical macroporous microcarriers made of gelatin. Other commercially available microcarriers, namely microporous, can also be successfully used to support the expansion of mESCs under stirred conditions (10).

6. Use prewarmed medium (37°C) when adding the microcarriers.

7. Before initiating ESC expansion on microcarriers under stirred conditions, cells are expanded under static conditions for at least two passages in order to obtain enough cells to inoculate the spinner flask. Moreover, if cells are previously expanded under serum-containing medium, it will be necessary to adapt cells to serum-free conditions during several passages before beginning expansion under stirred conditions.

8. The removing/replenishment of the culture medium is performed immediately after the spinner flasks have been removed from the stirring plate and placed under the laminar flow hood. After the microcarriers containing cells have sedimented, (10–15 Min) change the culture medium.

9. Cells can be released from the *Cultispher®* S microcarriers by completely digesting the gelatin matrix. To perform this digestion, a more concentrated 1% trypsin solution should be used. Occasional flicking must be performed in order to facilitate gelatin matrix digestion. When cells attain large densities on the microcarriers, it becomes difficult to dissolve the gelatin matrix even when using this concentrated trypsin solution. In this case, it may be necessary to use longer incubation times. An alternative microcarrier dissolution protocol in situ (i.e. inside the spinner flask) was successfully tested. After the microcarrier-containing cells settled down, the supernatant was removed and 30 mL of PBS was used to wash the cells/microcarriers. After the washing step, 15 mL of 1% trypsin

was added and the spinner flask was put into the incubator under agitation until the microcarriers were dissolved. After the dissolution of the microcarriers, 45 mL of medium-containing serum was added to deactivate the trypsin, and the cell suspension was removed from the spinner flask. A cell strainer may be necessary (e.g. 100-μm pore size) to separate the cells from the viscous gelatin matrix that did not dissolve completely.

10. If necessary, cells can be maintained during several days in 2% PFA at 4°C until performing flow cytometry acquisition.

11. For an alternative protocol on how to stain human ESC colonies for alkaline phosphatase see Chapter 4.

12. In order to improve cell attachment under serum-free conditions prior to the neural commitment protocol, gelatin coating should be performed for at least 1 h.

Acknowledgements

The authors wish to acknowledge Fundação para a Ciência e a Tecnologia (FCT), Portugal (grant POCTI/BIO/46695/2002 and fellowship SFRH/BD/36070/2007 to Ana M. Fernandes), Professor Austin Smith for providing the 46C ESC line, Domingos Henrique for helpful discussion, as well as Francisco Santos, Pedro Andrade, and Gemma Eibes for SEM of 46C ESCs.

References

1. Smith, A. G. (2001) Embryo-derived stem cells: of mice and men. *Annu. Rev. Cell. Dev. Biol.* **17**, 435–642.

2. Mukhida, K., Mendez, I., McLeod, M., Kobayashi, N., Haughn, C., Milne, B. et al. (2007) Spinal GABAergic transplants attenuate mechanical allodynia in a rat model of neuropathic pain. *Stem Cells.* **25**, 2874–2885.

3. Levenberg, S., Burdick, J. A., Kraehenbuehl, T., and Langer, R. (2005) Neurotrophin-induced differentiation of human embryonic stem cells on three-dimensional polymeric scaffolds. *Tissue Eng.* **11**, 506–512.

4. Thomson, H. (2007) Bioprocessing of embryonic stem cells for drug discovery. *Trends Biotechnol.* **25**, 224–230.

5. Cezar, G. G. (2007) Can human embryonic stem cells contribute to the discovery of safer and more effective drugs? *Curr. Opin. Chem. Biol.* **11**, 405–409.

6. Sun, Y., Pollard, S., Conti, L., Toselli, M., Biella, G., Parkin, G. et al. (2008) Long-term tripotent differentiation capacity of human neural stem (NS) cells in adherent culture. *Mol. Cell Neurosci.* **38**, 245–258.

7. Portner, R., Nagel-Heyer, S., Goepfert, C., Adamietz, P., and Meenen, N. M. (2005) Bioreactor design for tissue engineering. *J. Biosci. Bioeng.* **100**, 235–245.

8. Cabrita, G. J., Ferreira, B. S., da Silva, C. L., Goncalves, R., Almeida-Porada, G., and Cabral, J. M. (2003) Hematopoietic stem cells: from the bone to the bioreactor. *Trends Biotechnol.* **21**, 233–240.

9. Cormier, J. T., zur Nieden, N. I., Rancourt, D. E., and Kallos, M. S. (2006) Expansion of undifferentiated murine embryonic stem cells as aggregates in suspension culture bioreactors. *Tissue Eng.* **12**, 3233–3245.

10. Abranches, E., Bekman, E., Henrique, D., and Cabral, J. M. (2007) Expansion of mouse

embryonic stem cells on microcarriers. *Biotechnol. Bioeng.* **96**, 1211–1221.

11. Fok, E. Y., and Zandstra, P. W. (2005) Shear-controlled single-step mouse embryonic stem cell expansion and embryoid body-based differentiation. *Stem Cells.* **23**, 1333–1342.

12. Ying, Q. L., Nichols, J., Chambers, I., and Smith, A. (2003) BMP induction of Id proteins suppresses differentiation and sustains embryonic stem cell self-renewal in collaboration with STAT3. *Cell.* **115**, 281–292.

13. McDowell, C. L., and Papoutsakis, E. T. (1998) Serum increases the CD13 receptor expression, reduces the transduction of fluid-mechanical forces, and alters the metabolism of HL60 cells cultured in agitated bioreactors. *Biotechnol. Bioeng.* **60**, 259–628.

14. Ng, Y. C., Berry, J. M., and Butler, M. (1996) Optimization of physical parameters for cell attachment and growth on macroporous microcarriers. *Biotechnol. Bioeng.* **50**, 627–635.

15. Fernandes, A. M., Fernandes, T. G., Diogo, M. M., da Silva, C. L., Henrique, D., and Cabral, J. M. (2007) Mouse embryonic stem cell expansion in a microcarrier-based stirred culture system. *J. Biotechnol.* **132**, 227–236.

16. Ying, Q. L., Stavridis, M., Griffiths, D., Li, M., and Smith, A. (2003) Conversion of embryonic stem cells into neuroectodermal precursors in adherent monoculture. *Nat. Biotechnol.* **21**, 183–186.

17. Diogo, M. M., Henrique, D., and Cabral, J. M. (2008) Optimization and integration of expansion and neural commitment of mouse embryonic stem cells. *Biotechnol. Appl. Biochem.* **49**, 105–112.

18. Mygind, T., Stiehler, M., Baatrup, A., Li, H., Zou, X., Flyvbjerg, A., Kassem, M., and Bünger, C. (2007) Mesenchymal stem cell ingrowth and differentiation on coralline hydroxyapatite scaffolds. *Biomaterials.* **28**, 1036–1047.

Chapter 9

Embryoid Body Formation: Recent Advances in Automated Bioreactor Technology

Susanne Trettner, Alexander Seeliger, and Nicole I. zur Nieden

Abstract

While spontaneous differentiation is an undesired feature of expanding populations of embryonic stem cells, a variety of methods have been described for their intended differentiation into specialized cell types, such as the osteoblast or chondrocyte. Most commonly, differentiation initiation involves the aggregation of ESCs into a so-called embryoid body (EB), a sphere composed of approximately 15,000 differentiating cells. EB formation has been optimized through the years, for example through invention of the hanging drop protocol. Yet, it remains a highly laborious process.

Here we describe the use of computer-controllable suspension bioreactors to form EBs in an automated and highly reproducible process and their subsequent differentiation along the osteoblast lineage. The development of the differentiating cells taken from bioreactor EBs to EBs formed in static control cultures through the hanging drop method will be compared.

Key words: Embryoid body, Suspension, Bioreactor, Osteogenesis

1. Introduction

Expanding cultures of embryonic stem cells (ESCs) are typically maintained by the addition of so-called pluripotency factors to the culture medium. As described in Chapters 2, 3, and 5, these "pluripotency factors" differ between mouse and human or non-human primate ESCs. In reality, these pluripotency factors do not actively maintain the self-renewing capacity of the cells nor their capacity to differentiate into cell types of all three germ layers, but rather prevent that differentiation signals are being transduced and early specification events are initiated as adequately described by Austin Smith and colleagues (1). The willingness of ESCs to spontaneously give up their pluripotent status in culture is not surprising given the fact that the fate of the inner cell mass cells in the embryo is the differentiation into the embryo proper and not

Nicole I. zur Nieden (ed.), *Embryonic Stem Cell Therapy for Osteo-Degenerative Diseases*, Methods in Molecular Biology, vol. 690, DOI 10.1007/978-1-60761-962-8_9, © Springer Science+Business Media, LLC 2011

the continuous proliferation. For the purpose of this chapter, however, we will keep calling them pluripotency factors.

In contrast to maintenance cultures of ESCs, in which differentiation is not desired, a great effort has been put into devising strategies to direct differentiation of these cells into specific lineages or even particular cell types. This seems to be particularly difficult as differentiation occurs according to a stochastic event by which a percentage of the cells commits to the endodermal fate, whereas other portions of the cells specify to the mesodermal and ectodermal fate. Typically, differentiation initiation involves the removal of pluripotency factors and the formation of an embryoid body (EB), a small aggregate consisting of cells that loose the expression of pluripotency markers. At the rise of the field, a good 30 years ago, EB formation was first performed in nonadherent culture conditions with embryonal carcinoma cells (2) and then later with murine ESCs (3). Individual ESCs were plated as a suspension into nonadherent conditions and over a few days period, which may last anywhere from 2 to 6 days (3, 4), the single cells spontaneously grow into small balls of cells, caused by their natural desire and potency to proliferate and differentiate. For a detailed description of the static suspension method for murine ESCs, see the next chapter and for human ESCs see Chapter 4. The inner differentiating cells in these EBs are surrounded by an outer layer of endoderm-like cells and extracellular matrix (3, 5).

Although cells inside the EB expressed early differentiation markers, the size distribution of these EBs was rather inhomogeneous. As a result of this size inhomogeneity, differentiation processes and gene expression in different EBs were not synchronized. Consequently, researchers were searching for alternative ways to form EBs. Indeed, in 1991, Wobus et al. published the hanging drop protocol, which is nowadays widely used in ESC research (6). Compared to other processes of EB formation, the "hanging drop" method provides the best EBs in terms of reproducibility of the resulting aggregate sizes as well as synchrony of differentiation and gene expression, since the protocol starts out with a defined cell number per "drop" (Fig. 1). Briefly, the hanging drop (HD) protocol entails the pipetting of individual droplets of cell suspension with a defined number of cells. The drops are distributed on the inner side of the lid of a petri dish, which is then flipped into its regular position. Within 3 days, EBs are formed by gravitation, which is what forces the cells to aggregate at the bottom of the droplet (for an illustration see Fig. 2). In the second step of the protocol, the EBs are maintained in suspension culture dishes for an additional 2 days before they are finally plated. In those initial 5 days, the 750 cells, which are seeded per droplet, undergo approximately five population doublings resulting in a sphere composed of up to 15,000 cells. Depending on the desired differentiation outcome, the timing of the hanging drop protocol

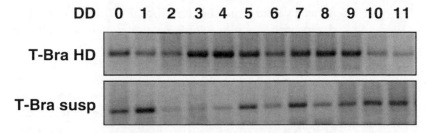

Fig. 1. Expression profile of T-Brachyury compared for hanging drop cultures (HD) and static suspension cultures (susp). Note that the expression profile is different. Since T-Brachyury is a pan-mesodermal marker, its expression pattern suggests that differentiation of ESCs into osteoblasts, which stem from the mesodermal lineage, would be accelerated in hanging drops versus static suspension cultures. *DD* differentiation day.

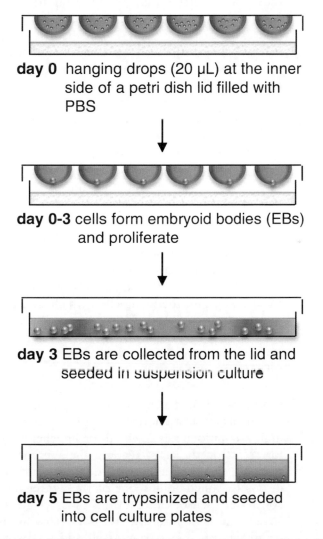

day 0 hanging drops (20 µL) at the inner side of a petri dish lid filled with PBS

day 0-3 cells form embryoid bodies (EBs) and proliferate

day 3 EBs are collected from the lid and seeded in suspension culture

day 5 EBs are trypsinized and seeded into cell culture plates

Fig. 2. The hanging drop protocol. Cells are cultivated in drops on the inner side of a petri dish lid. Within 3 days EBs are formed. On day 3 of differentiation, EBs are collected and cultured for additional 2 days in suspension. On day 5, EBs are trypsinized and seeded to cell culture plates. Alternatively, intact EBs may be plated onto tissue culture substrate. Timing among the individual steps may vary.

may be varied from the described $3+2$ day protocol (7) to a $2+4$ day protocol, which seems to support neural differentiation (8) or a $2+3$ day protocol, which has been described for myogenic differentiation (9).

Although tremendously advantageous over spontaneous EB formation, the hanging drop protocol is highly labor intensive. It poses particular difficulties when larger numbers of cells are needed, for example for the biochemical characterization of protein interactions and the interrogation of signaling pathways, which is typically accomplished by western blot analysis. In order to collect adequate amounts of protein, over 100 dishes of hanging drops have to be pipetted, sometimes more depending on the experimental setup. In addition to the low throughput, another concern about the HD protocol is its lack of automation. The fact that the HD protocol is difficult to automate, makes it impossible to standardize. The second concern thus applies especially to pharmaceutical screening assays, such as the EST assay (10), which makes use of differentiating ESCs. Not only does the laborious hanging drop protocol require cost intensive personnel resources, but by not being automatable human error could be introduced, a nondesirable feature of an in vitro screening assay.

Due to the mentioned concerns, alternative means to generate EBs have been investigated, among them suspension culture bioreactors (11–13) and microwell patterning (14), which are the focus of this and the next chapter. Suspension culture bioreactors have recently been employed for a higher throughput EB formation alternative. In contrast to the nonadherent suspension culture conditions described above and in the next chapter, suspension culture bioreactors offer constant stirring of the media. Stirring is achieved by a magnetic impeller, which is suspended into the culture medium and is run by a magnetic plate. The single cells, which are inoculated into the reactor, will grow into more or less perfectly round spheres. It is clear that the pO_2 within these spheres decreases rapidly with increased distance of the cells from the liquid interface (15, 16). While stirring results in continuous mixing of the media and homogeneous dissolving of oxygen into the media, it also controls the size of the resulting EBs. With higher agitation rates at which the impeller is moved, higher shear forces act on the cells, keeping the aggregate size in check and allowing the proper flow of nutrients into the aggregate and disposal of metabolites out of the aggregate. Truly computer-controlled sensor integrating bioreactors supply the added advantage of monitoring and controlled adjustment of culture parameters such as oxygen, pH, and temperature. Hence, with minimal labor, larger numbers of cells can be cultured in a more homogenous microenvironment under standardized and replicable conditions.

2. Materials

2.1. Preparation of Bioreactors and Inoculation

1. Waterbath (i.e. VWR) and ultrasonic waterbath.
2. Bioreactor system (DASGIP) or similar (see Note 1).
3. Distilled water (dH_2O).
4. Sigmacote SL2 (Sigma). Highly flammable, harmful, and dangerous for the environment. Wear protective equipment and store as appropriate.
5. Ethanol (70%). Highly flammable.
6. Bioreactor control unit (DASGIP) or similar (see Note 2).
7. Control Differentiation Medium (CDM): Dulbecco's Modified Eagle's Medium (DMEM) (1×), liquid (high-glucose) with L-glutamine, without pyruvate containing 15% fetal bovine serum (FBS, PAN, batch tested for osteogenic differentiation of ESCs), 1% nonessential amino acids (NEAA, 100×), 0.1 mM 2-mercaptoethanol (cell culture tested), and penicillin G/streptomycin sulfate (final conc. 50 U/mL penicillin and 50 µg/mL streptomycin – *optional*). Filter medium – use a sterile filter unit. Store the filtered medium at 4°C up to 2 weeks.
8. Embryonic stem cells. If using mouse ESCs for osteogenic differentiation, we suggest to use the D3 ESC line (American Type Culture Collection, see Note 3).
9. Trypsin, 0.25% (1×) with ethylenediaminetetraacetic acid (EDTA) 4Na, liquid (Invitrogen).
10. Hemocytometer (VWR).
11. 0.4% Trypan blue (i.e. Sigma).

2.2. Hanging Drop Culture

1. 1× Phosphate-buffered saline (PBS), without Ca^{2+} and Mg^{2+}, pH 7.4 (Invitrogen).
2. Plastic ware, such as bacteriological plates ($\varnothing$ 100 mm), i.e. Greiner Bio-One and 15 and 50 mL Falcon tubes.
3. Repeator® Pipettor plus combitips (Eppendorf).
4. Inverted (phase) microscope for checking EB formation, mineralization, and counting cells.
5. Leica DMI 400B microscope and Leica Application Suite, Version 2.8.1.

2.3. Differentiation of ESCs Toward Osteoblasts

1. Tissue culture-treated plates, 6-well and 15 and 50-mL Falcon tubes (i.e. Greiner Bio-One).
2. Trypsin, 0.25% (1×) with EDTA 4Na, liquid (Invitrogen).
3. Beta-glycerophosphate. Maintain a stock solution of 1 M in PBS, filter sterilize through a 0.2-µm filter and keep at –20°C.

4. Ascorbic acid. Make a stock of 50 mg/mL in PBS, filter sterilize (0.2-μm filter) and maintain at –20°C.

5. $1\alpha,25$-$(OH)_2$ vitamin D_3 (VD_3). Prepare a stock solution of 1.2×10^{-4} M in DMSO. Aliquot and store at –20°C.

6. Osteogenic Differentiation Medium (ODM): CDM supplemented with 10 mM beta-glycerophosphate, 25 μg/mL ascorbic acid, and 5×10^{-8} M VD_3. Filter media – use a sterile filter unit. Store the filtered medium at 4°C up to 2 weeks (see Note 4).

7. Inverted (phase) microscope for checking mineralization and counting cells.

3. Methods

3.1. Preparation of Bioreactors

Prepare bioreactors for the inoculation of cells. This includes the cleaning and sterilization of the bioreactors, the coating of the bioreactor vessels with sigmacote to prevent the adhesion of cells and the calibration of pO_2, temperature, and pH probes.

3.1.1. Cleaning and Sterilization of the Bioreactors

1. Disassemble the bioreactor and stirrer into its component parts.

2. Clean all components in an ultrasonic waterbath.

3. Fill the reactor vessel with approximately 2 mL of dH_2O in order to ensure that the pO_2 probe will be surrounded by water vapor during autoclaving.

4. Assemble all individual parts: Screw the closure head including the stirrer into place, slide the pO_2 probe, the pH probe, and the temperature probe into position (the depth and position have been marked beforehand); only loosely screw on sample port (provides venting during autoclaving) (see Note 5).

5. Wrap all sample ports as well as the gas in/out filters at the closure head in aluminum foil and autoclave at 121°C for 20 min.

3.1.2. Coating of the Bioreactor Vessels with Sigmacote

1. Place bioreactors under the laminar flow hood. Remove aluminum foil, screw of caps, deposit stirrer, and sample port on a petri dish.

2. Aspirate off all water that had condensed during autoclaving and let the vessels dry for at least 3 h.

3. Prior to coating, remove the pO_2, temperature, and pH probes or lift them so high that the coating solution cannot reach them.

4. Coat the vessel interior by adding 60 mL of sigmacote (calculated for a working medium volume of 45 mL) (see Note 6).

5. Decant the liquid and dry the vessels for 10 min. Then rinse twice with dH_2O.

3.1.3. Calibration of the Probes

1. Fill the coated reactor vessels with 42 mL of CDM, return all probes back to their original position and screw on all caps and the closure head including the stirrer.

2. Connect the bioreactor vessels to the bioreactor control unit.

3. Adjust the process conditions to working conditions directly on the control unit, for example $n = 110$ rpm, $T = 37°C$, $V_{Gas} = 1$ sl/h, $CO_2 = 5\%$, $O_2 = 21\%$ (see Note 7).

4. Calibrate the probes under stable process conditions according to manufacturer's information (see Note 8).

3.2. Cell Inoculation and EB Formation in Suspension Bioreactors

Murine ESCs are routinely cultivated as described in Chapter 2 (see Notes 9 and 10).

1. Trypsinize the ESCs into single cells. Decant the ESC medium or aspirate it off (see Note 11).

2. Add 1–3 mL of trypsin/EDTA to a T25-cm² culture flask of undifferentiated ESCs or 5 mL to a T75-cm². If you use dishes to cultivate your cells, adjust the volume of trypsin/EDTA accordingly. Stop the trypsin/EDTA after approximately 5 min with at least the same amount of OCM.

3. Depending on the total volume, transfer cells into a 15- or 50-mL Falcon tube.

4. Centrifuge the cells at $200 \times g$ for 5 min. Decant the supernatant or aspirate it and resuspend the cell pellet carefully in CDM by pipetting up and down (see Note 12).

5. Count cells in a hemocytometer with trypan blue exclusion. For that, add 20 µL of trypan blue solution to 180 µL of cells and count 10 µL of this cell solution. Prepare a dilution of 5.625×10^5 cells per mL in CDM only taking the viable cells into consideration for your calculation (see Note 13).

6. Disconnect the bioreactors from the control units and place them under the laminar flow hood. Add 3 mL of the ESC dilution per bioreactor vessel through the sample port using a sterile 5 mL serological pipette.

7. Connect the bioreactors to the control units, set the working conditions and start the experiment directly in the software ($n = 110$ rpm, $T = 37°C$, $V_{Gas} = 1$ sl/h, $CO_2 = 5\%$, $O_2 = 21\%$) (see Notes 14–16).

8. Let the bioreactors run for 5 days without changing the medium. The set parameters will be monitored over the entire runtime of the experiment. Screenshots are depicted in Fig. 3.

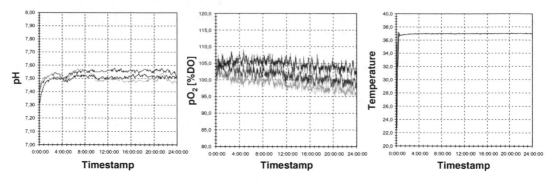

Fig. 3. Screenshots of the bioreactor control station. The pH, oxygen tension, and temperature were monitored over a bioreactor run with three vessels for 24 h. The pH and oxygen are just monitored and not regulated.

3.3. Harvesting EBs from the Suspension Bioreactors

1. In order to harvest the EBs on day 5 of the experiment from the bioreactors, stop the experiment in the bioreactor system software. Disconnect the bioreactors from the control system and transfer the vessels to the laminar flow hood.

2. Remove the closure head including the stirrer from the bioreactor vessels and let them stand for 5 min to allow the EBs to settle down.

3. Aspirate half of the medium without dispersing the EBs.

4. Transfer the rest of the medium with the EBs to a bacteriological petri dish. Check the size and shape of EBs under a microscope.

3.4. Embryoid Body Formation in Static Culture: The Hanging Drop Method

1. Prepare a solution of ESCs of 3.75×10^4 cells per mL in CDM following exactly the same steps as described in Subheading 3.2, steps 1–7.

2. Fill a bacteriological dish ($\varnothing$ 100 mm) with 10 mL of sterile PBS.

3. Using a manual or digital dispenser pipettor, dispense 20 µL of cell suspension (750 cells) on the inner side of the lid of the petri dish. 50–80 drops are pipetted per lid (see Note 17).

4. Turn the lid carefully into its regular position and put it on the top of the petri dish filled with PBS.

5. Place in incubator as soon as you have finished one dish, then pipette the next.

6. Incubate the "hanging drops" for 3 days in a humidified atmosphere with 5% CO_2 at 37°C (see Note 18).

7. After 3 days of incubation collect drops from the lid. In order to do so, hold the lid at approximately a 45° angle and rinse the EBs down to the bottom using a pipette with a sterile 1 mL tip.

8. Gently transfer the total suspension to a bacteriological dish filled with approximately 5 mL of CDM.

9. Cultivate this EB suspension culture for 2 days in a humidified atmosphere with 5% CO_2 at 37°C.

3.5. Analyzing Embryoid Body Formation

On day 5 of differentiation, the phenotype and size of the EBs is analyzed.

1. Take your dishes with EBs that you have generated either in the bioreactor or in hanging drops to a microscope.

2. Take good quality pictures with a magnification that allows you to see entire EBs.

3. You want to take as many pictures as you need in order to analyze 200–300 EBs.

4. Use a software program that allows you to measure the EB diameters (see Note 19).

Figure 4 shows an example of the size distribution of EBs that you are getting with both methods. EBs formed in suspension bioreactors under the conditions mentioned in Subheading 3.2 are typically smaller than EBs generated with the hanging drop method. In the experiment shown, hanging drop EBs had a mean diameter of 320 μm and EBs generated in suspension bioreactors (110 rpm) had a mean diameter of 137 μm. However, EBs generated in suspension bioreactors are more equal in size compared to EBs generated with the hanging drop protocol. Most EBs (37.2%) formed in suspension culture had a diameter of 125–150 μm and only 22.3% of EBs generated with the hanging drop protocol had a common size of 325–350 μm. Differentiation in stirred suspension furthermore seems to favor some lineages over others (Fig. 5).

3.6. Differentiation of ESCs Toward Osteoblasts

After 5 days of differentiation either in suspension bioreactors (see Subheadings 3.2 and 3.3) or hanging drop culture (see Subheading 3.4), EBs are trypsinized and seeded into cell culture plates. From day 5 to day 30 cells are differentiated in ODM.

1. On day 5 collect EBs in the center of the petri dish by gently rotating the dish in one direction.

2. Fill a new bacteriological petri dish (∅ 100 mm) with approximately 5 mL of prewarmed (37°C) trypsin/EDTA.

3. Transfer EBs into the petri dish filled with trypsin/EDTA. Try to transfer as less medium as you can. Incubate for 5 min at 37°C.

4. Mechanically disperse the EBs in the petri dish by pipetting up and down a few times. Check under the microscope, if cells are single cells (see Note 20).

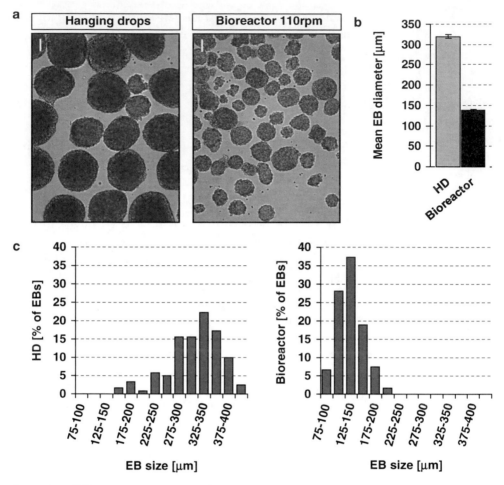

Fig. 4. Comparison of EBs between static culture (HD) and bioreactors. (**a**) The pictures show EBs on day 5 of differentiation generated with the hanging drop method (*left*) and in suspension bioreactors with a stable agitation rate of 110 rpm (*right*). Scale bar 100 μm. (**b**) The diagram shows the mean diameters of EBs on day 5 of differentiation, which were generated with the hanging drop method and in suspension bioreactors with an agitation rate of 110 rpm. EBs generated in suspension bioreactors are significantly smaller than EBs formed with the hanging drop protocol ($P < 0.001$). (**c**) The diagram show the frequency distribution of EB diameters for EBs generated with the hanging drop method (*left*) and in suspension bioreactors (*right*). *HD* hanging drops.

5. Transfer the cell solution to a Falcon tube and centrifuge at $200 \times g$ for 5 min. Resuspend in ODM for cells, which should be differentiated toward the osteogenic lineage and CDM for the control cells.

6. Count the cells and plate them at a density of 5×10^4 per cm^2 onto cell culture plates coated with 0.1% gelatin (see Note 21).

7. Cultivate the cells for 25 days in humidified atmosphere with 5% CO$_2$ at 37°C.

8. Change medium every second day.

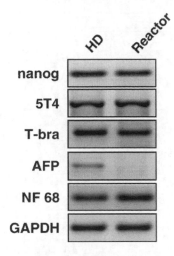

Fig. 5. Expression profiling for lineage markers comparing EBs generated through hanging drops (HD) and in bioreactors. Agitation speed was set to 130 rpm and samples taken on differentiation day 5. RT–PCR was performed for nanog, 5T4 (18), a general differentiation marker, T-brachyury (T-bra), alpha-fetoprotein (AFP), neurofilament 68 (NF68) and compared to the housekeeper GAPDH.

3.7. Endpoint Analysis of Differentiated Osteoblasts

On day 30 of differentiation, cells can be analyzed for osteoblast markers such as mineral deposition, calcium content or osteocalcin expression (see Chapter 17). In addition, the cultures should be checked for their morphology under a microscope. Successful osteogenic differentiation has taken place when a black calcified deposit is noticeable in the cultures (17). Pictures of EBs, which were generated with the HD protocol or in suspension and subsequently differentiated into osteoblasts, are shown in Fig. 6.

4. Notes

1. Bioreactor system consisting of Mini-Spinner suspension bioreactor vessels with stirrer (35–60 mL working volume), adapted pO_2 probe, pH probe, heating station, temperature probe, magnetic stirrer, and gas connection system.

2. Bioreactor control unit consisting of a temperature control station, a magnetic stirrer control station, a gas mixing station, and a personal computer (PC) with control software.

3. D3 mESCs possess a higher capability to differentiate into mesodermal cell types, among them cardiomyocytes and osteoblasts, than other murine ESC lines, such as R1.

4. Vitamin D_3 is light sensitive. You want to wrap your medium bottle in aluminum foil when it is in use. Moreover, always take pictures of your cells before you change the medium.

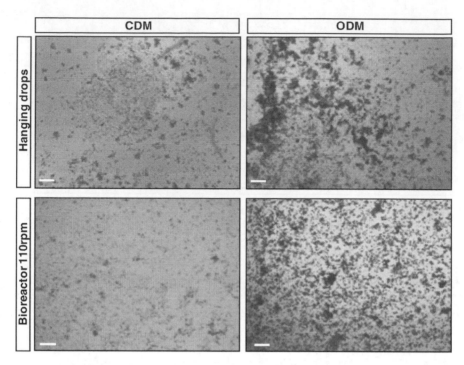

Fig. 6. Mineral deposition of ESC-derived osteoblasts. Mature osteoblasts are characterized by a mineralized matrix. This mineralization appears black in phase contrast microscopy without any additional staining (16). The photographs compare the phenotype between control cells cultivated in control culture medium without osteogenic factors (CDM) and osteogenic cultures cultivated in ODM. EBs were formed either in hanging drops (*top panel*) or in the bioreactor for 5 days at 110 rpm. EBs were then trypsinized and plated as monolayers into adherent conditions.

5. The positions of the probes in the bioreactor vessels should be equal in each experiment you run to have stable stirring conditions. Often pH probes are fragile for interferences when autoclaved too frequently, in this case it is also possible to sterilize pH probes by using 70% ethanol instead of autoclaving. For this, close the port for the pH probe with aluminum foil before autoclaving the bioreactors and assemble the ethanol cleaned pH probe in the laminar hood after autoclaving the bioreactors.

6. Vessels are coated immediately after the addition of sigmacote. The solution may be transferred into the next vessel without any incubation time. After all bioreactor vessels have been coated, you may return the solution into the flask. It can be reused up to 1,000 times.

7. Conditions can be different in different systems and different runs according to the wishes of the researcher. However, it is important to calibrate the different probes under the actual working conditions.

8. The calibration process differs from company to company. For the DASGIP bioreactor system, the temperature probe

calibration is done every few years by the company. The pH probe is calibrated before each run by a three-point calibration, placing the probe into a pH 4 solution and then into one of pH 7. The pO_2 probe is calibrated by placing it into an atmosphere of 21% O_2 (100% DO) and then the probes are disconnected from the system to calibrate for 0% O_2 (0% DO).

9. Routinely check your cells for mycoplasma contamination, as mycoplasma may change the behavior of your cells.

10. ESCs may accumulate karyotypical changes. To maintain a good quality of cells, do not use them for more than 25 passages after thawing. Thaw a new batch whenever you detect a change in proliferation rates, even when the 25 passages have not been reached.

11. If the medium is decanted, wash your cells twice with PBS. This step is not necessary, when the medium is aspirated.

12. This is best accomplished with a 2-mL serological pipette or a 2.5-mL Eppendorf pipettor.

13. By adding 3 mL of cell suspension (5.625×10^5 cells per mL) to the 42 mL of media, the final cell concentration in the bioreactors is 3.75×10^4 cells per mL. This cell concentration was tested to show best results for the hanging drop method and for expanding cultures of ESCs in bioreactors (13).

14. As described in the bioreactor system handbook, the bioreactor system software has to be adjusted. Parameters like pH, O_2 concentration, temperature, agitation rate, and gas flow rate have to be set for every bioreactor vessel separately before an experiment is to be initiated. By starting an actual experimental run, the software begins to record and safe all of the described parameters every minute.

15. The O_2 concentration is regulated to a constant 21% by regulating the mixture of the inflowing gas, which contains N_2, CO_2, and air. When the O_2 concentration drops, the concentration of O_2 in the inflowing gas mixture will be increased by reducing the N_2 concentration. After it has reached the initial set value, the O_2 concentration in the gas mixture will be decreased, respectively.

16. Depending on the system used and the data to be recorded, all of these parameters can also be regulated, if desired. This is of particular interest, if experiments under low oxygen shall be performed.

17. A yellow pipette tip should be used in addition to the dispenser combitip. This allows easy change of the pipette tip in case you touch the outside of the dish without having to throw out the entire dispenser combitip.

18. By gravity force cells collect in the bottom center of the drop. Within the first 3 days, they stick together and proliferate. The 750 cells per drop are an optimized cell number for the EB formation of D3 ESCs and the osteogenic differentiation process.

19. EBs were photographed on a Leica DMI 400B microscope. EB sizes were measured with a special Leica software tool, the Leica Application Suite.

20. This step is critical for the success of the differentiation. If the EBs have not been digested into single cells completely after the set time, only place the dish back into the incubator for a very short period of time. Longer incubation will result in the generation of strings of DNA, which will lead to total clumping of your cell suspension. If you are uncertain about your technique, rather increase your initial incubation before you disaggregate the EBs with the pipette.

21. You can either use noncoated tissue culture plates (and coat them with gelatin yourself, see Chapter 5) or alternatively, you may use BD Falcon primaria plates without further coating.

Acknowledgement

This work was supported by a German Ministry for Science, Education, and Research (BMBF) grant to NzN. The authors would like to express their sincerest thanks to Dr. Matthias Arnold and Dr. Christoph Bremus for their continuous technical support and helpful discussions on bioreactor technology.

References

1. Ying, Q.L., Wray, J., Nichols, J., Batlle-Morera, L., Doble, B., Woodgett, J., et al. (2008) The ground state of embryonic stem cell self-renewal. *Nature* **453**, 519–523.

2. Martin, G.R. and Evans, M.J. (1975) Differentiation of clonal lines of teratocarcinoma cells: formation of embryoid bodies in vitro. *Proc. Natl. Acad. Sci. U.S.A.* **72**, 1441–1445.

3. Martin, G.R. (1981) Isolation of a pluripotent cell line from early mouse embryos cultured in medium conditioned by teratocarcinoma stem cells. *Proc. Natl. Acad. Sci. U.S.A.* **78**, 7634–7638.

4. Lin, R.Y., Kubo, A., Keller, G.M., and Davies, T.F. (2003) Committing embryonic stem cells to differentiate into thyrocyte-like cells in vitro. *Endocrinology* **144**, 2644–2649.

5. Bratt-Leal, A.M., Carpenedo, R.L., and McDevitt, T.C. (2009) Engineering the embryoid body microenvironment to direct embryonic stem cell differentiation. *Biotechnol. Prog.* **25**, 43–51.

6. Wobus, A.M., Wallukat, G., and Hescheler, J. (1991) Pluripotent mouse embryonic stem cells are able to differentiate into cardiomyocytes expressing chronotropic responses to adrenergic and cholinergic agents and Ca2+ channel blockers. *Differentiation* **48**, 173–182.

7. zur Nieden, N.I., Kempka, G., Rancourt, D.E., and Ahr, H.J. (2005) Induction of chondro-, osteo- and adipogenesis in embryonic stem cells by bone morphogenetic protein-2: effect of cofactors on differentiating lineages. *BMC Dev. Biol.* **5**, 1.

8. Strübing, C., Ahnert-Hilger, G., Shan, J., Wiedenmann, B., Hescheler, J., and Wobus, A.M. (1995) Differentiation of pluripotent embryonic stem cells into the neuronal lineage in vitro gives rise to mature inhibitory and excitatory neurons. *Mech. Dev.* **53**, 275–287.

9. Rose, O., Rohwedel, J., Reinhardt, S., Bachmann, M., Cramer, M., Rotter, M., et al. (1994) Expression of M-cadherin protein in myogenic cells during prenatal mouse development and differentiation of embryonic stem cells in culture. *Dev. Dyn.* **201**, 245–259.

10. Scholz, G., Pohl, I., Genschow, E., Klemm, M., and Spielmann, H. (1999) Embryotoxicity screening using embryonic stem cells in vitro: correlation to in vivo teratogenicity. *Cells Tissues Organs* **165**, 203–211.

11. Fok, E.Y.L. and Zandstra, P.W. (2005) Shear-controlled single-step mouse embryonic stem cell expansion and embryoid body-based differentiation. *Stem Cells* **23**, 1333–1342.

12. Cormier, J.T., zur Nieden, N.I., Rancourt, D.E., and Kallos, M.S. (2006) Expansion of undifferentiated murine embryonic stem cells as aggregates in suspension culture bioreactors. *Tissue Eng.* **12**, 3233–3245.

13. zur Nieden, N.I., Cormier, J.T., Rancourt, D.E., and Kallos, M.S. (2007) Embryonic stem cells remain highly pluripotent following long term expansion as aggregates in suspension bioreactors. *J. Biotechnol.* **129**, 421–432.

14. Karp, J.M., Yeh, J., Eng, G., Fukuda, J., Blumling, J., Suh, K.Y., et al. (2007) Controlling size, shape, and homogeneity of embryoid bodies using poly(ethylene glycol) microwells. *Lab Chip* **7**, 786–794.

15. Youn, B.S., Sen, A., Behie, L.A., Girgis-Gabardo, A., and Hassell, J.A. (2006) Scaleup of breast cancer stem cell aggregate cultures to suspension bioreactors. *Biotechnol. Prog.* **22**, 801–810.

16. Zhao, F., Pathi, P., Grayson, W., Xing, Q., Locke, B.R., and Ma, T. (2005) Effects of oxygen transport on 3-D human mesenchymal stem cell metabolic activity in perfusion and static cultures: experiments and mathematical model. *Biotechnol. Prog.* **21**, 1269–1280.

17. zur Nieden, N.I., Kempka, G., and Ahr, H.J. (2003) In vitro differentiation of embryonic stem cells into mineralized osteoblasts. *Differentiation* **71**, 18–27.

18. Ward, C.M., Barrow, K., Woods, A.M., and Stern, P.L. (2003) The 5T4 oncofoetal antigen is an early differentiation marker of mouse ES cells and its absence is a useful means to assess pluripotency. *J. Cell. Sci.* **116**, 4533–4542.

Chapter 10

Methods for Embryoid Body Formation: The Microwell Approach

Dawn P. Spelke, Daniel Ortmann, Ali Khademhosseini, Lino Ferreira, and Jeffrey M. Karp

Abstract

Embyroid body (EB) formation is a key step in many embryonic stem cell (ESC) differentiation protocols. The EB mimics the structure of the developing embryo, thereby providing a means of obtaining any cell lineage. Traditionally, the two methods of EB formation are suspension and hanging drop. The suspension method allows ESCs to self aggregate into EBs in a nonadherent dish. The hanging drop method suspends ESCs on the lid of a dish and EBs form through aggregation at the bottom of the drops. Recently, alternative methods of EB formation have been developed that allow for highly accurate control of EB size and shape, resulting in reproducibly produced homogeneous EBs. This control is potentially useful for directed differentiation, as recent studies have shown that EB size may be a useful determinant of the resulting differentiated cell types. One particular approach to generate homogeneous EBs utilizes nonadhesive microwell structures. The methodology associated with this technique, along with the traditional approaches of suspension and hanging drop, is the focus of this chapter.

Key words: Embryoid body, Embryonic stem cell differentiation, Suspension method, Hanging drop method, Microwell fabrication

1. Introduction

The therapeutic potential of embryonic stem cells (ESCs) lies in their ability to differentiate into a variety of clinically useful cell types, such as hepatocytes (1, 2), cardiomyocytes (3, 4), osteoblasts (5, 6), and neural cells (7, 8). The key step in many ESC differentiation protocols is the formation of embryoid bodies (EBs), achieved by removing environmental cues that maintain ESCs in an undifferentiated state. EBs are three-dimensional cell aggregates that can mimic some structure of the developing embryo and can differentiate into cells of all three germ layers (9, 10).

Nicole I. zur Nieden (ed.), *Embryonic Stem Cell Therapy for Osteo-Degenerative Diseases*, Methods in Molecular Biology, vol. 690, DOI:10.1007/978-1-60761-962-8_10, © Springer Science+Business Media, LLC 2011

Inside the embryo are spatially distinct microenvironments, which may regulate how a cell is differentiated. Similar to these *in vivo* instructive mechanisms, cells at the periphery of EBs, for example, tend to form primitive endoderm (11). Further environmental cues and cell–cell interactions can subsequently induce increasing specification. This ability of EBs to recapitulate the developing embryo makes them, in addition to their therapeutic usage, a valuable tool for studying embryonic development (9, 12, 13).

The ability to direct the differentiation of ESCs down particular lineages and with high efficiency is necessary for therapeutic applications, which require only one or a few specific cell populations in the highest yield possible. Numerous studies have examined parameters that control ESC differentiation in an attempt to develop optimized protocols for specific cell types (14, 15). Most of these efforts have focused on soluble factors; however, recently it was found that EB size is also a strong indicator of the resulting differentiated cell types due to the relationship between size and cell–cell interactions (16, 17). For example, Ng et al. found that EBs formed with about 1,000 ESCs show optimal hematopoietic differentiation compared to EBs of different sizes. The number of EBs containing erythroid cells nearly doubled when 1,000 ESCs were used versus 500 (18).

The standard methods of EB formation are static suspension culture (19) and hanging drop (12). These techniques are successful in generating EBs that produce various differentiated cell types, but they allow only limited control over aggregate size and shape and present practical difficulties with scale-up. In suspension culture, ESCs are cultured on dishes that prevent cell attachment. As a result, the cells will self-aggregate into masses that form EBs. A detailed method on how to generate static suspension EBs from human ESCs is described in Chapter 3 and example pictures from murine EBs can be found in Fig. 1. However, due to the random aggregation of ESCs and the combination of multiple EBs that can occur, this method produces a large variety in EB sizes (20). In the hanging drop method, ESCs are suspended in drops on a surface turned upside down. The shape of the drops and gravity will promote cell aggregation into EBs at

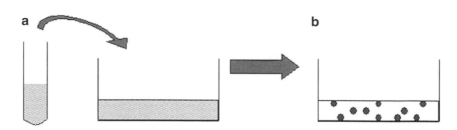

Fig. 1. Static suspension culture method. (a) A suspension of ESCs in media without pluripotency-associated factors (e.g. LIF) is transferred to a nonadherent petri dish. (b) The ESCs will self-aggregate into EBs.

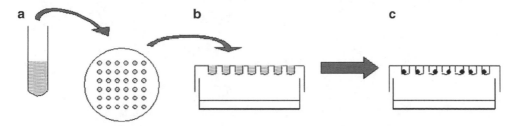

Fig. 2. Hanging drop method. (**a**) A suspension of ESCs in media without pluripotency associated factors (e.g. LIF) is plated in drops onto the lid of a petri dish. (**b**) The bottom of the dish is filled with PBS and then the lid is reversed and placed over the bottom. (**c**) The ESCs will self-aggregate into EBs.

the bottom of the drop, as shown in Fig. 2. Forming the EBs in hanging drops allows initial control of the aggregate size through the cell number in each drop. However, shape cannot be controlled and after harvesting the EBs from the drops, the size homogeneity is lost similar to the situation with the suspension method described above. Additionally, as also described in Chapter 9, the hanging drop method cannot easily be scaled up for large EB production, limiting its practical usefulness.

To overcome the limitations of existing methods of EB formation, new techniques have been developed, such as the use of microfluidic devices (21) and stirred vessel bioreactors (22, 23), a technique described in the last two chapters.

One of the most successful approaches utilizes nonadhesive microwell structures to control EB size (24, 25); this method will be the focus of this chapter. In this technique, small wells (in the submillimeter range) are produced and cells are seeded inside, as shown in Fig. 3. The cells then aggregate and grow until they are limited by the size of the microwells. This leads to accurate control of EB size and produces homogeneous EBs, as shown in Figs. 4 and 5. Furthermore, it can be readily scaled-up, particularly in comparison to the hanging drop technique. It must be noted, however, that the microwell approach requires additional instrumentation and knowledge compared with the traditional methods. For example, the fabrication of microwells requires a clean room and specialized equipment, while EB formation using suspension or hanging drop techniques requires only materials routinely found in a biology lab. Thus, those not familiar with these techniques are encouraged to purchase reagent grade microwells (see Note 1), rather than producing them in the lab as described in this chapter. The choice of EB formation method, then, depends on the goals of the particular study and the balance between EB control and costs that is desired. In the following, the two traditional methods of suspension culture and hanging drop (see also Chapters 4 and 9) and the new method of microwells are described in detail.

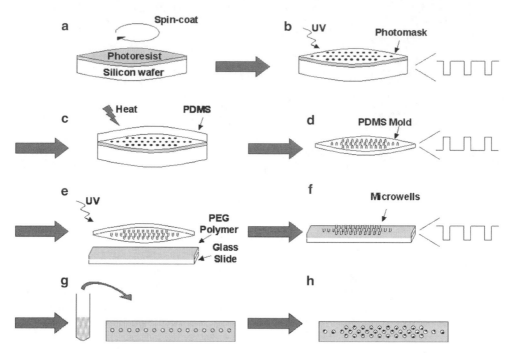

Fig. 3. Microwell method. (**a**) A silicon wafer is spin-coated with photoresist to create a film with a thickness equal to the desired microwell depth. (**b**) A printed photomask foil is placed on top of the coated wafer and then exposed to UV light. The uncovered areas of the photoresist polymerize, while the covered areas do not, leaving a pattern of indentations. (**c**) A PDMS solution is poured onto the patterned wafer and then cured in an oven. (**d**) The PDMS is now a solid mold with a raised pattern that can be peeled from the wafer. (**e**) A solution of PEG and photoinitiator is coated on a treated glass slide. The PDMS mold is then placed on top of the PEG film and exposed to UV light. (**f**) The PDMS mold is then removed, leaving behind fully formed PEG microwells. (**g**) A suspension of ESCs in media without pluripotency associated factors (e.g. LIF) is divided among the microwells. (**h**) The ESCs will self-aggregate into uniform EBs.

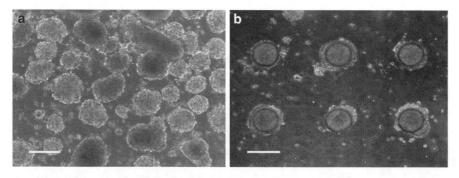

Fig. 4. EBs formed using different methods. Scale bars correspond to 200 μm. (**a**) EBs formed in static suspension culture are of a wide range of sizes and shapes. (**b**) EBs formed in microwells with a diameter of 150 μm exhibit highly defined dimensions and low variability. Reproduced from ref. (24).

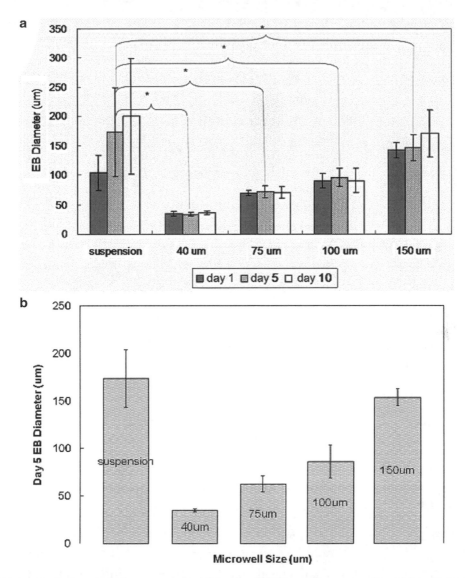

Fig. 5. EB size distributions with different methods. (a) The size of ESC aggregates grown without LIF in 40, 75, 100, and 150 μm microwells quantified 1, 5, and 10 days after seeding compared with suspension culture. Data are shown here as average ± standard deviation for $n = 50$ samples within one experiment. (b) The size of ESC aggregates grown without LIF in 40, 75, 100, and 150 μm microwells after 5 days in culture analyzed across three independent experiments and compared with suspension culture. Data are shown here as average ± standard deviation for $n = 3$ experiments. Reproduced from ref. (24).

2. Materials

2.1. Static Suspension Culture

1. ESC medium: Knockout™ Dulbecco's Modified Eagle's Medium (K-DMEM) (Invitrogen) containing 15% fetal bovine serum (FBS, ESC qualified, Invitrogen) with or without 1,000 U/mL leukemia inhibitory factor (LIF) from mouse as denoted below.

2. Phosphate-buffered saline (PBS) without calcium and magnesium (Sigma-Aldrich).

3. Trypsin, 0.25% (10×) with ethylenediaminetetraacetic acid (EDTA, 4Na, Invitrogen).

4. Conical tube.

5. Ultra Low Attachment culture dish, 100 mm (Corning) (see Note 2).

2.2. Hanging Drop Culture

1. ESC medium: Knockout™ Dulbecco's Modified Eagle's Medium (K-DMEM) (Invitrogen) containing 15% FBS (ESC qualified).

2. Polystyrene petri dish, 100 mm.

3. Multichannel pipette.

4. Phosphate-buffered saline (PBS) without calcium and magnesium.

2.3. Microwells

2.3.1. Photomask and Wafer Production

1. Photomask foil ordered from Art Services Inc. The desired photomask pattern is sent to the company for printing on foil.

2. Silicon wafer (Wafernet, Inc.).

3. Piranha solution: Mix 70% sulfuric acid (H_2SO_4) and 30% hydrogen peroxide (H_2O_2).

4. SU-8 2050 photoresist (MicroChem Corporation).

5. Heat plate adjustable to 65 and 95°C.

6. UV light source.

7. Developer solution.

8. Acetone.

2.3.2. PDMS Mold Fabrication

1. Polydimethylsiloxane (PDMS) elastomer composed of pre-polymer and curing agent (Sylgard 184, Essex Chemical).

2. Vacuum chamber (i.e. ThermoScientific, 01-060A).

2.3.3. Formation of PEG Microwells

1. Deionized (DI) water.

2. Trimethylsilyl methacrylate (TMSMA).

3. Poly(ethylene glycol) dimethacrylate (PEGDM, MW 330, see Note 3).

4. Glass slides.

5. Irgacure 2959 photoinitiator (Ciba Corporation).

6. OmniCure® S2000 UV/Visible Spot Curing System (EXFO Life Sciences & Industrial Division).

7. Ethanol.

8. 6-well cell culture plate, nontreated polystyrene.

<table>
<tr><td>2.3.4. EB Formation in PEG Microwells</td><td>

1. Phosphate-buffered saline (PBS) without calcium and magnesium.
2. ESC medium: Knockout™ Dulbecco's Modified Eagle's Medium (K-DMEM) containing 15% FBS (ESC qualified) with or without 1,000 U/mL leukemia inhibitory factor (LIF) from mouse as denoted below.
</td></tr>
</table>

3. Methods

All three methods of EB formation assume the same initial protocol for culturing ESCs in an undifferentiated state. Mouse and human ESCs should be cultured on a feeder layer, such as inactivated mouse embryonic fibroblasts (MEFs), and/or in the presence of stem cell renewal factors, such as mouse leukemia inhibitory factor (see Chapter 2). In this state, the ESCs will grow in clusters. To differentiate the ESCs by forming EBs, the feeder layer and/or stem cell renewal factors are removed, and the ESC clumps must be dissociated (into a single cell suspension for mouse ESCs or suspensions of small cell aggregates for human ESCs). The protocol below is specifically defined for mouse ESCs and can be easily adopted for human ESCs.

3.1. Static Suspension Culture

1. Remove the medium from the ESC culture and wash with PBS.
2. Add 3–5 mL of trypsin/EDTA for T75-cm² flasks and return to incubator at 37°C for 3–5 min (see Note 4).
3. Once the cells have detached from the flask, as determined by checking under a microscope, inactivate the trypsin by adding 3–5 mL of ESC medium.
4. Transfer cells into a Falcon tube and centrifuge the cells at $270 \times g$ for 5 min.
5. Remove the supernatant and resuspend the pellet in ESC medium without LIF at a concentration of 10^3–10^6 cells/mL.
6. Seed 10 mL of the cell solution onto a 100-mm ultra low attachment culture dish (see Note 5).
7. Culture in a humidified incubator at 37°C and 5% CO_2 and feed with ESC medium without LIF every 2 days.

3.2. Hanging Drop Culture

1. Trypsinize the ESCs (as described in Subheading 3.1, steps 1–4) and resuspend the pellet in ESC medium without LIF at a concentration of 10^4–5×10^4 cells/mL. The concentration of the ESCs and the size of the drops will initially affect the size of the EBs that will form.

2. Plate 20–30 µL drops of the cell solution onto the lid of a 100-mm petri dish using a multichannel pipette to achieve a regular pattern. About 30–40 drops should fit on the lid.

3. Fill the bottom of the dish with PBS to avoid evaporation.

4. Reverse the lid and place it on the bottom of the dish. Now the ESC drops are hanging and the rounded bottom of the drops will promote ESC aggregation and EB formation.

5. Culture in a humidified incubator at 37°C and 5% CO_2.

6. Harvest the EBs after 2–3 days of culture by flushing with about 5 mL of ESC medium without LIF over the lid of the dish (see Note 6).

7. Plate the EBs on culture dishes or transfer them to suspension culture. For a variation of the hanging drop protocol see Chapter 9.

3.3. Microwells

3.3.1. Photomask and Wafer Production

1. To create a photomask with the desired pattern, use a computer program, such as AutoCad, or draw freehand. For microwells, the pattern should be filled black dots with the configuration and diameter of the desired microwells. The photomask is then sent to a company that can print the photomask on foil.

2. Clean the silicon wafer with piranha solution for 5 min and then wash with DI water.

3. Spin-coat the silicon wafer with SU-8 2050 photoresist at $14 \times g$ for 5 s and then at $160 \times g$ for 30 s. Make sure to center the wafer on the spin-coater. The wafer is then fixed on the machine by a vacuum pump. This combination of photoresist and spinning will result in a 100 µm film, which translates to microwells with a depth of 100 µm (see Note 7).

4. Preincubate the wafer on a heat plate at 65°C for 5 min and then at 95°C for 15 min. The photoresist should now be solid.

5. Place the photomask foil on top of the photoresist-coated wafer and expose to UV-light for 2 min. This step will polymerize the photoresist not covered by the black dots of the photomask foil; the covered areas will remain unpolymerized.

6. Postincubate at 65°C for 4 min and then at 95°C for 9 min.

7. Put the wafer in developer solution for 8 min and then wash with acetone. After these steps, an indented pattern of the microwells should be visible as the unpolymerized photoresist will have been washed away.

8. Hard bake the wafer by putting it at 180–210°C for at least 1 h to ensure permanent binding of the photomask to the silicon wafer.

3.3.2. PDMS Mold
Fabrication

1. To create the PDMS prepolymer, mix the silicon elastomer base solution and the curing agent at a 10:1 ratio using a balance. Make sure to mix the solution very well by stirring.

2. Remove all gas from the solution by placing it in a vacuum chamber until no bubbles are visible.

3. Pour the PDMS solution on the silicon wafer patterned with photoresist. Be careful to avoid bubble formation.

4. Again, remove all gas created by pouring using a vacuum chamber.

5. Cure the wafer plus PDMS solution by placing in an oven at 70°C for 2 h (or as long as overnight).

6. Carefully peel the PDMS mold from the silicon wafer and clean with ethanol and acetone. The depressed features of the wafer will result in corresponding PDMS replicas with protruding features.

7. Cut out the desired part of the mold that will be used in the fabrication of the PEG microwells.

3.3.3. Formation
of PEG Microwells

1. To create surfaces capable of PEG attachment, treat glass slides with TMSMA for 5 min and cure at 100°C for 10 min. Following curing, wash two times with DI water.

2. Prepare a solution of 99.5 wt% PEGDA and 0.5 wt% Irgacure 2959 photoinitiator just before use. Evenly distribute a few drops of this solution on the treated glass slides resulting in a PEG polymer film.

3. Place the PDMS mold directly on the polymer film and expose it to 365 nm, 300 mW/cm² UV-light using a UV spot lamp for 30 s (see Note 8). The PEG microwells are now formed.

4. Remove the PDMS mold and place the glass slide with the microwells in 75% ethanol for 30 min to ensure sterility.

5. Aspirate the ethanol, replace with PBS, and place in the incubator over night (see Note 9).

6. Cut the PEG microwell substrate into pieces approximately 24.5 × 24.5 mm in size and place them within the wells of a 6-well plate immersed in PBS. The number of microwells per substrate will depend on the size and patterning of the microwells.

3.3.4. Formation
of Embryoid Bodies

1. Just before seeding the ESCs onto the microwells, aspirate the PBS completely so that there is no liquid film on the gels. To remove any liquid that may remain, let the microwells dry for 5–10 min.

2. Trypsinize the ESCs (as described in Subheading 3.1, steps 1–4) and resuspend the pellet in ESC medium with LIF at a concentration of ~4 × 10⁶ cells/mL.

3. Add 150–200 μL of the cell solution directly to the microwells in each well of a 6-well plate and let settle for 30 min.

4. Wash the cultures with a gentle flow of PBS to remove cells that have not docked within the microwells.

5. Return the cultures to a humidified incubator at 37°C and 5% CO_2 and feed with ESC medium without LIF every 2 days.

6. Use a 1-mL pipette to retrieve the EBs mechanically by dispensing medium directly on the microwell structures until all the EBs are flushed out.

4. Notes

1. A microwell platform is commercially available and can be purchased from StemCell Technologies. These microwells would be useful if only short-term work with microwells is intended or if the required materials and equipment are unavailable.

2. An alternative to the ultra low attachment culture dish is any nontreated polystyrene culture dish. The lack of treatment also reduces cell attachment, but to a lesser degree than the ultra low attachment dish, which is specifically designed for this purpose.

3. A variety of molecular weight PEGDAs have been used to form microwells. In general, using a higher molecular weight leads to reduced cell-adhesion and an increased retrieval of embryoid bodies (25).

4. The amounts of trypsin/EDTA needed for detachment and media needed for inactivation depend upon the size of the tissue culture flask or petri dish this is being used. The quantities given are for a T75-cm^2 tissue culture flask and can be scaled up or down as necessary.

5. This hydrogel coated polystyrene dish is hydrophilic and nonionic so the ESCs cannot adhere to the surface, resulting in ESC aggregation and EB formation.

6. The EBs have to be harvested after a short time because the drops will dry out and some of the components of the medium will become depleted. If the EBs are not transferred after this time, they will begin to fall apart.

7. If microwells of a different depth are desired, the speed and length of spin-coating can be varied.

8. Crosslinking the PEGDA solution can also be achieved with other UV lamps at a lower intensity (i.e. 1–30 mW/cm^2), if a spot lamp is not available. The photocrosslinking process

should then be optimized for this setting. The calculation of the dose (exposure time multiplied by intensity) gives a hint for this.

9. This washing step is important as the photoinitiator is toxic to the cells and must be removed completely.

References

1. Ishii, T., Yasuchika, K., Fujii, H., Hoppo, T., Baba, S., Naito, M., et al. (2005) In vitro differentiation and maturation of mouse embryonic stem cells into hepatocytes. *Exp. Cell. Res.* **309**, 68–77.

2. Lavon, N., Yanuka, O., and Benvenisty, N. (2004) Differentiation and isolation of hepatic-like cells from human embryonic stem cells. *Differentiation* **72**, 230–238.

3. Mummery, C., Ward-van Oostwaard, D., Doevendans, P., Spijker, R., van den Brink, S., Hassink, R., et al. (2003) Differentiation of human embryonic stem cells to cardiomyocytes: role of coculture with visceral endoderm-like cells. *Circulation* **107**, 2733–2740.

4. Kehat, I., Kenyagin-Karsenti, D., Snir, M., Segev, H., Amit, M., Gepstein, A., et al. (2001) Human embryonic stem cells can differentiate into myocytes with structural and functional properties of cardiomyocytes. *J. Clin. Invest.* **108**, 407–414.

5. Karp, J. M., Ferreira, L. S., Khademhosseini, A., Kwon, A. H., Yeh, J., and Langer, R. (2006) Cultivation of human embryonic stem cells without the embryoid body step enhances osteogenesis in vitro. *Stem Cells* **24**, 835–843.

6. Bielby, R. C., Boccaccini, A. R., Polak, J. M., and Buttery, L. D. (2004) In vitro differentiation and in vivo mineralization of osteogenic cells derived from human embryonic stem cells. *Tissue Eng.* **10**, 1518–1525.

7. Kim, J. H., Auerbach, J. M., Rodriguez-Gomez, J. A., Velasco, I., Gavin, D., Lumelsky, N., et al. (2002) Dopamine neurons derived from embryonic stem cells function in an animal model of Parkinson's disease. *Nature* **418**, 50–56.

8. Schuldiner, M., Eiges, R., Eden, A., Yanuka, O., Itskovitz-Eldor, J., Goldstein, R. S., et al. (2001) Induced neuronal differentiation of human embryonic stem cells. *Brain Res.* **913**, 201–205.

9. Desbaillets, I., Ziegler, U., Groscurth, P., and Gassmann, M. (2000) Embryoid bodies: an in vitro model of mouse embryogenesis. *Exp. Physiol.* **85**, 645–651.

10. Itskovitz-Eldor, J., Schuldiner, M., Karsenti, D., Eden, A., Yanuka, O., Amit, M., et al. (2000) Differentiation of human embryonic stem cells into embryoid bodies comprising the three embryonic germ layers. *Mol. Med.* **6**, 88–95.

11. Hamazaki, T., Oka, M., Yamanaka, S., and Terada, N. (2004) Aggregation of embryonic stem cells induces Nanog repression and primitive endoderm differentiation. *J. Cell. Sci.* **117**, 5681–5686.

12. Keller, G. M. (1995) In vitro differentiation of embryonic stem cells. *Curr. Opin. Cell. Biol.* **7**, 862–869.

13. Smith, A. G. (2001) Embryo-derived stem cells: of mice and men. *Annu. Rev. Cell. Dev. Biol.* **17**, 435–462.

14. Nakayama, T., Momoki-Soga, T., Yamaguchi, K., and Inoue, N. (2004) Efficient production of neural stem cells and neurons from embryonic stem cells. *Neuroreport* **15**, 487–491.

15. Xu, C., Police, S., Rao, N., and Carpenter, M. K. (2002) Characterization and enrichment of cardiomyocytes derived from human embryonic stem cells. *Circ. Res.* **91**, 501–508.

16. Bauwens, C. L., Peerani, R., Niebruegge, S., Woodhouse, K. A., Kumacheva, E., Husain, M., et al. (2008) Control of human embryonic stem cell colony and aggregate size heterogeneity influences differentiation trajectories. *Stem Cells* **26**, 2300–2310.

17. Park, J., Cho, C. H., Parashurama, N., Li, Y., Berthiaume, F., Toner, M., et al. (2007) Microfabrication-based modulation of embryonic stem cell differentiation. *Lab Chip* **7**, 1018–1028.

18. Ng, E. S., Davis, R. P., Azzola, L., Stanley, E. G., and Elefanty, A. G. (2005) Forced aggregation of defined numbers of human embryonic stem cells into embryoid bodies fosters robust, reproducible hematopoietic differentiation. *Blood* **106**, 1601–1603.

19. Martin, G. R. and Evans, M. J. (1975) Differentiation of clonal lines of teratocarcinoma cells: formation on embryoid bodies in vitro. *Proc. Natl. Acad. Sci. U.S.A.* **72**, 1441–1445.

20. Dang, S. M., Kyba, M., Perlingeiro, R., Daley, G. Q., and Zandstra, P. W. (2002) Efficiency of embryoid body formation and hematopoietic development from embryonic stem cells in different culture systems. *Biotechnol. Bioeng.* **78**, 442–453.

21. Torisawa, Y. S., Chueh, B. H., Huh, D., Ramamurthy, R., Roth, T. M., Barald, K. F., et al. (2007) Efficient formation of uniform-sized embryoid bodies using a compartmentalized microchannel device. *Lab Chip.* **7**, 770–776.

22. Schroeder, M., Niebruegge, S., Werner, A., Willbold, E., Burg, M., Ruediger, M., et al. (2005) Differentiation and lineage selection of mouse embryonic stem cells in a stirred bench scale bioreactor with automated process control. *Biotechnol. Bioeng.* **92**, 920–933.

23. zur Nieden, N. I., Cormier, J. T., Rancourt, D. E., and Kallos, M. S. (2007) Embryonic stem cells remain highly pluripotent following long term expansion as aggregates in suspension bioreactors. *J. Biotechnol.* **129**, 421–432.

24. Karp, J. M., Yeh, J., Eng, G., Fukuda, J., Blumling, J., Suh, K. Y., et al. (2007) Controlling size, shape, and homogeneity of embryoid bodies using poly(ethylene glycol) microwells. *Lab Chip.* **7**, 786–794.

25. Moeller, H. C., Mian, M. K., Shrivastava, S., Chung, B. G., Khademhosseini, A. (2008) A microwell array system for stem cell culture. *Biomaterials* **29**, 752–763.

Chapter 11

Human Embryonic Stem Cell-Derived Mesenchymal Progenitors: An Overview

Peiman Hematti

Abstract

Mesenchymal stromal/stem cells (MSCs) were originally isolated from bone marrow (BM), but are now known to be present in all fetal and adult tissues. These multipotent cells can be differentiated into at least three downstream mesenchymal lineages that include bone, cartilage, and fat. However, under some experimental conditions, these cells can differentiate into nonmesenchymal cell types and/or participate in regeneration of damaged tissues through a variety of mechanisms. Most recently, MSCs have been derived from human embryonic stem cells (hESCs) through several different methodologies. Human MSCs derived from hESCs have been shown to possess characteristics very similar to BM-derived MSCs. Thus, the generation of MSCs from hESCs provides an opportunity to study the developmental biology of cells of mesenchymal lineages in an appropriate in vitro model. Furthermore, MSCs from different adult tissue sources are being actively investigated in a multitude of clinical trials; therefore, hESCs could provide an unlimited source of MSCs for potential clinical applications in the future. Such MSCs could be used without further differentiation for regeneration of tissues, or they could be directed towards specific lineage pathways, such as bone and cartilage, for reconstruction of tissues. Finally, immuno-modulatory properties of hESC-derived MSCs are likely to prove valuable for inducing immune tolerance toward other cells or tissues derived from the same hESC lines.

Key words: Embryonic stem cell, Mesenchymal stromal cell, Mesenchymal stem cell

1. Introduction

Mesenchymal stromal/stem cells (MSCs) are multipotent cells capable of differentiating into cells of mesenchymal lineage such as bone, cartilage, and fat (1, 2). Although MSC preparations generated ex vivo have a homogenous appearance under light microscopy, they presumably consist of a heterogeneous group of progenitor cells that do not fulfill strict criteria for true stem cells at a single cell level (i.e. self-renewal and multilineage

Nicole I. zur Nieden (ed.), *Embryonic Stem Cell Therapy for Osteo-Degenerative Diseases*, Methods in Molecular Biology, vol. 690, DOI:10.1007/978-1-60761-962-8_11, © Springer Science+Business Media, LLC 2011

differentiation capacity). MSCs were originally isolated from bone marrow (BM), although similar populations have also been isolated from a wide variety of other adult tissues such as adipose tissue (3), skeletal muscle (4), synovium (5), dental pulp (6), heart, and spleen (7); from neonatal tissues such as the placenta (8), amniotic fluid (9), and umbilical cord blood (10); and from fetal liver, blood, and BM (11). Most recently, several groups have reported directed differentiation of human embryonic stem cells (hESCs) into cells with characteristics very similar to adult tissue-derived MSCs. In addition to providing new platforms to study basic human developmental processes, such as investigating the development of mesodermal tissues in vitro, hESCs could also potentially provide an unlimited source of MSCs for a wide variety of clinical applications.

Due to inconsistencies regarding how to define MSCs, a group of experts recently set forth criteria for MSCs based on a combination of specific culture properties, as well as phenotypic and functional characteristics (12). According to these widely accepted criteria, culture-expanded MSCs do not express hematopoietic markers, such as CD34 and CD45, and, although variation exists, they typically express a number of cell surface molecules, including CD29, CD44, CD73, CD90, CD105, and CD166. Finally, the biological property that most uniquely identifies MSCs is their capacity for trilineage differentiation potential in vitro into bone, cartilage, and fat upon the addition of necessary exogenous growth factors. The differentiation potential of MSCs into myogenic, or other cell types of mesenchymal and/or nonmesenchymal origin, is not part of these criteria.

2. Generation of MSCs from Human ESCs

Interestingly, although many investigators reported generation of mesenchymal-like cells from hESCs that likely fit the above criteria for MSCs, in many of the original studies, such cells were not fully characterized and thus were not classified as such. Xu et al. were the first to report derivation of fibroblast-like cells from hESCs (13). These cells were derived from hESCs (H1 cell line) and induced to differentiate using embryoid body (EB) formation, followed by culture on gelatin-coated plates. After a few passages, the cultured cells became homogenous and assumed a fibroblast/mesenchymal-like morphology, and were designated human embryonic fibroblast-like cells (HEF1). After infection with a retrovirus containing human telomerase reverse transcriptase (hTERT), flow cytometry analysis showed that HEF1-hTERT cells expressed MSC markers, including CD29, CD44, CD71, and CD90, and were able to differentiate into an osteogenic lineage, but not chondrogenic or adipogenic lineages. Since the authors did not examine HEF1 cell characteristics prior to retroviral transduction, it is not clear whether transduction with hTERT was necessary for their differentiation into MSC-like phenotype cells. Stojkovic et al. analyzed fibroblast-like cells that spontaneously formed in ESC cultures (designated hESC-derived fibroblasts or hESC-dF) and showed the expression of MSC cell surface markers, including CD44 and CD90. These cells could support growth of hESCs in the undifferentiated state. Although these cells were probably MSCs, their differentiation potential into mesenchymal lineages was not further tested (14). Wang et al. (15) and Yoo et al. (16) also reported generation of fibroblast-like cells from hESCs capable of supporting growth of hESCs in the undifferentiated state. Although the mesenchymal characteristics of these cells were not evaluated, the similarity of their culture derivation to that reported by other investigators makes it likely these cells also possessed MSC characteristics.

Barberi et al. were the first to fully characterize mesenchymal-looking cells developed in tissue culture after 40 days of coculture of hESCs with OP9 cells, which are murine BM stromal cells, and show that these cells had characteristics of MSCs (17). These cells expressed many markers of adult-derived MSCs, including cell surface markers CD44, CD73, CD105, CD166, CD106, CD29, and STRO-1, and were capable of differentiating into osteogenic, chondrogenic, adipogenic, and myogenic lineages. Furthermore, gene expression analysis of MSCs derived from H1 and H9 hESC lines and human adult MSCs showed 579 transcripts in common. Our group later showed that pure MSC populations can be derived within the first 2 weeks of coculturing hESCs with OP9 cells (18). In contrast, Olivier et al. reported that the Raclure method generated MSCs independent of coculture with OP9 or any other feeder cells (19). In this method, differentiating cells around the hESC colonies are scraped, replated, then differentiated into cells that possess MSC markers and are capable of differentiation into osteogenic and adipogenic lineages. This was followed by a report from Lian et al., who used fluorescence-activated cell sorting (FACS) to sort and expand MSCs from the CD105+/CD24– population of cells differentiated from hESCs (20), a technique described in Chapter 12 of this book. These cells had cell surface antigens, differentiation potential, and gene expression profiles similar to BM-derived MSCs. Like Olivier et al., these authors did not utilize any cocultured feeder layers for their derivation of MSCs. A lack of animal-derived feeder cells could be advantageous if these cells are designed to be subsequently used in clinical applications. Our laboratory has also shown that coculturing undifferentiated hESCs with OP9 cells is not necessary for MSC generation (21). In contrast to these mainly non-EB methods, Hwang et al. reported generation of

MSCs after transferring EBs into gelatin-coated plates, followed by subculturing for four passages in MSC media (22). These cells were positive for MSC markers CD29, CD44, and CD105 and could be further differentiated into chondrogenic tissues. More recently, the same group, using an articular cartilage defect model in athymic nude rats, showed that transplantation of chondrogenically committed hESC-derived MSCs could reestablish normal cartilage architecture in vivo (23).

In an elegant study, Bendall et al. reported the important observation that in hESC cultures, cells surrounding and generated directly from undifferentiated hESC colonies provide a regulatory stem cell niche for hESCs (24). These authors propose that hESCs maintain their culture homeostasis within an autologously generated niche by spontaneously and continuously differentiating into hESC fibroblast-like cells. These fibroblast-like cells, designated as hdFs, are presumably very similar to MSCs reported by other investigators. Based on the similar methodology reported for these studies, we propose generation of MSCs might be the default pathway during differentiation of hESCs.

In addition to these in vitro results, researchers recently investigated the in vivo engraftment potential of hESC-derived MSCs. Barberi et al. generated myoblasts from ESC-derived MSCs and showed viability after transplantation into the tibialis anterior muscle of immunodeficient mice in the absence of teratoma formation (25). The same group later showed that neural crest stem cells derived from hESCs are capable of directed differentiation toward mesenchymal lineages of smooth muscle, osteogenic, chondrogenic, and adipogenic cells, in addition to peripheral nervous system lineages (26). The main difference between mesenchymal precursors derived from neural crest stem cells and their previously published mesenchymal precursor populations was the lack of skeletal muscle production and decreased efficiency of adipocytic differentiation.

Not all cells reported by different investigators have the exact same phenotype and/or genotype. Small permutations in the culture methodology are likely to lead to either subtle or major differences in the phenotype and/or genotype of differentiated cells. For example, cells reported by Olivier et al. (19) had many MSC characteristics, including expression of CD44, CD73, CD105, and CD166. However, unlike MSCs derived from hESCs by our group (18, 21), their hESC-derived MSCs also expressed SSEA4, a marker present in hESCs. Similarly, MSCs generated by Hwang et al. (22) exhibited many markers of MSCs including CD29, CD44, and CD105, but lacked CD73, a marker widely believed to be expressed on MSCs (12). This is consistent with the wide heterogeneity of cell surface characteristics reported for MSCs derived from BM or other adult or fetal tissues. This variability possibly arises from differences in the derivation methodology, passage number, or other ill-defined factors, such as type of serum or culture media used. For example, the expression of SSEA4 in MSCs derived from adult BM has been reported by some (27), but not all (28, 29), investigators. Characterization of cells as MSCs depends on a combination of factors and not a single phenotypic marker; and it is not known whether subtle differences in cell surface markers have any significant impact on the functional properties of MSCs.

Finally, although not discussed in this overview, many investigators have reported generation of bone and cartilage directly from ESCs and not from ESC-derived MSCs (30–35). It is not clear whether, in these studies, bone or cartilage lineages were generated through non-MSC pathways, through potentially redundant differentiation paths toward osteogenic and chondrogenic cells, or if MSCs were only transiently present in the cultures and thus not identified.

3. Potential Applications of ESC-Derived MSCs

The generation of different types of cells from hESCs allows a unique opportunity to study developmental biology of the earliest stages of body formation processes in vitro, a technically impossible task using human embryos. Therefore, generation of mesodermal progenitors from hESCs could provide a novel in vitro model to study the earliest stages of embryonic mesodermal development. Nevertheless, the biggest excitement surrounds the potential transplantation of cells derived from hESCs to repair damaged or diseased tissues. One straightforward potential application of ESC-derived MSCs would be directed differentiation into osteogenic or chondrogenic lineages for orthopedic applications, including bone or cartilage repair. MSCs may not need to be fully differentiated into osteogenic or chondrogenic lineages before implantation if the host tissue environment provides the necessary signals to direct terminal differentiation of either ESC-derived MSCs or their semidifferentiated progenies (36). Alternatively, as mentioned above, it may be possible to generate fully differentiated bone or cartilage tissues bypassing a MSC phase. Such cells could also be used for the reconstruction of bone and cartilage tissues. However, there has been no direct comparison of the functional capacity of cells generated using these various methodologies.

In addition to potential applications in orthopedic and reconstructive surgery, MSCs are being actively investigated in the much wider field of regenerative medicine. A number of initial studies suggested that MSCs do not exclusively differentiate into other types of cells of mesodermal lineage, but they also differentiate into cells of endodermal and ectodermal lineages, including cardiomyocytes (37), endothelial cells (38), lung epithelial cells (39), hepatocytes (40), neurons (41), and pancreatic islets (42). Although these cells have been tested in a variety of injury models, the degree of contribution to different tissues through *trans*-differentiation is now a matter of strong debate (43, 44). Nevertheless, new functional mechanisms for these cells are being discovered, such as migration to the inflammatory site, stimulation of proliferation and differentiation of resident progenitor cells, secretion of growth factors, and matrix remodeling to promote recovery of injured cells. All these potential roles individually and collectively provide the rationale to continue investigating the regenerative potential of MSCs (45–50). Culture expanded BM-derived MSCs have been used in several small phase I–II trials for a variety of diseases, including metachromatic leukodystrophy and Hurler's disease (51), osteogenesis imperfecta (52), myocardial infarction (53), amyotrophic lateral sclerosis (54), graft versus host disease (55), and Crohn's disease (56), among others. In many of these trials, MSCs from third-party donors were used without any HLA-typing; thus, it is theoretically feasible that a few clinical grades of validated hESCs could provide sufficient MSCs for clinical applications in a large number of patients.

4. Immunomodulatory Properties of MSCs

Clinical exploitation of MSCs derived primarily from adult tissues, typically BM, has been greatly facilitated by understanding the immune-privileged nature of MSCs. MSCs possess exceptional immunosuppressive and/or immunomodulatory properties (57) that make their use in the allogeneic setting very attractive and practical. The emerging body of data indicates that MSCs derived from BM and other tissues modulate the immune system through interaction with a broad range of immune cells, including T-lymphocytes, B-lymphocytes, natural killer cells, and dendritic cells (58–60). These immunomodulatory properties of MSCs, and not their *trans*-differentiation potential, have been the basis for their use in treating conditions, such as Crohn's disease and graft versus host disease after allogeneic hematopoietic stem cell transplantation (61).

Human MSCs express human leukocyte antigen (HLA) class I on their cell surface but not HLA class II (62); human MSCs also do not express costimulatory molecules CD80, CD86, or CD40. Our laboratory was able to show (21) that ESC-derived MSCs also possess immunological properties very similar to those of BM-derived MSCs, including the expression of HLA-I but not

HLA-II or costimulatory molecules, lack of immunogenicity when cocultured with third-party lymphocytes, and that they have immunosuppressive effects in mixed lymphocyte culture assays. These observations have now been confirmed by other investigators (63).

We predict that the inclusion of human ESC-derived MSCs could provide protective immunomodulatory functions toward cotransplanted cells or tissues derived from the same ESC lines. Despite the huge potential of induced pluripotent (iPS) cells (64) to generate patient-specific ESCs, there will still be instances in which the use of allogeneic cells is preferable to autologously derived cells. For instance, in type-1 diabetes, autoantibodies in the recipient could prove to be harmful to autologously generated iPS-derived pancreatic islets, but not to allogeneic islet cells (65, 66). In such situations, the immunomodulatory properties of MSCs derived from the same hESC lines used for pancreatic islet generation could be potentially useful for promoting engraftment of the transplanted ESC-derived islet cells.

5. Safety Concerns

Infusion of ex vivo expanded MSCs derived from adult tissues is considered relatively safe based on the assumption that these cells are not immunogenic and actually elicit an immunomodulatory effect in the recipient. This has been substantiated by the impressive safety record of MSCs in numerous clinical trials thus far (61). Therefore, it can be assumed that MSCs derived from hESCs under clinically acceptable conditions are likely to provide an equally safe alternative to adult tissue-derived MSCs. However, in the case of MSCs derived from hESCs, there are two major concerns. ESCs, by definition, generate teratomas when transplanted into the immunodeficient host (67), and this potential complication has been the biggest concern hindering the application of hESCs into clinical use (68). However, in contrast to other cells and tissues derived from hESCs (69–72), MSCs have been generated with a very high level of purity. It is unlikely that ESCs can survive the repeated passaging that is a necessary part of MSC derivation. Furthermore, the resilience of MSCs in culture is sufficiently robust to allow additional FACS sorting prior to their transplantation. Such precautionary strategies will further enhance culture purity without major effects on functionality. The second safety concern surrounding the clinical use of MSCs is potential tumor formation by MSCs themselves, whether derived from adult tissues or ESCs, and their potential in promoting growth of other malignancies. Some recent studies suggest that BM-derived MSCs may become neoplastic after long term in vitro culture (73–75); however, this phenomenon has not been observed in vivo yet. Furthermore, Djouad et al. demonstrated that MSCs prevent rejection of allogeneic tumor cells in immunocompetent mice and promote tumor formation when injected locally or systemically (76), and Karnoub et al. showed, in a murine xenograft model, that BM-derived human MSCs, when mixed with otherwise weakly metastatic human breast carcinoma cells and injected into a subcutaneous site, cause cancer cells to greatly increase their metastatic potency (77). As a result, although it has not been seen in clinical trials yet, it is theoretically possible for MSCs to generate tumors or promote growth of other malignancies, either through a direct effect on tumor growth or via immunosuppression of antitumor responses. This potential risk is not limited to ESC-derived MSCs and can happen with any type of MSCs used. Finally, it is important to recognize that ESC-derived MSCs are not immortalized cell lines, but have a limited passaging capability similar to BM-derived MSCs.

6. Future Directions

Much of our knowledge of ESC-derived MSCs is based on in vitro experiments, and much more research is needed before such cells can be used clinically. There are important questions to be answered: With the availability of MSCs derived easily from BM and other adult tissues such as fat, do we need a new source of MSCs for clinical applications? How can generation of MSCs from hESCs be standardized? What differences exist between ESC-derived MSCs and BM, or other adult tissue-derived MSCs, and are these differences sufficient to make one preferable over others? What, if any, are the differences between osteogenic and chondrogenic cells generated from ESC-derived MSCs versus BM-derived MSCs? Despite all the above mentioned uncertainties regarding the potential clinical applications of ESC-derived cells in the near future, there is no doubt that ESCs have opened wide-ranging opportunities for further exploration in the field of regenerative medicine.

Acknowledgment

I thank Dr. Laura H. Hogan for her critical review of the manuscript. The author is the recipient of NIH/NHLBI HL081076 K08 award.

References

1. Friedenstein, A. J., Petrakova, K. V., Kurolesova, A. I., and Frolova, G. P. (1968) Heterotopic of bone marrow. Analysis of precursor cells for osteogenic and hematopoietic tissues. *Transplantation*. **6**, 230–247.

2. Pittenger, M. F., Mackay, A. M., Beck, S. C., Jaiswal, R. K., Douglas, R., Mosca, J. D., et al. (1999) Multilineage potential of adult human mesenchymal stem cells. *Science*. **284**, 143–147.

3. Zuk, P. A., Zhu, M., Mizuno, H., Huang, J., Futrell, J. W., Katz, A. J., et al. (2001) Multilineage cells from human adipose tissue: implications for cell-based therapies. *Tissue Eng*. **7**, 211–228.

4. Williams, J. T., Southerland, S. S., Souza, J., Calcutt, A. F., and Cartledge, R. G. (1999) Cells isolated from adult human skeletal muscle capable of differentiating into multiple mesodermal phenotypes. *Am. Surg*. **65**, 22–26.

5. De Bari, C., Dell'Accio, F., Tylzanowski, P., and Luyten, F. P. (2001) Multipotent mesenchymal stem cells from adult human synovial membrane. *Arthritis Rheum*. **44**, 1928–1942.

6. Gronthos, S., Mankani, M., Brahim, J., Robey, P. G., and Shi, S. (2000) Postnatal human dental pulp stem cells (DPSCs) in vitro and in vivo. *Proc. Natl. Acad. Sci. USA*. **97**, 13625–13630.

7. Hoogduijn, M. J., Crop, M. J., Peeters, A. M., Van Osch, G. J., Balk, A. H., Ijzermans, J. N., et al. (2007) Human heart, spleen, and perirenal fat-derived mesenchymal stem cells have immunomodulatory capacities. *Stem Cells Dev*. **16**, 597–604.

8. In 't Anker, P. S., Scherjon, S. A., Kleijburg-van der Keur, C., de Groot-Swings, G. M., Claas, F. H., Fibbe, W. E., et al. (2004) Isolation of mesenchymal stem cells of fetal or maternal origin from human placenta. *Stem Cells*. **22**, 1338–1345.

9. In 't Anker, P. S., Scherjon, S. A., Kleijburg-van der Keur, C., Noort, W. A., Claas, F. H., Willemze, R., Fibbe, W. E., et al. (2003) Amniotic fluid as a novel source of mesenchymal stem cells for therapeutic transplantation. *Blood*. **102**, 1548–1549.

10. Bieback, K., Kern, S., Kluter, H., and Eichler, H. (2004) Critical parameters for the isolation of mesenchymal stem cells from umbilical cord blood. *Stem Cells*. **22**, 625–634.

11. Campagnoli, C., Roberts, I. A., Kumar, S., Bennett, P. R., Bellantuono, I., and Fisk, N. M. (2001) Identification of mesenchymal stem/progenitor cells in human first trimester fetal blood, liver, and bone marrow. *Blood*. **98**, 2396–2402.

12. Dominici, M., Le Blanc, K., Mueller, I., Slaper-Cortenbach, I., Marini, F., Krause, D., et al. (2006) Minimal criteria for defining multipotent mesenchymal stromal cells. The International Society for Cellular Therapy position statement. *Cytotherapy*. **8**, 315–317.

13. Xu, C., Jiang, J., Sottile, V., McWhir, J., Lebkowski, J., and Carpenter, M. K. (2004) Immortalized fibroblast-like cells derived from human embryonic stem cells support undifferentiated cell growth. *Stem Cells*. **22**, 972–980.

14. Stojkovic, P., Lako, M., Stewart, R., Przyborski, S., Armstrong, L., Evans, J., et al. (2005) An autogeneic feeder cell system that efficiently supports growth of undifferentiated human embryonic stem cells. *Stem Cells*. **23**, 306–314.

15. Wang, Q., Fang, Z. F., Jin, F., Lu, Y., Gai, H., and Sheng, H. Z. (2005) Derivation and growing human embryonic stem cells on feeders derived from themselves. *Stem Cells*. **23**, 1221–1227.

16. Yoo, S. J., Yoon, B. S., Kim, J. M., Song, J. M., Roh, S., You, S., and Yoon, H. S. (2005) Efficient culture system for human embryonic stem cells using autologous human embryonic stem cell-derived feeder cells. *Exp. Mol. Med*. **37**, 399–407.

17. Barberi, T., Willis, L. M., Socci, N. D., and Studer, L. (2005) Derivation of multipotent mesenchymal precursors from human embryonic stem cells. *PLoS Med*. **2**, e161.

18. Trivedi, P. and Hematti, P. (2007) Simultaneous generation of CD34(+) primitive hematopoietic cells and CD73(+) mesenchymal stem cells from human embryonic stem cells cocultured with murine OP9 stromal cells. *Exp. Hematol*. **35**, 146–154.

19. Olivier, E. N., Rybicki, A. C., and Bouhassira, E. E. (2006) Differentiation of human embryonic stem cells into bipotent mesenchymal stem cells. *Stem Cells*. **24**, 1914–1922.

20. Lian, Q., Lye, E., Suan Yeo, K., Khia Way Tan, E., Salto-Tellez, M., Liu, T. M., et al. (2007) Derivation of clinically compliant MSCs from CD105+, CD24– differentiated human ESCs. *Stem Cells*. **25**, 425–436.

21. Trivedi, P. and Hematti, P. (2008) Derivation and immunological characterization of mesenchymal stromal cells from human embryonic stem cells. *Exp. Hematol*. **36**, 350–359.

22. Hwang, N. S., Varghese, S., Zhang, Z., and Elisseeff, J. (2006) Chondrogenic differentiation of human embryonic stem cell-derived cells in arginine-glycine-aspartate-modified hydrogels. *Tissue Eng.* **12**, 2695–2706.

23. Hwang, N. S., Varghese, S., Lee, H. J., Zhang, Z., Ye, Z., Bae, J., et al. (2008) In vivo commitment and functional tissue regeneration using human embryonic stem cell-derived mesenchymal cells. *Proc. Natl. Acad. Sci.USA.* **105**, 20641–20646.

24. Bendall, S. C., Stewart, M. H., Menendez, P., George, D., Vijayaragavan, K., Werbowetski-Ogilvie, T., et al. (2007) IGF and FGF cooperatively establish the regulatory stem cell niche of pluripotent human cells in vitro. *Nature.* **448**, 1015–1021.

25. Barberi, T., Bradbury, M., Dincer, Z., Panagiotakos, G., Socci, N. D., and Studer, L. (2007) Derivation of engraftable skeletal myoblasts from human embryonic stem cells. *Nat. Med.* **13**, 642–648.

26. Lee, G., Kim, H., Elkabetz, Y., Al Shamy, G., Panagiotakos, G., Barberi, T., et al. (2007) Isolation and directed differentiation of neural crest stem cells derived from human embryonic stem cells. *Nat. Biotechnol.* **25**, 1468–1475.

27. Gang, E. J., Bosnakovski, D., Figueiredo, C. A., Visser, J. W., and Perlingeiro, R. C. (2007) SSEA-4 identifies mesenchymal stem cells from bone marrow. *Blood.* **109**, 1743–1751.

28. Cheng, L., Hammond, H., Ye, Z., Zhan, X., and Dravid, G. (2003) Human adult marrow cells support prolonged expansion of human embryonic stem cells in culture. *Stem Cells.* **21**, 131–142.

29. Wagner, W., Wein, F., Seckinger, A., Frankhauser, M., Wirkner, U., Krause, U., et al. (2005) Comparative characteristics of mesenchymal stem cells from human bone marrow, adipose tissue, and umbilical cord blood. *Exp. Hematol.* **33**, 1402–1416.

30. Karp, J. M., Ferreira, L. S., Khademhosseini, A. H., Kwon, A., Yeh, J., and Langer, R. S. (2006) Cultivation of human embryonic stem cells without the embryoid body step enhances osteogenesis in vitro. *Stem Cells.* **24**, 835–843.

31. Cao, T., Heng, B. C., Ye, C. P., Liu, H., Toh, W. S., Robson, P., et al. (2005) Osteogenic differentiation within intact human embryoid bodies result in a marked increase in osteocalcin secretion after 12 days of in vitro culture, and formation of morphologically distinct nodule-like structures. *Tissue Cell.* **37**, 325–334.

32. Woll, N. L., Heaney, J. D., and Bronson, S. K. (2006) Osteogenic nodule formation from single embryonic stem cell-derived progenitors. *Stem Cells Dev.* **15**, 865–879.

33. Bielby, R. C., Boccaccini, A. R., Polak, J. M., and Buttery, L. D. (2004) In vitro differentiation and in vivo mineralization of osteogenic cells derived from human embryonic stem cells. *Tissue Eng.* **10**, 1518–1525.

34. Toh, W. S., Yang, Z., Liu, H., Heng, B. C., Lee, E. H., and Cao, T. (2007) Effects of culture conditions and bone morphogenetic protein 2 on extent of chondrogenesis from human embryonic stem cells. *Stem Cells.* **25**, 950–960.

35. Sottile, V., Thomson, A., and McWhir, J. (2003) In vitro osteogenic differentiation of human ES cells. *Cloning Stem Cells.* **5**, 149–155.

36. Alsberg, E., von Recum, H. A., and Mahoney, M. J. (2006) Environmental cues to guide stem cell fate decision for tissue engineering applications. *Expert Opin. Biol. Ther.* **6**, 847–866.

37. Makino, S., Fukuda, K., Miyoshi, S., Konishi, F., Kodama, H., Pan, J., et al. (1999) Cardiomyocytes can be generated from marrow stromal cells in vitro. *J. Clin. Invest.* **103**, 697–705.

38. Oswald, J., Boxberger, S., Jorgensen, B., Feldmann, S., Ehninger, G., Bornhauser, M., et al. (2004) Mesenchymal stem cells can be differentiated into endothelial cells in vitro. *Stem Cells.* **22**, 377–384.

39. Spees, J. L., Olson, S. D., Ylostalo, J., Lynch, P. J., Smith, J., Perry, A., et al. (2003) Differentiation, cell fusion, and nuclear fusion during ex vivo repair of epithelium by human adult stem cells from bone marrow stroma. *Proc. Natl. Acad. Sci. USA.* **100**, 2397–2402.

40. Schwartz, R. E., Reyes, M., Koodie, L., Jiang, Y., Blackstad M., Lund, T., et al. (2002) Multipotent adult progenitor cells from bone marrow differentiate into functional hepatocyte-like cells. *J. Clin. Invest.* **109**, 1291–1302.

41. Woodbury, D., Schwarz, E. J., Prockop, D. J., and Black, I. B. (2000) Adult rat and human bone marrow stromal cells differentiate into neurons. *J. Neurosci. Res.* **61**, 364–370.

42. Tang, D. Q., Cao, L. Z., Burkhardt, B. R., Xia, C. Q., Litherland, S. A., Atkinson, M. A., et al. (2004) In vivo and in vitro characterization of insulin-producing cells obtained from murine bone marrow. *Diabetes.* **53**, 1721–1732.

43. Phinney, D. G. and Prockop, D. J. (2007) Concise review: mesenchymal stem/multipotent stromal cells: the state of transdifferentiation and modes of tissue repair–current views. *Stem Cells.* **25**, 2896–2902.

44. Prockop, D. J. (2007) "Stemness" does not explain the repair of many tissues by mesenchymal stem/multipotent stromal cells (MSCs). *Clin. Pharmacol. Ther.* **82**, 241–243.

45. Chamberlain, G., Fox, J., Ashton, B., and Middleton, J. (2007) Concise review: mesenchymal stem cells: their phenotype, differentiation capacity, immunological features, and potential for homing. *Stem Cells.* **25**, 2739–2749.

46. Caplan, A. I. (2007) Adult mesenchymal stem cells for tissue engineering versus regenerative medicine. *J. Cell Physiol.* **213**, 341–347.

47. Keating, A. (2006) Mesenchymal stromal cells. *Curr. Opin. Hematol.* **13**, 419–425.

48. Deans, R. J. and Moseley, A. B. (2000) Mesenchymal stem cells: biology and potential clinical uses. *Exp. Hematol.* **28**, 875–884.

49. Dazzi, F. and Horwood, N. J. (2007) Potential of mesenchymal stem cell therapy. *Curr. Opin. Oncol.* **19**, 650–655.

50. Uccelli, A., Pistoia, V., and Moretta, L. (2007) Mesenchymal stem cells: a new strategy for immunosuppression? *Trends Immunol.* **28**, 219–226.

51. Koc, O. N., Day, J., Nieder, M., Gerson, S. L., Lazarus, H. M., and Krivit, W. (2002) Allogeneic mesenchymal stem cell infusion for treatment of metachromatic leukodystrophy (MLD) and Hurler syndrome (MPS-IH). *Bone Marrow Transplant.* **30**, 215–222.

52. Horwitz, E. M., Prockop, D. J., Fitzpatrick, L. A., Koo, W. W., Gordon, P. L., Neel, M., et al. (1999) Transplantability and therapeutic effects of bone marrow-derived mesenchymal cells in children with osteogenesis imperfecta. *Nat. Med.* **5**, 309–313.

53. Chen, S. L., Fang, W. W., Ye, F., Liu, Y. H., Qian, J., Shan, S. J., et al. (2004) Effect on left ventricular function of intracoronary transplantation of autologous bone marrow mesenchymal stem cell in patients with acute myocardial infarction. *Am. J. Cardiol.* **94**, 92–95.

54. Mazzini, L., Mareschi, K., Ferrero, I., Vassallo, E., Oliveri, G., Boccaletti, R., et al. (2006) Autologous mesenchymal stem cells: clinical applications in amyotrophic lateral sclerosis. *Neurol. Res.* **28**, 523–526.

55. Le Blanc, K., Frassoni, F., Ball, L., Locatelli, F., Roelofs, H., Lewis, I., et al. (2008) Mesenchymal stem cells for treatment of steroid-resistant, severe, acute graft-versus-host disease: a phase II study. *Lancet.* **371**, 1579–1586.

56. Taupin, P. (2006) OTI-010 Osiris Therapeutics/JCR Pharmaceuticals. *Curr. Opin. Invest. Drugs.* **7**, 473–481.

57. Le Blanc, K. and Ringden, O. (2005) Immunobiology of human mesenchymal stem cells and future use in hematopoietic stem cell transplantation. *Biol. Blood Marrow Transplant.* **11**, 321–334.

58. Stagg, J. and Galipeau, J. (2007) Immune plasticity of bone marrow-derived mesenchymal stromal cells. *Handb. Exp. Pharmacol.* **180**, 45–66.

59. Noel, D., Djouad, F., Bouffi, C., Mrugala, D., and Jorgensen, C. (2007) Multipotent mesenchymal stromal cells and immune tolerance. *Leuk. Lymphoma.* **48**, 1283–1289.

60. Nauta, A. J. and Fibbe, W. E. (2007) Immunomodulatory properties of mesenchymal stromal cells. *Blood.* **110**, 3499–3506.

61. Giordano, A., Galderisi, U., and Marino, I. R. (2007) From the laboratory bench to the patient's bedside: an update on clinical trials with mesenchymal stem cells. *J. Cell Physiol.* **211**, 27–35.

62. Le Blanc, K., Tammik, C., Rosendahl, K., Zetterberg, E., and Ringden, O. (2003) HLA expression and immunologic properties of differentiated and undifferentiated mesenchymal stem cells. *Exp. Hematol.* **31**, 890–896.

63. Yen, B. L., Chang, C. J., Liu, K. J., Chen, Y. C., Hu, H. I., Bai, C. H., and Yen, M. L. (2009) Brief report – human embryonic stem cell-derived mesenchymal progenitors possess strong immunosuppressive effects toward natural killer cells as well as T lymphocytes. *Stem Cells.* **27**, 451–456.

64. Yamanaka, S. (2007) Strategies and new developments in the generation of patient-specific pluripotent stem cells. *Cell Stem Cell.* **1**, 39–49.

65. Dieterle, C. D., Hierl, F. X., Gutt, B., Arbogast, H., Meier, G. R., Veitenhansl, M., Hoffmann, J. N., and Landgraf, R. (2005) Insulin and islet autoantibodies after pancreas transplantation. *Transpl. Int.* **18**, 1361–1365.

66. Laughlin, E., Burke, G., Pugliese, A., Falk, B., and Nepom, G. (2008) Recurrence of autoreactive antigen-specific CD4+ T cells in autoimmune diabetes after pancreas transplantation. *Clin. Immunol.* **128**, 23–30.

67. Thomson, J. A. and Odorico, J. S. (2000) Human embryonic stem cell and embryonic germ cell lines. *Trends. Biotechnol.* **18**, 53–57.

68. Baker, M. (2008) FDA to vet embryonic stem cells' safety. *Nature.* **452**, 670.

69. Dihne, M., Bernreuther, C., Hagel, C., Wesche, K. O., and Schachner, M. (2006) Embryonic stem cell-derived neuronally committed precursor cells with reduced teratoma formation after transplantation into the lesioned adult mouse brain. *Stem Cells.* **24**, 1458–1466.

70. Leor, J., Gerecht, S., Cohen, S., Miller, L., Holbova, R., Ziskind, A., Shachar, M., Feinberg, M. S, Guetta, E., and Itskovitz-Eldor, J. (2007) Human embryonic stem cell transplantation to repair the infarcted myocardium. *Heart.* **93**, 1278–1284.

71. Cai, J., Yi, F. F., Yang, X. C., Lin, G. S., Jiang, H., Wang, T., et al. (2007) Transplantation of embryonic stem cell-derived cardiomyocytes improves cardiac function in infarcted rat hearts. *Cytotherapy.* **9**, 283–291.

72. Arnhold, S., Klein, H., Semkova, I., Addicks, K., and Schraermeyer, U. (2004) Neurally selected embryonic stem cells induce tumor formation after long-term survival following engraftment into the subretinal space. *Invest. Ophthalmol. Vis. Sci.* **45**, 4251–4255.

73. Rubio, D., Garcia-Castro, J., Martin, M. C., de la Fuente, R., Cigudosa, J. C., Lloyd, A. C., et al. (2005) Spontaneous human adult stem cell transformation. *Cancer Res.* **65**, 3035–3039.

74. Wang, Y., Huso, D. L., Harrington, J., Kellner, J., Jeong, D. K., Turney, J., et al. (2005) Outgrowth of a transformed cell population derived from normal human BM mesenchymal stem cell culture. *Cytotherapy.* **7**, 509–519.

75. Tolar, J., Nauta, A. J., Osborn, M. J., Panoskaltsis Mortari, A., McElmurry, R. T., Bell, S., et al. (2007) Sarcoma derived from cultured mesenchymal stem cells. *Stem Cells.* **25**, 371–379.

76. Djouad, F., Plence, P., Bony, C., Tropel, P., Apparailly, F., Sany, J., et al. (2003) Immunosuppressive effect of mesenchymal stem cells favors tumor growth in allogeneic animals. *Blood.* **102**, 3837–3844.

77. Karnoub, A. E., Dash, A. B., Vo, A. P., Sullivan, A., Brooks, M. W., Bell, G. W., et al. (2007) Mesenchymal stem cells within tumour stroma promote breast cancer metastasis. *Nature.* **449**, 557–563

Chapter 12

Derivation of Mesenchymal Stem Cells from Human Embryonic Stem Cells

Andre Choo and Sai Kiang Lim

Abstract

Mesenchymal stem cells (MSCs) have been isolated from many tissues including differentiating human embryonic stem cells (hESCs). Derivation of MSCs from hESCs consists of two major steps: differentiation and isolation. In our hands, differentiation of hESCs towards MSC-enriched culture can be induced by trypsinizing hESCs into single cells and plating them on gelatin-coated plates in a culture condition that enhances survival of hESC-derived MSCs and not hESCs. The trypsinized hESCs were grown with feeder support and the medium was supplemented with basic fibroblast growth factor (FGF2) and platelet-derived growth factor (PDGF)-AB. A highly enriched MSC culture could be obtained by repeated passaging by trypsinization. The enriched MSC cultures could be further purified by limiting dilution or FACS sorting for CD105$^+$ or CD73$^+$ and CD24$^-$.

Key words: Mesenchymal stem cells, Human embryonic stem cells, CD105, CD73, CD24, Flow cytometry

1. Introduction

Mesenchymal stem cells (MSCs) are one of the easiest adult stem cells to isolate and propagate in culture. As such, they are highly amenable to cell-based therapies where they can be expanded ex vivo for autologous cell therapy. These cells have been isolated from many adult, fetal, and embryonic tissues such as bone marrow, fats, and skin (1). MSCs are known to be multipotent and have been reported to differentiate into an amazing array of cell types, such as osteoblasts, chondrocytes, adipocytes, and endothelial cells leading to many potential therapeutic applications (2). However, MSCs have been isolated using different approaches and are characterized using different parameters. Therefore, to facilitate comparing of results from different MSC studies, the

Nicole I. zur Nieden (ed.), *Embryonic Stem Cell Therapy for Osteo-Degenerative Diseases*, Methods in Molecular Biology, vol. 690, DOI 10.1007/978-1-60761-962-8_12, © Springer Science+Business Media, LLC 2011

International Society for Cellular Therapy has issued a position statement on a minimal criterion for defining multipotent MSCs (3). First, MSCs must be plastic-adherent when maintained in standard culture conditions. Second, MSCs must express CD105, CD73, and CD90 and lack expression of CD45, CD34, CD14 or CD11b, CD79α or CD19, and HLA-DR surface molecules. Thirdly, MSCs must be capable of differentiating into osteoblasts, adipocytes, and chondrocytes in vitro. Based on this criterion, several groups have reported the derivation of MSCs from human ESCs (4–6). In contrast to other derivation protocols, our process does not require the use of viral vectors, DNA transformation, or the use of feeder cells (5). It therefore circumvents the exposure to mouse feeder cells or other animal products and minimizes the risk of exposure to xenozootic infectious agents. A further advantage of the protocol described below is that the derivation of MSCs abstains from using serum. In summary, this is an easy method to derive clinically compliant MSCs.

It is generally observed that hESCs undergo spontaneous differentiation after trypsinization into single cells or when cultured on gelatinized plates without feeder cells or medium conditioned by feeder cells. By repeated trypsinization of hESCs into single cells and then plating them on gelatinized plates in culture medium supplemented with basic fibroblast growth factor (FGF2) and platelet-derived growth factor (PDGF) to promote MSC growth, a highly enriched polyclonal MSC population could be obtained. These enriched MSC cultures could be further purified by sorting for cells bearing MSC-associated markers such as CD105 and against cells bearing hESC-, but not MSC- associated markers such as CD24. In addition, the surface antigen profile of these enriched hESC-MSCs is positive for CD29, CD44, CD49, CD105, and CD166, and negative for CD34 and CD45, which characterizes them as being similar to bone marrow MSCs and adipose-derived MSCs (5). Although MSCs derived from hESCs exhibit many characteristics of MSCs derived from adult bone marrow, they also exhibit some distinct biological differences. For example, they express higher levels of genes that are associated with early embryonic processes and, unlike bone marrow-derived MSCs, differentiate more efficiently into adipocytes than osteoblasts or chondrocytes (5).

MSCs are also known to mediate tissue repair by secreting paracrine factors that promote cell growth and reduce tissue injury (7). More specifically, we have demonstrated that hESC-derived MSCs secrete more than 201 unique gene products (8) and that these secreted factors reduce reperfusion injury in a pig model of myocardial ischemia/reperfusion (9). We also demonstrated that these hESC-derived MSCs can be used as an autogenic feeder layer for hESCs and that biological factors secreted by them support the propagation of hESCs in an undifferentiated state (10).

For secretion-based and not cell-based applications of MSCs, hESC-derived MSCs offer several advantages over bone marrow-derived MSCs. The most distinct advantage is the almost infinitely expandable source of the parent hESCs for generating MSCs. This virtually ensures that highly uniformed batches of MSCs can be generated from the same hESC source, and therefore, minimizes batch to batch variations in large-scale production of secretion factors. In contrast, large-scale production of secretion factors from adult tissue-derived MSCs such as bone marrow or adipose tissues will require multiple donors and expensive testing for each individual donor.

2. Materials

2.1. Cell Culture

1. Trypsin, 0.05% (1×) with Ethylenediaminetetraacetic acid (EDTA), 4Na, liquid (Invitrogen).
2. 0.1% gelatin in water.
3. Phosphate-buffered saline (PBS, 1×), pH 7.4 (Invitrogen).
4. Basic fibroblast growth factor (bFGF), human, recombinant (Invitrogen). Prepare a 500 µg/mL stock solution with sterile PBS, aliquot, and freeze at −20°C.
5. Platelet-derived growth factor (PDGF)-AB (Cytolab). Make a 500 µg/mL stock solution and store at −20°C.
6. MSC derivation medium: Combine 435 mL Knockout Dulbecco's Modified Eagle's Medium (K-DMEM) (Invitrogen), 50 mL Knockout Serum Replacement (KSR, Invitrogen, final concentration of 10%), 5 mL Penicillin–streptomycin–glutamine (PSG, Invitrogen), 5 mL nonessential amino acids (100×), 500 µL 2-mercaptoethanol, 5 mL sodium pyruvate, 10 µL bFGF (final concentration of 10 ng/mL), 5 µL PDGF, and 5 mL Insulin–Transferrin–Selenium-X Supplement (Invitrogen).
7. Dimethyl sulfoxide (DMSO).
8. 60- and 100-mm tissue culture treated plates, i.e. Corning, Costar, BD Biosciences.

2.2. Cell Sorting

1. 100-mm bacterial dish, i.e. Greiner Bio-One.
2. Orbital shaker, such as OS-20 orbital shaker from Boeco, Germany.
3. 15-mL Falcon tubes (BD Biosciences).
4. FACS dilution solution: 1× PBS (Invitrogen, see above) with 2% fetal bovine serum (FBS, Invitrogen).
5. Antihuman CD105, FITC-conjugated (Serotec).

6. Antihuman CD24, PE-conjugated (BD Biosciences).

7. FACS Aria Flow Cytometer and FACS Diva software (BD Biosciences).

2.3. Freezing hESC-Derived MSCs

1. Prepare 2× Freezing Medium containing 80% (v/v) FBS (Invitrogen) and 20% (v/v) DMSO. Mix well. Keep medium at 4°C until use within 24 h.

2. Cryo 1° freezing container or "Mr Frosty" (Nalgene).

3. Methods

3.1. Derivation of MSCs from hESCs

The efficiency in deriving MSCs from hESCs using the method described depends on the hESC lines. In our hands, HuES9 (11) is most amenable to generating MSCs using this method and we have also derived MSCs from the WiCell H1 ESC line (12):

1. HuES9 human ESCs are maintained on mouse embryonic fibroblast as previously described (11). For a detailed protocol on how to expand hESCs, the reader is referred to Chapters 2, 3, 5, and 6.

2. To generate MSCs, it is recommended to start with a confluent 60-mm plate of HuES9 hESCs.

3. Aspirate the culture medium and rinse cells with 5 mL of PBS.

4. Add 1 mL of trypsin/EDTA and return the plate to the 37°C CO_2 incubator for 8 min (see Note 1).

5. After 8 min, remove the plate and tilt the plate from side to dislodge and disperse the cells into a cell suspension.

6. Neutralize trypsin with 5 mL of MSC derivation medium. Pipet cell suspension gently up and down about five times to break up any cell clumps (see Note 2).

7. Centrifuge the cell suspension for 5 min at $800 \times g$ at 4°C.

8. Discard the supernatant and loosen the cell pellet by flicking the side of the tube with a finger.

9. Add 5 mL of MSC derivation medium to resuspend the cells.

10. Plate the cell suspension on a 60-mm gelatinized plate. The 60-mm gelatinized plate is prepared by adding 1–2 mL of gelatin to completely cover the surface and leaving the plate to stand for at least 15 min or longer at room temperature. Just before use, remove the gelatin.

11. After plating, return the cell suspension to the CO_2 incubator for 48 h to allow cells to adhere to the plate.

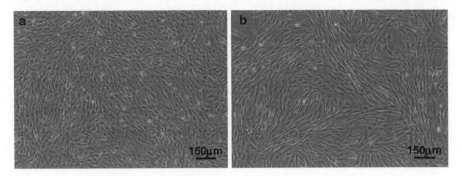

Fig. 1. Morphology of mesenchymal stem cells during derivation. (a) Morphology of cell culture during the early stages of differentiation when the cells are starting to acquire the fingerprint whorl morphology. (b) A typical hESC-derived mesenchymal stem cell culture with the characteristic fingerprint whorl.

12. After 48 h, remove culture medium and cell debris, wash the cell culture with 5 mL of PBS, and replenish cells with fresh MSC derivation medium. There should be extensive cell death and differentiation of hESCs (see Note 3).

13. Repeat trypsinization when the culture is confluent.

14. After the third or fourth trypsinization, the characteristic fingerprint whorl of confluent MSC cultures starts to become evident (Fig. 1a, see Note 4).

15. When the fingerprint whorl of confluent MSC cultures forms (Fig. 1b), trypsinize the cell culture and plate cells on a gelatinized 100-mm plate. MSC cultures should not be expanded at more than 1:3–1:4 per split.

3.2. Cell Sorting

1. The trypsinized cells can be sorted for positive expression of CD105 and negative expression of CD24 as early as 1 week after hESCs have been trypsinized.

2. Trypsinize the differentiating hESCs with 1 mL of trypsin/ EDTA for 8 min at 37°C. Tilt plate from side to side to ensure that all cells are lifted off the plates and there are no visible cell clumps.

3. Neutralize the trypsin with an equal volume of MSC derivation medium. Gently pipet the cells up and down about five times to break up any cell clumps.

4. Centrifuge the cell suspension for 5 min at $800 \times g$ (4°C), discard the supernatant, and resuspend the cell pellet in 10 mL of MSC derivation medium.

5. Plate the cell suspension on a 100-mm bacterial culture dish and place the plate on an orbital shaker with gentle shaking for 2 h at 37°C in a CO_2 incubator.

6. After 2 h at 37°C in the CO_2 incubator, collect the cells into a 15-mL Falcon tube.

7. Centrifuge the cell suspension for 5 min at $800 \times g$ at 4°C. Wash the cells twice with PBS by resuspending the cell pellet in 10 mL of PBS each time followed by centrifuging the cell suspension for another 5 min at $800 \times g$ at 4°C.

8. Count cells using trypan blue exclusion or using a Coulter counter (see Chapter 8) and prepare 1×10^6 aliquots in 0.5 mL FACS dilution solution. Aliquot into microcentrifuge tubes.

9. Add 10 μL of FITC-conjugated antihuman CD105 and 10 μL of PE-conjugated antihuman CD24 to the samples.

10. Incubate with gentle shaking for 40 min at room temperature.

11. Centrifuge the cell suspension for 5 min at $800 \times g$ (4°C). Wash twice with PBS.

12. Resuspend cells in 500 μL of MSC derivation medium and sort for CD105+, CD24− cells on a FACSAria using the FACS Diva software.

13. After sorting, plate cells at a density of about 500 cells/cm² on a gelatinized plate (see Note 5).

3.3. Passaging hESC-Derived CD105+/ CD24− MSCs

1. Passage hESC-derived MSCs at a 1:3 or 1:4 split.

2. Aspirate medium and rinse cells with PBS.

3. For a 150-mm plate, add 5 mL of trypsin/EDTA to the cells. Leave plate in the incubator for 8 min.

4. Upon removal of plates from the incubator, tilt plate from side to side to ensure that all cells are lifted off the plates and there are no visible cell clumps.

5. Add 5 mL of MSC derivation medium to the plate to neutralize trypsin.

6. Gently pipette the cells up and down about five times to break up any cell clumps.

7. Transfer cell suspension into a 50-mL Falcon tube.

8. Wash the plate that was trypsinized with another 10 mL of fresh MSC derivation medium to collect cells that were left behind. Transfer this medium/cell suspension into the same Falcon tube.

9. Centrifuge at $800 \times g$ for 3 min in a cooled centrifuge.

10. Aspirate supernatant, dislodge cell pellet by tapping on the outside of the tube, and resuspend cells with 40 mL of fresh MSC derivation medium. Pipette up and down until cells are evenly dispersed.

11. Distribute 10 mL of the cell suspension onto a new gelatinized 150-mm plate followed by 10 mL of pure MSC derivation medium.

12. Feed cells every 48 h until the culture is ~60–75% confluent and thereafter every 24 h. Cells reach confluency in about 7–9 days (see Note 6).

3.4. Freezing/Thawing hESC-Derived CD105+/ CD24− MSCs

1. Trypsinize a 100-mm confluent plate with CD105+/CD24− MSCs with 2 mL of trypsin/EDTA for 8 min at 37°C.

2. Upon removal of plates from the incubator, tilt plate from side to side to ensure that all cells are lifted off the plates and there are no visible cell clumps.

3. Neutralize the trypsin/EDTA with an equal volume of MSC derivation medium. Gently pipette the cells up and down about five times to break up any cell clumps.

4. Centrifuge the cell suspension for 5 min at $200 \times g$ (4°C), discard the supernatant, and resuspend the cell pellet in 2 mL of MSC derivation medium.

5. Add 2 mL of 2× freezing medium slowly while shaking tube to mix the freezing medium with the cell suspension.

6. When addition is completed, pipette the cell suspension up and down a few times to ensure complete and even distribution of the freezing medium.

7. Once freezing medium is added, work quickly to get cells into a −80°C freezer.

8. Aliquot 1 mL of the suspension into prelabeled cryovials.

9. Transfer the cryovials to a cryo 1°C freezing container or "Mr Frosty" and allow the cells to freeze for 48 h at −80°C before transferring to long-term storage in liquid nitrogen or at −150°C.

4. Notes

1. Dissociation of hESCs to derive MSCs by Invitrogen's TrypLE Express or collagenase does not induce differentiation of human ESCs.

2. Dissociation of hESCs into single cells is critical in the derivation.

3. Human ESCs can be propagated on MSCs. During the early stages of MSC derivation, poorly dissociated hESC colonies will continue to propagate with the newly derived MSCs.

4. Repeated trypsinization of hESCs more than five times as described generally generates a fairly homogenous MSC culture.

5. Sorting of cells by FACS significantly reduces the viability of cells. An alternative method to enhance homogeneity of the

MSC cultures is to use the more laborious method of limiting dilution.

6. At the 68th population doubling, we began to observe random chromosomal aberrations.

Acknowledgments

This work was supported by funding from A*STAR. We thank members of our laboratories for their contributions to this work.

References

1. Wagner, W., and Ho, A.D. (2007) Mesenchymal stem cell preparations - comparing apples and oranges. *Stem Cell Rev.* **3**, 239–248.

2. Brooke, G., Cook, M., Blair, C., Han, R., Heazlewood, C., Jones, B., et al. (2007) Therapeutic applications of mesenchymal stromal cells. *Semin. Cell. Dev. Biol.* **18**, 846–858.

3. Dominici, M., Le Blanc, K., Mueller, I., Slaper-Cortenbach, I., Marini, F., Krause, D., et al. (2006) Minimal criteria for defining multipotent mesenchymal stromal cells. The International Society for Cellular Therapy position statement. *Cytotherapy* **8**, 315–317.

4. Barberi, T., Willis, L.M., Socci, N.D., and Studer, L. (2005) Derivation of multipotent mesenchymal precursors from human embryonic stem cells. *PLoS Med.* **2**, e161.

5. Lian, Q., Lye, E., Suan Yeo, K., Khia Way Tan, E., Salto-Tellez, M., Liu, T.M., et al. (2007) Derivation of clinically compliant MSCs from CD105+, CD24– differentiated human ESCs. *Stem Cells* **25**, 425–436.

6. Olivier, E.N., Rybicki, A.C., and Bouhassira, E.E. (2006) Differentiation of human embryonic stem cells into bipotent mesenchymal stem cells. *Stem Cells* **24**, 1914–1922.

7. Caplan, A.I., and Dennis, J.E. (2006) Mesenchymal stem cells as trophic mediators. *J. Cell. Biochem.* **98**, 1076–1084.

8. Sze, S.K., de Kleijn, D.P., Lai, R.C., Khia Way Tan, E., Zhao, H., Yeo, K.S., et al. (2007) Elucidating the secretion proteome of human embryonic stem cell-derived mesenchymal stem cells. *Mol. Cell. Proteomics* **6**, 1680–1689.

9. Timmers, L., Lim, S.K., Arslan, F., Armstrong, J.S., Hoefer, I.E., Doevendans, P.A., et al. (2007) Reduction of myocardial infarct size by human mesenchymal stem cell conditioned medium. *Stem Cell Res.* **1**, 129–137.

10. Choo, A., Ngo, A.S., Ding, V., Oh, S., and Kiang, L.S. (2008) Autogeneic feeders for the culture of undifferentiated human embryonic stem cells in feeder and feeder-free conditions. *Methods Cell Biol.* **86**, 15–28.

11. Cowan, C.A., Klimanskaya, I., McMahon, J., Atienza, J., Witmyer, J., Zucker, J.P., et al. (2004) Derivation of embryonic stem-cell lines from human blastocysts. *N. Engl. J. Med.* **350**, 1353–1356.

12. Thomson, J.A., Itskovitz-Eldor, J., Shapiro, S.S., Waknitz, M.A., Swiergiel, J.J., Marshall, V.S., et al. (1998) Embryonic stem cell lines derived from human blastocysts. *Science* **282**, 1145–1147.

Chapter 13

Differentiation of Human Embryonic Stem Cells into Mesenchymal Stem Cells by the "Raclure" Method

Emmanuel N. Olivier and Eric E. Bouhassira

Abstract

Mesenchymal stem cells also called mesenchymal stromal cells (MSCs) are multipotent progenitors that can be found in many connective tissues including fat, bone, cartilage, and muscle. We report here a simple method to reproducibly differentiate human embryonic stem cells (hESCs) into MSCs that does not require the use of any feeder layers or exogenous cytokines. The cells obtained with this procedure have a normal karyotype, are morphologically similar to bone marrow MSCs, are contact-inhibited, can be grown in culture for about 20–25 passages, exhibit an immuno-phenotype similar to bone marrow MSCs (negative for CD34 and CD45, but positive for CD44, CD71, CD73, CD105, CD166, HLA ABC, and SSEA-4), and can differentiate into osteocytes and adipocytes. They are also a very useful source of autogenic feeder cells to support the growth of undifferentiated hESCs. The ability to produce MSCs from hESCs should prove useful in obtaining large amounts of genetically identical and genetically modifiable MSCs that can be subsequently used to study the biology of MSCs as well as possible therapeutic applications.

Key words: Embryonic stem cells, Mesenchymal stem cells, Adipocytes, Osteocyte

1. Introduction

Mesenchymal stromal cells (MSCs) are multipotent progenitors that are thought to be essential for the formation of many connective tissues including fat, bone, cartilage, and muscle (1). Because of their ability to differentiate, MSCs have a large therapeutic potential for cell therapy, regenerative or reconstructive medicine and have already been used clinically, for instance to treat diseases such as *osteogenesis imperfecta* (2–4).

MSCs can be isolated from several adult tissues, including bone marrow and fat, as well as from fetal tissues. Interestingly, the proliferation rate of MSCs depends on their developmental

Nicole I. zur Nieden (ed.), *Embryonic Stem Cell Therapy for Osteo-Degenerative Diseases*, Methods in Molecular Biology, vol. 690, DOI 10.1007/978-1-60761-962-8_13, © Springer Science+Business Media, LLC 2011

age, since MSCs derived from adults grow slowly in culture (one or two divisions per week), while MSCs derived from fetal sources grow at a faster rate (two to three divisions per week).

The particular protocol described here has the potential to produce MSCs from human embryonic stem cells (hESCs) that divide up to once a day, two to three times faster than fetal MSCs. This is an important practical difference, considering this rapid rate of proliferation allows the rapid production of large amounts of cells.

Adult and fetal MSCs can differentiate in vitro into multiple lineages including adipocytes, osteocytes, and chondrocytes. The described protocol, which produces MSCs from hESCs, yields cells that can reproducibly differentiate into adipocytes and osteocytes. There is a likelihood that these hESC-derived MSCs can also produce chondrocytes, but we have not yet fully explored this line of research. We therefore described our cells as bipotent, although they may have a broader potential. The fact that hESCs can give rise to multipotential MSCs has been demonstrated in a study in which hESC-derived MSCs were shown to have the potential to differentiate into adipocytes, osteocytes, chondrocytes, and myocytes (5, 6).

In recent studies, we have compared erythroid cells produced from hESCs to their in vivo counterparts at different stages of development and found that hESCs differentiate into red blood cells that are most similar to embryonic or fetal cells (6, 7). Other groups have reported similar findings in other lineages (i.e. cardiomyogenic, hepatogenic, pancreatic) (8–10). To our knowledge, adipocytes, osteocytes, and other cells that can be obtained from MSCs have not been compared in detail to cells obtained from fetal or adult MSCs. However, it is likely that hESC-derived MSCs give rise to embryonic adipocytes and osteocytes that are subtly different from their adult counterparts. The optimal developmental age of MSCs for the purpose of regenerative medicine is unknown and might vary for different applications.

The techniques used to derive MSCs from hESCs are probably directly adaptable to induced pluripotent stem cells (iPSCs) (11) and will probably prove invaluable to produce almost infinite amounts of donor-specific cells to repair connective tissues or for other applications.

MSCs arise from the mesoderm and can be induced from hESCs by the activation of the Wnt pathways and the expression of transcription factors involved in epithelial–mesenchymal transition (12, 13). Human ESCs grown on mouse embryonic fibroblasts (MEFs) are exposed to high levels of Wnt signaling. This might explain why MSCs appear to be a default differentiation pathway for hESCs. This hypothesis is supported by the fact that most of the publications reporting MSC differentiation from hESCs, including our own, use either hESCs grown on feeder layers, medium conditioned by MEF, or co-culture with fibroblastic cells (5, 14–17).

2. Materials

<table>
<tr>
<td valign="top">

2.1. Human ESC
Culture and Raclure
Derivation

</td>
<td valign="top">

1. hESCs: H1 and H9 from Wicell have been tested.

2. MEFs: CF-1 mice have been used, but the origin of MEFs does not seem to matter. See Chapters 2 or 6 for preparation of feeder cells.

3. Fetal bovine serum (FBS): Sera from Invitrogen and Atlanta Biologicals have been used, which were originally selected on their ability to generate hematopoietic stem cells from hESCs. Freeze upon arrival.

4. Knockout serum replacement (KSR) (Invitrogen): Keep at −20°C.

5. High-glucose Dulbecco's Modified Eagle's Medium (DMEM) with L-glutamine: Keep refrigerated.

6. DMEM/F12 with L-glutamine: Store at 4°C.

7. 1× Dulbecco's phosphate-buffered saline (DPBS) without Ca^{2+}/Mg^{2+} (Cellgro or Hyclone).

8. L-glutamine 200 mM (100×) or GlutaMAX I (100×) (Invitrogen): Store at 4°C.

9. Nonessential amino acids (NEAA; 100×): Store at 4°C.

10. Penicillin (50 U/mL)/streptomycin (50 μg/mL): Store at −20°C.

11. Collagenase type IV, trypsin/EDTA, Dispase, and TrypLE Express.

12. Basic fibroblast growth factor (bFGF, R&D Systems): Lyophilized samples are stable for up to 12 months at −20°C. Prepare a 100 μg/mL stock solution in sterile PBS containing 0.1% bovine serum albumin. Upon reconstitution, store at 4°C for 1 month or at −20 to −70°C in a manual defrost freezer for 3 months. Avoid repeated freeze–thaw cycles. Add freshly to medium before use.

13. Gelatin, 2% in water, tissue culture grade, sterile, Type B, cell culture tested: Store at 4°C.

14. Dimethyl sulfoxide (DMSO): Store at room temperature. Wear gloves as harmful if absorbed through skin.

15. Tissue culture ware, tissue culture treated (i.e. Nunc, Falcon or Corning).

16. Modified Pasteur pipette: Elongate and twist on a Bunsen burner, so the end is curve and closed. Do so under a stereoscopic binocular in a custom made hood (a Plexiglas box with doors, which is sterilized with UV light before use).

</td>
</tr>
</table>

17. Human ESC medium: Combine DMEM/F12, 20% KSR, 2 mM L-glutamine, or 1× GlutaMAX I as well as 0.1 mM NEAA solution and 4 ng/mL bFGF.

18. Freezing medium: 80% DMEM supplemented with 10% FBS, and 10% DMSO.

2.2. Flow Cytometric Analysis

1. Antibodies used for CD13, CD71, CD105, HLA-ABC, and isotypes control for IgG1, IgG2a, IgG3, and FITC Rat anti-mouse IgG (H+L) are from eBioscience. CD34 FITC, CD45PE, CD73PE, and isotypes control for IgG_1K FITC, IgG_1K PE, and FITC rat anti-mouse IgG_1 are from BD Biosciences, CD44 and SSEA-4 are from Developmental Studies Hybridoma Bank.

2. 1× DPBS without Ca^{2+}/Mg^{2+} (Cellgro or Hyclone): Prepare staining buffer with 5% KSR in DPBS.

3. Formaldehyde (37%): Toxic and combustible. Wear appropriate protection. Make a 10% formalin in DPBS solution. The final concentration of formaldehyde is 3.7%, as formalin is generally provided as an aqueous solution of 37% formaldehyde.

2.3. Osteogenic Differentiation

1. D10 medium: DMEM supplemented with 10% of FBS, 1× NEAA, and 1% penicillin/streptomycin (ingredients see Subheading 2.1).

2. Dexamethasone: Store powder at 4°C. Use personal protective equipment when weighing in and avoid breathing dust. Prepare a stock solution by adding 1 mL absolute ethanol per milligram product (2.55 mM). Add 25.5 mL sterile D10 medium per milliliter of ethanol added to achieve the final concentration of 1 μM. Sterile filter if necessary. Freeze working aliquots, avoid repeated freeze–thaw cycles.

3. Ascorbic acid-2-phosphate: Prepare a 50 mM stock solution in sterile water or DPBS without Ca^{2+}/Mg^{2+} (Hyclone or Cellgro). Store aliquots at –20°C.

4. β-Glycerophosphate: Prepare a 1 M stock solution by adding 10 mL of DPBS per 2.16 g of β-glycerophosphate.

5. 70% Ethanol: Highly flammable. Store at room temperature.

6. Alizarin Red S: Prepare a 0.5% Alizarin Red S staining solution in water. Adjust pH to 4. Store at room temperature.

2.4. Adipogenic Differentiation

1. D10 medium: DMEM supplemented with 10% of FBS, 1× NEAA, and 1% penicillin/streptomycin (ingredients see Subheading 2.1).

2. Dexamethasone: Prepare a 1 μM stock solution as described above.

3. Indomethacin: Cyclooxygenase inhibitor with specificity selective for COX-1. Substance is practically insoluble in

water, but soluble in ethanol, ether, and acetone. Not stable in alkaline solutions, therefore adjust pH to 7.

4. Insulin: Store powder at 0°C. To prepare 10 mg/mL stock solution, add 10 mL of acidified H_2O. Acidified H_2O is made by approximately 0.1 mL of glacial acetic acid. The pH is ≤2. Stock solution is stable for 1 year at –20°C.

5. 3-Isobutyl-1-methylxanthine (IBMX): Considered stable at room temperature, but freezing is recommended. Nonspecific inhibitor of cAMP and cGMP phosphodiesterases. Dissolve in DMSO and store aliquots at –20°C. Stable for several months.

6. 1× DPBS without Ca^{2+}/Mg^{2+} (Cellgro or Hyclone).

7. Formaldehyde (37%), see above.

8. 70% Ethanol (see above).

9. Oil Red-O: Make a 0.3% Oil Red-O staining solution in isopropanol and store at room temperature.

3. Methods

3.1. The "Raclure" Method

In most cultures of undifferentiated hESCs, a small amount of differentiation occurs as the colonies increase in size. These differentiated cells, which are easily recognizable under a microscope at low power, must be removed by mechanical scraping, otherwise they invade the whole culture. The "raclure" method, which produces MSCs from hESCs, is based on observations that these spontaneously differentiated cells give rise to homogenous cultures of MSCs when placed in appropriate growth conditions.

1. Culture hESCs on feeder cells in hESC medium. Feeder cells were either MEFs or previously derived ESC-MSCs, γ-irradiated (80 Gy), and plated at 75,000 cells/cm² in gelatinized (0.1%) six-well plates. For details on how to gelatinize plates and generate feeder layers see Chapter 2. Human ESC-derived MSCs can be used up to passage 15.

2. Scrape differentiating colonies or chunks of colonies from plates of hESCs ready to be passaged (6–7 days after they are plated) using one of the modified Pasteur pipette. Figure 1a illustrates the morphology of a differentiated colony (see Fig. 1a).

3. Afterwards, collect the scraps (or raclures in French) in a 15-mL tube and leave them to sediment at the bottom of the tube for at least 5 min (see Note 1).

4. Discard the supernatant and plate the "raclures," a mixture of irradiated feeder cells and pieces of hESC colonies at different stages of differentiation. Plate raclures from one six-well plate

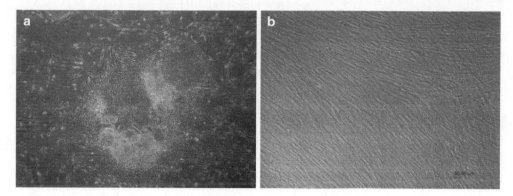

Fig. 1. Morphological appearance of hESCs and hESC-derived MSCs. (a) Micrograph of a large differentiated colony bordered by a small undifferentiated one. (b) Micrograph of hESC-derived MSCs.

of hESCs that contains no more than 10–15% of differentiated cells (see Note 2), in a T25-cm² flask (Corning) in 8–10 mL of D10 medium at 7.5–10% CO_2 (see Note 3).

5. Incubate the raclures for about 2 months (see Note 4) and feed once every 1–2 weeks by completely changing the medium (D10) (see Note 5). At this point, the hardest part is to not completely forget about the cells (see Notes 6 and 7).

6. After about 2 months, the epithelium structure can be dissociated using a mixture of collagenase type IV (200 U/ mL) and dispase (50 U/mL) for 1 h (2 mL of each), followed by the addition of 4 mL of trypsin and 3–5 more hours of incubation, or simply by using TrypLe Express for 2–3 h.

7. Plate the dissociated cells including the remaining clumps of cells in T75-cm² (Corning) in D10 medium (see Note 8). Two to three days later, the cells should attach and form a layer of cells.

8. After 2–3 days, (see Note 9), the culture should be almost uniform and exhibit a regular fingerprint-like pattern (see Fig. 1b).

9. After 2–3 additional passages, the morphology of the cell population should be completely homogenous and characterization by flow cytometry and functional differentiation can be carried out as described below (see Fig. 2a, b).

3.2. Flow Cytometric Analysis

1. Harvest cells using TripLE Express, then wash and resuspend the cells in staining buffer at a concentration of 10^6 cells/mL.

2. Stain 10^5 cells with each antibody or antibody combinations. Use antibody concentrations according to the manufacturer's instructions and dilute in DPBS.

3. Incubate for 20 min at 4°C.

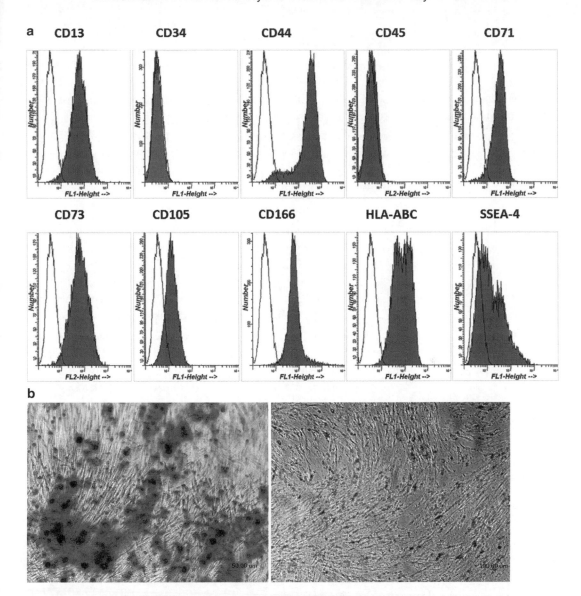

Fig. 2. Marker expression and differentiation capacity of hESC-derived MSCs. (a) Antigenic profile of hESC-derived MSCs. As for MSCs from other origin, expression of CD13 and SSEA4 may vary. (b) Micrograph of hESC-derived MSCs after osteogenic and adipogenic differentiation.

4. After staining, wash cells twice with staining buffer or DPBS and resuspended in 500 µL of staining buffer.

5. At least 10,000 events are acquired for each sample and distinction between viable and nonviable cells is estimated based on the granularity of the cells (side scatter).

6. Cells can be fixed after staining. To do this, rinse the cells twice with DPBS only (no protein) after they have been stained, resuspend in 10% formalin in DPBS solution, and

store at 4°C in a dark environment until they are ready to be analyzed. The cells may be stored for up to 96 h without noticeable decay of the fluorescent signals.

3.3. Osteogenic Differentiation Assay

1. Seed cells at 1,000–3,000 cells/cm² in D10 medium in a 6- or 12-well plate (see Note 10).

2. When the highest plated density reaches 50–70% confluence, supplement the growth medium with 100 nM dexamethasone, 50 μM ascorbic acid-2-phosphate, and 10 mM β-glycerophosphate.

3. Replace the medium every 3–4 days for 21 days.

4. Then wash cultures twice with DPBS, fix them in a solution of ice-cold 70% ethanol for 1 h, and stain the cultures for 10 min with 1 mL of 40 mM Alizarin Red S (18).

3.4. Adipogenic Differentiation

1. Cells are seeded at 0.5×10^4 to 10^4 cells/cm² and two methods can be used to induce adipogenic differentiation.

2. In the first method, which is widely used to differentiate adult MSCs into adipocytes, place the confluent MSCs in D10 medium supplemented with 1 μM dexamethasone, 0.2 mM indomethacin, 10 μg/mL insulin, and 0.5 mM IBMX.

3. In the second method, termed SWH for serum withdrawal/hypoxia method, place the confluent MSCs in hESC medium in partial hypoxia (5% O_2) (see Note 11).

4. In both cases, replace the medium every 3–4 days for 21 days.

5. Then wash cells three times with DPBS, fix in 10% formalin in DPBS for 1–2 h, and rinse three times with water and once with 70% ethanol. Afterwards, stain cells for 15 min with fresh Oil Red-O staining solution, rinse several times with 70% ethanol to clear the diffusing dye, and cover with water (19).

4. Notes

1. Instead of using scraps of cells, MSCs can also be produced by mechanically dissociating slightly overgrown undifferentiated hESCs (2 days more than a regular passage to capitalize on the spontaneous differentiation at the edge of the colonies) with a 5- or 10-mL glass pipette and by following the same protocol. If this alternate procedure is used, it is important not to completely dissociate the cells, since at this stage, clumps of cells appear essential to initiate the formation of the epithelium, and single cells would float and die. The feeders coming along within the scraps do not have to be removed

and, in fact, may help the clumps to settle down. Since they are mitotically inactivated, they will be eliminated from the differentiation culture during the medium changes.

2. Initial seeding density must be kept low to allow the clumps to become isolated adherent colonies. High seeding density seems to favor the apparition of fast growing, short living fibroblasts, and impair the differentiation into MSCs.

3. A relatively high CO_2 concentration (at least 7.5%) is important as the increased acidity improves the differentiation toward MSCs. Production of MSCs in the presence of 5% CO_2 was not successful in our hands.

4. Important: Early dissociation of the epithelium-like structure yields cells that do not have the typical characteristics of MSCs. We hypothesize that the differentiating hESCs undergo an epithelial–mesenchymal transition, and this process happens pretty late in our conditions. It might be possible to recover MSCs at earlier stages by sorting them using CD73, CD271 (LNGF receptor), or CD349 (Frizzled 9) (20), but we have not fully explored this protocol.

5. We have not extensively tested alternative medium formulation. It is therefore quite possible that more elaborate or specialized media (low-glucose DMEM or serum-free formulation) may yield hESC-derived MSCs with a broader differentiation potential.

6. A thick epithelium-like structure should progressively form and cover the whole culture area. During the 2-month incubation, layer of cells often detach from the borders of the plate and curl. New cells generally start growing underneath the detaching cells (see Fig. 3). The detached layer of cells can therefore be discarded. Alternatively, the detached layer of cells can be plated on a new flask since they will also give rise to MSCs.

7. The optimal time of differentiation is between 45 and 60 days. Before that, the cells obtained are not MSCs in majority, although MSCs can be sorted. Incubation longer than 2 months does not seem to add anything to the protocol.

8. At the end of the initial long step of differentiation, the ability of the cells to attach to plastic, one of the characteristics of MSCs, is a natural way to select them when the epithelium-like structure is dissociated and replated.

9. After the first dissociation, the entire contents of a T25-cm² flask are plated in a T75-cm² flask. The flasks should be checked frequently and cells should be passed as soon as confluence is reached to keep dynamic culture and to avoid loss of potential. Time to confluence varies between 2 and 10 days.

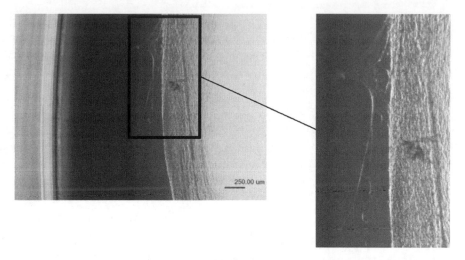

Fig. 3. Raclure cells during establishment of cultures. Micrograph of a cell layer detaching from the culture plate. The *panel* shows cells starting to grow at the interface of the plate and the detaching layer.

10. Rapid cell proliferation seems to impair the differentiation if confluence is reached too soon. We therefore recommend starting with a range of cell concentrations.

11. The SWH method is used preferentially as it is simpler and less dependent on the cell concentration.

Acknowledgements

The monoclonal antibodies for CD44 and SSEA-4 developed by J.T. August and J.E.K. Hildreth, and D. Solter, respectively, were obtained from the Developmental Studies Hybridoma Bank, developed under the auspices of the NICHD, and maintained by the University of Iowa, Department of Biological Sciences, Iowa City, IA 52242.

References

1. Pittenger, M. F., Mackay, A. M., Beck, S. C., Jaiswal, R. K., Douglas, R., Mosca, J. D., et al. (1999) Multilineage potential of adult human mesenchymal stem cells. *Science* **284**, 143–147.

2. Horwitz, E. M., Prockop, D. J., Fitzpatrick, L. A., Koo, W. W., Gordon, P. L., Neel, M.D., et al. (1999) Transplantability and therapeutic effects of bone marrow-derived mesenchymal cells in children with osteogenesis imperfecta. *Nat. Med.* **5**, 309–313.

3. Horwitz, E. M., Prockop, D. J., Gordon, P. L., Koo, W. W., Fitzpatrick, L. A., Neel, M. D., et al. (2001) Clinical responses to bone marrow transplantation in children with severe osteogenesis imperfecta. *Blood* **97**, 1227–1231.

4. Le Blanc, K., Gotherstrom, C., Ringden, O., Hassan, M., McMahon, R., Horwitz, E., et al. (2005) Fetal mesenchymal stem-cell engraftment in bone after in utero transplantation in a patient with severe osteogenesis imperfecta. *Transplantation* **79**, 1607–1614.

5. Barberi, T., Willis, L. M., Socci, N. D., and Studer, L. (2005) Derivation of multipotent mesenchymal precursors from human embryonic stem cells. *PLoS Med.* **2**, e161.

6. Olivier, E. N., Qiu, C., Velho, M., Hirsch, R. E., and Bouhassira, E. E. (2006) Large-scale production of embryonic red blood cells from human embryonic stem cells. *Exp. Hematol.* **34**, 1635–1642.

7. Qiu, C., Olivier, E. N., Velho, M., and Bouhassira, E. E. (2008) Globin switches in yolk sac-like primitive and fetal-like definitive red blood cells produced from human embryonic stem cells. *Blood* **111**, 2400–2408.

8. Rolletschek, A., Blyszczuk, P., and Wobus, A. M. (2004) Embryonic stem cell-derived cardiac, neuronal and pancreatic cells as model systems to study toxicological effects. *Toxicol. Lett.* **149**, 361–369.

9. Kehat, I. and Gepstein, L. (2003) Human embryonic stem cells for myocardial regeneration. *Heart Fail. Rev.* **8**, 229–236.

10. Kehat, I., Amit, M., Gepstein, A., Huber, I., Itskovitz-Eldor, J., and Gepstein, L. (2003) Development of cardiomyocytes from human ES cells. *Methods Enzymol.* **365**, 461–473.

11. Takahashi, K., Tanabe, K., Ohnuki, M., Narita, M., Ichisaka, T., Tomoda, K., et al. (2007) Induction of pluripotent stem cells from adult human fibroblasts by defined factors. *Cell* **131**, 861–872.

12. Lindsley, R. C., Gill, J. G., Kyba, M., Murphy, T. L., and Murphy, K. M. (2006) Canonical wnt signaling is required for development of embryonic stem cell-derived mesoderm. *Development* **133**, 3787–3796.

13. Lindsley, R. C., Gill, J. G., Murphy, T. L., Langer, E. M., Cai, M., Mashayekhi, M.,

et al. (2008) Mesp1 coordinately regulates cardiovascular fate restriction and epithelial–mesenchymal transition in differentiating ESCs. *Cell Stem Cell* **3**, 55–68.

14. Olivier, E. N., Rybicki, A. C., and Bouhassira, E. E. (2006) Differentiation of human embryonic stem cells into bipotent mesenchymal stem cells. *Stem Cells* **24**, 1914–1922.

15. Trivedi, P. and Hematti, P. (2007) Simultaneous generation of CD34+ primitive hematopoietic cells and CD73+ mesenchymal stem cells from human embryonic stem cells cocultured with murine OP9 stromal cells. *Exp. Hematol.* **35**, 146–154.

16. Lian, Q., Lye, E., Suan Yeo, K., Khia Way Tan, E., Salto-Tellez, M., Liu, T. M., et al. (2007) Derivation of clinically compliant MSCs from CD105+, CD24- differentiated human ESCs. *Stem Cells* **25**, 425–436.

17. Trivedi, P. and Hematti, P. (2008) Derivation and immunological characterization of mesenchymal stromal cells from human embryonic stem cells. *Exp. Hematol.* **36**, 350–359.

18. Colter, D. C., Sekiya, I., and Prockop, D. J. (2001) Identification of a subpopulation of rapidly self-renewing and multipotential adult stem cells in colonies of human marrow stromal cells. *Proc. Natl. Acad. Sci. USA* **98**, 7841–7845.

19. Kasturi, R. and Joshi, V. C. (1982) Hormonal regulation of stearoyl coenzyme A desaturase activity and lipogenesis during adipose conversion of 3T3-L1 cells. *J. Biol. Chem.* **257**, 12224–12230.

20. Buhring, H. J., Battula, V. L., Treml, S., Schewe, B., Kanz, L., and Vogel, W. (2007) Novel markers for the prospective isolation of human MSC. *Ann. N. Y. Acad. Sci.* **1106**, 262–271.

Chapter 14

Improved Media Compositions for the Differentiation of Embryonic Stem Cells into Osteoblasts and Chondrocytes

Beatrice Kuske, Vuk Savkovic, and Nicole I. zur Nieden

Abstract

Differentiation procedures leading to osteogenic and chondrogenic differentiation of embryonic stem cells (ESCs) have been established and well upgraded over the past decade. Novel cell-culture conditions, signaling inducers, and chemical modifications of cellular environment have been found and optimized for use as steering or supporting modules in ESC differentiation.

While most of the novel studies of osteoblasts or chondrocytes differentiated from ESCs deal with their regenerative potential, the "childhood diseases" of basic differentiation have not yet been quite solved. Purification procedures are still facing a lack of exclusive markers for osteogenic progenitors and a collateral development of other cell types at the end points of differentiation that possibly lead to teratomas.

This chapter discusses the role of novel markers and inducers in osteogenic and chondrogenic differentiation, their effect on signaling pathways, particularly on that of Wnt/beta-catenin, and the time-specific manner of their action. We present an improved osteogenic differentiation protocol based on the hanging drop method and a time-optimized use of $1\alpha,25\text{-(OH)}_2$ vitamin D_3, all-trans retinoic acid, and bone morphogenetic protein 2 (BMP-2) with an end point efficiency increased up to 90% and a protocol for chondrogenic differentiation, which employs BMP-2 and transforming growth factor $\beta 1$ as chondrogenic inducers, with 60% chondrogenic end point efficiency.

Key words: Osteogenesis, Vitamin D3, ATRA, Hanging drop protocol, Morphometric image analysis, Chondrogenesis, Expression analysis

1. Introduction

1.1. Osteogenesis in Embryonic Stem Cells

Differentiation procedures leading to osteogenic differentiation of embryonic stem cells (ESCs) have been constantly upgraded over the past decade. Novel additives have been used, and once elucidated, time points of their use have been specified. In its present status, in vitro differentiation of ESCs into osteoblasts demands addition of several factors, which are normally released by the in vivo surroundings of the osteoblasts. The prerequisite

Nicole I. zur Nieden (ed.), *Embryonic Stem Cell Therapy for Osteo-Degenerative Diseases*, Methods in Molecular Biology, vol. 690, DOI:10.1007/978-1-60761-962-8_14, © Springer Science+Business Media, LLC 2011

additives are β-glycerophosphate as a source of phosphate, ascorbic acid, and the active form of vitamin D_3 (VD_3) or dexamethasone, which has been acknowledged as a differentiation factor and a potentiator of mineralization (reviewed in (1)).

The potential of dexamethasone to convert a portion of differentiating murine ESCs into matrix-secreting osteoblasts has first been shown in 2001 by Buttery and coworkers (2). The following three publications on osteogenic differentiation of ESCs concentrated on evaluating different inducers in mice (3, 4) and the same inducer in man (5). While the field has been relatively slow in describing optimized techniques for the steered differentiation of both murine and human ESCs (hESCs), novel reports focus on evaluating the repair capabilities of such ESC-derived osteoblasts and/or their precursors (6, 7). Yet, the application of ESC-derived osteoblasts in transplantation models faces the same challenges as other ESC-derived cell types: the impure differentiation outcome and the lack of proper purification favor the growth of teratomas when such mixtures of cells are transplanted. Therefore, this chapter briefly assesses studies that discuss how such differentiations could be effectively enhanced, while it suggests a particular protocol to steer differentiation outcome in osteoblasts, in the methods section to follow.

Osteogenic induction is a very time-specific process, and an accurate use of each inducer appears to be essential; vice versa, wrong timing can be very counterproductive. The initial publications already showed that osteogenic induction of ESCs is a highly time-specific and time-regulated process, as osteogenic inducers were only effective at specific differentiation periods as assessed by the number of mineralized nodules and expression of the bone-specific genes osteocalcin and bone sialoprotein (2, 3). For example, controversial reports exist that *all-trans* retinoic acid (ATRA), commonly known as a regulator of neurogenesis, cardiogenesis, body axis extension, and development of the forelimb buds, foregut, and eye (8), may enhance or decrease the number of mineralized nodules (4, 9). Ultimately, it has been suggested that both ATRA and bone morphogenetic protein 2 (BMP-2) aid the osteogenic differentiation process at specific stages of the in vitro developmental program (4, 9–11).

Staging of the progression of the osteogenic developmental program is not an easy task, first and foremost, due to the lack of appropriate and exclusive markers for osteogenic progenitors. With the aim of identifying novel markers for progenitors with osteogenic potential, gene microarray analyses have been carried out simultaneously by our group and by Bourne and coworkers (9, 12). Bourne and coworkers identified cadherin-11, an osteoblast-specific molecule of the cadherin family (13), as a useful cell surface marker for purifying an osteogenic progenitor population (12). Cadherin-11-positive cells not only differentiated into a population of cells that formed a higher number of mineralized nodules, but also responded to BMP-2 and dexamethasone, whereas the cadherin-11-negative cells remained undifferentiated and nonresponsive.

In contrast, our group utilized these microarray gene profiles to devise novel osteogenic induction schemes, increasing osteogenic efficiency to over 90% (9). We succeeded by closely mimicking the endogenous expression of members of specific signaling pathways by adding ATRA and BMP-2 as exogenous activators of such pathways, finding that these would modulate the expression and localization of beta-catenin (CatnB), a key molecule in the Wnt signaling cascade. These findings suggest an existence of a common mechanism for VD_3 and ATRA that mediates reduction of canonical Wnt signaling through CatnB at particular differentiation time points, while BMP-2 induced nuclear CatnB activity at other time points.

Despite the controversial reports in regard to various time points, a time-optimized use of VD_3, ATRA, and BMP-2 has helped develop a highly efficient standard operating procedure for osteogenic differentiation of ESCs. This improved osteogenic differentiation medium composition is presented here. Although still suboptimal, since it relies on medium supplemented with undefined fetal bovine serum (FBS), this protocol could represent a large step forward in osteoblast tissue engineering.

1.2. Chondrogenesis in Embryonic Stem Cells

The chondrogenic differentiation potential of a stem cell immensely depends on its origin, but can be manipulated with certain media additives and the right kind of microenvironment. For example, coculture conditions have been employed to enhance chondrogenic differentiation from ESCs and to promote growth. As such, fully differentiated chondrocytes provide conditioning of the microenvironment and are able to direct ESCs into chondrogenic differentiation. Coculturing of human ESCs (hESCs) with limb bud progenitor cells or primary chondrocytes (14, 15) supports cell growth and a mature chondrogenic phenotype.

Furthermore, chondrogenic growth factors such as insulin-like growth factor 1 (IGF-1) (15) or combinations of growth factors (16, 17) represent other options for inducing chondrogenesis not only in pellet micromass cultures, but also in monolayer approaches (18). Early studies have suggested that chondrogenic differentiation of mesenchymal progenitor cells is modulated by numerous cytokines of the transforming growth factor beta (TGF-β) superfamily (19). Involvement of TGF-β isoforms 1, 2, and 3, as well as BMPs, in chondrogenic differentiation has been clearly shown in terms of affecting the expression of collagen type IIA and aggrecan, the two most specific chondrocyte markers (16, 20–22).

Mostly, the pellet culture system is used for chondrogenic differentiation of stem cells, for example, mesenchymal stem cells (MSCs), which is an established model for evaluating chondrogenesis (23–25). Pellets are made from single-cell suspensions of MSCs by placing them in nonadherent conditions and subsequently subjecting them to forced aggregation, induced by centrifugation. However, the high cell density and extensive cell–cell contact within the pellet may not fully represent the original cartilage repair strategies. Alternatively, tissue engineering strategies using matrices target a lower density of cells distributed within a 3D environment.

High cell density cultures, such as embryoid body (EB) direct-plating outgrowth, EB-derived high-density micromass and pelleted mass, vigorously increase expression levels of cartilage markers in hESC-derived EBs and mesenchymal precursor cells (26–28). Apparently, a high-density 3D environment favors cell-cell interactions, which mimic those of precartilage condensations and hereby promote chondrogenic differentiation of ESCs. Polyethylene glycol (PEG) hydrogels, for example, provide such a three-dimensional environment, which can mimic a chondrogenic environment in vitro. Still, in the absence of proper growth factors, the initial alginate or PEG gel encapsulation studies with ESC-derived MSCs have shown no difference in chondrogenic commitment between encapsulated and monolayer cultures of ESCs.

Addition of TGF-β_1 to PEG gels induces neocartilage formation from encapsulated ESC-derived MSCs, as observed through high levels of cartilage markers and deposition of extracellular matrix (29). Chemical modification of PEG gels further improved mimicking of the condensing mesenchyme not only by its mechanistic but also by its signaling properties. Arg-Gly-Asp-(RGD)-modified polyethylene glycol-diacrilate hydrogel (PEGDA), enriched with collagen type I and hyaluronic acid, actively induced differentiation of hESCs into a chondrogenic phenotype (30). Due to a limited diffusion of TGF-β_1 through the gel, only a restricted matrix production has been observed but it is still sufficient to claim a better support of chondrogenic differentiation than the encapsulation in alginate. Although alginate alone is expected to be instructive as a scaffold for the development and maintenance of chondrogenic phenotype, alginate-encapsulated intact EBs as well as monolayer-grown disrupted EB-derived cells display a very modest, if any, chondrogenic phenotype (27).

We present here a hanging drop approach to induce chondrogenic differentiation in intact adherently grown EBs, using some of the typical chondrogenic inducers, such as BMP-2 and TGF-β1, which can produce roughly 60% of chondrogenic cells in the mixture of differentiating ESCs (22).

2. Materials

2.1. Osteogenic Differentiation

1. 1× Phosphate-buffered saline (PBS), without Ca^{2+} and Mg^{2+}, pH 7.4 (Invitrogen).

2. Murine embryonic stem cells, preferably the D3 ESC line (American Type Culture Collection, see Note 1).

3. Control differentiation medium (CDM): High-glucose Dulbecco's Modified Eagle's Medium (DMEM) (1×), with L-glutamine, without pyruvate (Invitrogen). Prepare complete medium by adding 15% FBS (PAN, batch-tested for osteogenic and chondrogenic differentiation of ESCs, respectively), 1% nonessential amino acids (NEAA, 100×), 0.1 mM 2-mercaptoethanol (cell culture tested), and penicillin G/streptomycin sulfate (final concentration 50 U/mL penicillin and 50 µg/mL streptomycin). Store the complete medium at 4°C for up to 2 weeks.

4. Beta-glycerophosphate: maintain a stock solution of 1 M in PBS, filter-sterilize through a 0.2-µm filter, and keep at –20°C.

5. Ascorbic acid: make a stock of 50 mg/mL in PBS, filter-sterilize (0.2-µm filter), and maintain at –20°C.

6. 1α, 25-$(OH)_2$ vitamin D_3: prepare a stock solution of 1.2×10^{-4} M in DMSO. Aliquot and store at –20°C.

7. All-trans retinoic acid (ATRA): dissolve vial content in DMSO. Store this stock at –20°C. Predilute upon need freshly 1:1,000 in tissue culture medium. Use a final dilution of 1:1,000 of this predilution in your culture.

8. Bone morphogenetic protein 2 (BMP-2, R&D Systems): reconstitute 100 µg BMP-2 in 10 mL of sterile PBS/0.1% BSA to obtain a 10 µg/mL stock solution. Aliquots can be stored at –20°C for 4 weeks or at –80°C for up to 3 months.

9. Osteogenic differentiation medium (ODM1): CDM supplemented with 10 mM beta-glycerophosphate, 25 µg/mL ascorbic acid, and 5×10^{-8} M 1α, 25-$(OH)_2$ vitamin D_3. Use a sterile filter unit to filter the media. Store the filtered medium at 4°C for up to 2 weeks (see Note 2).

10. Osteogenic differentiation medium (ODM2): CDM supplemented with 10 mM beta-glycerophosphate, 25 µg/mL ascorbic acid, 1×10^{-7} M ATRA, and 5×10^{-8} M 1α, 25-$(OH)_2$ vitamin D_3. Filter-sterilize as described for ODM1 and store at 4°C.

11. Osteogenic differentiation medium (ODM3): CDM supplemented with 10 mM beta-glycerophosphate, 25 µg/mL ascorbic acid, 10 ng/mL BMP-2, and 5×10^{-8} M 1α, 25-$(OH)_2$ vitamin D_3. Filter-sterilize and store as described for ODM1 and ODM2.

12. Trypsin, 0.25% (1×) with ethylenediaminetetraacetic acid (EDTA) · 4Na, liquid (Invitrogen).

13. 0.4% trypan blue.

14. Automatic cell counter (i.e. CASY model TTC) or hemocytometer (VWR).

15. Plasticware such as bacteriological petri dishes (100 mm $\varnothing \times 20$ mm), i.e. Greiner Bio-One, tissue culture-treated plates, 24-well or 6-well plates (Greiner Bio-One), and 15- and 50-mL Falcon tubes (BD Biosciences).

2.2. Chondrogenic Differentiation

1. T-flasks 25 cm² and 24-well tissue culture plates, i.e. Corning or BD Bioscience (see Note 2).

2. Bacterial petri dishes 100 mm $\varnothing \times 20$ mm for the hanging drop culture, such as Greiner Bio-One and bacterial petri dishes 60 mm $\varnothing \times 15$ mm for EB suspensions (Greiner Bio-One).

3. 1× PBS, without Ca^{2+} and Mg^{2+}, pH 7.4 (Invitrogen).

4. Transforming growth factor β1 (TGF-$β_1$), human, recombinant (Sigma): dissolve the vial content in sterile 4 mM HCl (Merck)/0.1% bovine serum albumin (BSA) to obtain a 2 μg/mL stock solution. Aliquots can be stored at –20°C.

5. Insulin: dissolve the sterile vial content in 10 mL of sterile PBS to prepare a stock solution of 10 mg/mL. Aliquot into cryotubes and store at –20°C.

6. L-ascorbic acid: prepare a stock solution of 5 mg/mL in PBS. Aliquot into cryotubes and store at –20°C.

7. BMP-2 (R&D Systems): reconstitute 100 μg BMP-2 in 10 mL of sterile PBS/0.1% BSA (Sigma) to obtain a 10 μg/mL stock solution. Aliquots can be stored at –20°C for 4 weeks or at –80°C for up to 3 months.

8. Chondrogenic control medium (CCM): Same as CDM (see above).

9. Chondrocyte differentiation medium 1 (CCDM1): High-glucose DMEM (Invitrogen) containing 2 mM L-glutamine supplemented with 20% FBS (Invitrogen), 1% MEM nonessential amino acids, and 0.1 mM 2-mercaptoethanol. Add TGF-$β_1$ and BMP-2 to final concentrations of 10 ng/mL. Store at 4°C for 2 weeks only.

10. Chondrocyte differentiation medium 2 (CCDM2): High-glucose DMEM with 2 mM L-glutamine (Invitrogen) containing 20% FBS (Invitrogen), 1% MEM nonessential amino acids, and 0.1 mM 2-mercaptoethanol. Supplement with 1 μg/mL insulin and 50 μg/mL ascorbic acid, as well as

BMP-2, to a final concentration of 10 ng/mL. Store at 4°C and use for not longer than 2 weeks.

2.3. RT-PCR and Quantitative PCR Analysis

1. 1× PBS, without Ca^{2+} and Mg^{2+}, pH 7.4 (Invitrogen).

2. RNeasy Mini Kit, including lysis buffer. Prepare a working solution of RNA lysis buffer by adding 10 µL 2-mercaptoethanol (Sigma) per mL RLT-buffer (provided with the kit).

3. Qiashredder homogenization columns (Qiagen).

4. 21-G needle and syringe.

5. TE buffer, pH 7.5–8.0: 10 mM Tris base/1 mM EDTA in distilled H_2O.

6. NanoDrop or regular spectrophotometer to measure RNA concentration.

7. Random hexamer primers (Invitrogen).

8. Deoxynucleotide triphosphates (dNTPs): Make a 10 mM stock solution by combining one part of each 100 mM dNTP stock plus six parts of TE buffer.

9. RevertAid reverse transcriptase (Fermentas), provided with 5× first strand (FS) buffer.

10. RiboLock RNase inhibitor (Fermentas).

11. RNase H (i.e. Invitrogen).

12. DEPC-treated RNase-free water.

13. Gene-specific primers and primers for housekeeping genes (Table 1) at a concentration of 20 mM each forward and reverse primer [RT-polymerase chain reaction (RT-PCR)] and at 2.5 mM for quantitative PCR. Dilute primers in TE buffer.

14. Taq Polymerase, provided with $MgCl_2$ and 10× PCR buffer.

15. ABGene qPCR Kit.

16. qPCR plates (i.e. Eppendorf).

17. Thermocycler, i.e. Mastercycler (Eppendorf) and qPCR cycler, i.e. Lightcycler (Roche).

18. Agarose (i.e. Peqlab).

19. 1% EtBr solution.

20. GelDoc System (gel scanning and documentation), i.e. Kodak Gel Logic 100 (Raytest) or similar ones.

2.4. IMAGE Analysis

1. Inverted microscope with 4× and 10× objectives, i.e. Olympus IX70 with SPOT Advanced Imaging system (Diagnostic Instruments).

2. IMAGE J 1.33u image analysis program. Freely available from the National Institutes of Health at http://rsbweb.nih.gov/ij/download.html.

Table 1
Primer sequences to characterize osteogenic and chondrogenic gene expression in differentiating ESCs by RT-PCR and qPCR

Gene	Tissue specificity	Forward primer	Reverse primer	qPCR	RT-PCR	T_a in °C	Amplicon size
18S rRNA	Housekeeper	CGC-GGT-TCT-ATT-TTG-TTG-GT	AGT-CGG-CAT-CGT-TTA-TGG-TC	Yes	3	60	218
GAPDH	Housekeeper	GCA-CAG-TCA-AGG-CCG-AGA-AT	GCC-TTC-TCC-ATG-GTG-GTG-AA	Yes	2/3	60	151
aggrecan	Chondrocyte	GAT-CTG-GCA-TGA-GAG-AGG-CG	GCC-ACG-GTG-CCC-TTT-TTA-C	Yes	2	61	81
biglycan	Connective tissue	CAT-GAC-AAC-CGT-ATC-CGC-AA	ATT-CCC-GCC-CAT-CTC-AAT-G	Yes	2	60	81
brachyury	Mesoderm	CCC-TGC-ACA-TTA-CAC-ACC-AC	GTC-CAC-GAG-GCT-ATG-AGG-AG	Yes	3	60	150
Alk Phos	Osteoblast	GTG-CCC-TGA-CTG-AGG-CTG-TC	GGA-TCA-TCG-TGT-CCT-GCT-CAC	Yes	2	60	81
bone sialoprotein	Osteoblast	AAA-GTG-AAG-GAA-AGC-GAC-GA	GTT-CCT-TCT-GCA-CCT-GCT-TC	No	3	60	215
bone sialoprotein	Osteoblast	CAG-AGG-AGG-CAA-GCG-TCA-CT	CTG-TCT-GGG-TGC-CAA-CAC-TG	Yes	2	62	81
Cbfa1 (runx2)	Osteoblast	CCG-AGT-CAT-TTA-AGG-CTG-CAA	TGC-GCT-GAA-GAG-GCT-GTT-T	Yes	2	58	101

Gene	Cell type	Primer 1	Primer 2				Ref
collagen type I	Connective tissue	GCA-TGG-CCA-AGA-AGA-CAT-CC	CCT-CGG-GTT-TCC-ACG-TCT-C	Yes	2/3	60	83
collagen type II A	Chondrocyte	GCT-GCT-GAC-GCT-GCT-CAT-C	GGT-TCT-CCT-TTC-TGC-CCC-TT	Yes[a]	2	60	295
collagen type II B	Chondrocyte	GCT-GCT-GAC-GCT-GCT-CAT-C	GGT-TCT-CCT-TTC-TGC-CCC-TT	Yes[a]	2	60	87
collagen type X	Chondrocyte	CAA-GCC-AGG-CTA-TGG-AAG-TC	AGC-TGG-GCC-AAT-ATC-TCC-TT	N/A	2/3	60	154
decorin	Connective tissue	ATG-ACC-CTG-ACA-ATC-CCC-TG	CCC-AGA-TCA-GAA-CAC-TGC-ACC	Yes	2	60	82
link protein	Chondrocyte	TTC-TGG-GCT-ATG-ACC-GCT-G	AGC-GCC-TTC-TTG-GTC-GAG-A	Yes	2	60	81
M-CSF	Osteoblast	AGT-CAA-CAG-AGC-AAC-CAA	CTT-CCT-GGG-TCA-AAA-ATC	N/A	3	60	178
osteocalcin	Osteoblast	CCG-GGA-GCA-GTG-TGA-GCT-TA	TAG-ATG-CGT-TTG-TAG-GCG-GTC	Yes	2	60	81
osteonectin	Osteoblast	ATC-CAG-AGC-TGT-GGC-ACA-CA	GGA-AAG-AAA-CGC-CCG-AAG-A	Yes	2	61	81
osteopontin	Osteoblast	GAT-GCC-ACA-GAT-GAG-GAC-CTC	CTG-GGC-AAC-AGG-GAT-GAC-AT	Yes	2	60	81
osterix (SP7)	Osteoblast	GGA-AAG-GAG-GCA-CAA-AGA-AG	TGA-GGG-AAG-GGT-GGG-TAG-TC	Yes	2	60	163
RANKL	Osteoblast	ACC-AGC-ATC-AAA-ATC-CCA-AG	TTT-GAA-AGC-CCC-AAA-GTA-CG	N/A	3	60	204

(continued)

Table 1
(continued)

Gene	Tissue specificity	Forward primer	Reverse primer	qPCR	RT-PCR	T_a in °C	Amplicon size
scleraxis	Chondrocyte	GGA-CCG-CAA-GCT-CTC-CAA-G	ACC-CAC-CAG-CAG-CAC-ATT-G	Yes	2	62	82
Sox9	Chondrocyte	GCA-GAC-CAG-TAC-CCG-CAT-CT	CTC-GCT-CTC-GTT-CAG-CAG-C	Yes	2	62	81

[a]For quantitative PCR, a TaqMan approach with hybridization probes must be used to distinguish between the two isoforms. The probes are as follows: collagen type IIA 5′-CGAGATCCCTTCGGAGAGTGCTGT-3′ and collagen type IIB 5′-CCAGGATGCCCGAAAATTAGGGCCAA-3′. Number in column RT-PCR denotes whether a 2-step or a 3-step cycle program is to be used for PCR.

3. Methods

3.1. Differentiation Initiation

Murine, human, and primate embryonic stem cells are routinely cultivated as described in Chapters 2 and 5, with routine checks for karyotypic stability and mycoplasma contamination. Examine the cells daily using a phase contrast microscope. Any changes in morphology or their adhesive properties should be noted. Prior to use, all media should be prewarmed to 37°C in a water bath or incubator, and trypsin/EDTA should be brought to room temperature. When the cells approach 80% confluency, they should be removed from the culture flask by trypsinization as follows:

1. Decant the ESC medium or aspirate it off. When decanting, wash cells twice with 1× PBS.

2. Trypsinize murine ESCs into single cells (see Note 3). Add 1–5 mL of trypsin/EDTA to the cells depending on the surface area of your dish or flask (1 mL for a T25-cm² culture flask, 5 mL for a T75-cm² flask) and incubate at 37°C.

3. Quench the trypsin/EDTA after approximately 5 min with at least the same amount of CDM. Break murine ESC colonies into single cells by repeatedly pipetting up and down (see Note 4).

4. Depending on the total volume, transfer cells into a 15-mL or 50-mL Falcon tube.

5. Centrifuge the cells at $200 \times g$ for 5 min.

6. While the cells are being pelleted, remove the 100-mm dishes needed for your experiment (see Subheading 3.2, step 6) out of the plastic sleeve. Add 10 mL of sterile PBS into each dish.

7. Decant or aspirate the supernatant from the pelleted cells and resuspend the cell pellet carefully and thoroughly in CDM by repeatedly pipetting up and down with a 2-mL serological pipette or a 2.5-mL Eppendorf pipettor.

8. Count cells using an automatic cell counter, which distinguishes between viable and dead cells, or a hemocytometer by trypan blue exclusion. For the latter, add 20 µL of trypan blue solution to 180 µL of cells and use 10 µL of this cell solution for counting. Multiply cell count by 10 to factor in dilution with trypan blue.

9. Prepare a dilution of 3.75×10^4 cells per mL in a 50-mL Falcon tube in CDM, taking only the viable cells into consideration for your calculation (see Note 5).

10. Using a Repeator® pipettor with a 2-mL combitip attached, pipette 20 µL droplets of cell suspension onto the lid of the petri dish (see Note 6).

11. Turn the lid over carefully into its regular position (the droplets should now be hanging from the inner side of the

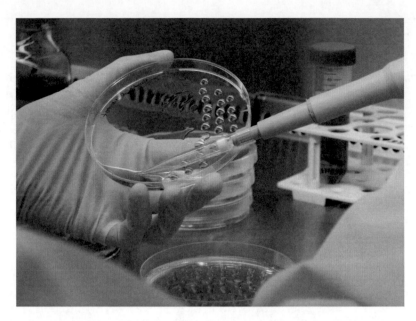

Fig. 1. Illustration of the hanging drop culture procedure on day three of differentiation. EBs that have been generated in hanging drops during the first three days of differentiation are washed off the slightly elevated petri dish lid with a 1-mL pipette tip.

lid) and put on top of the petri dish filled with PBS. Place the "hanging drops" in the incubator (5% CO_2, 37°C).

12. On day three of this protocol, pipette 5 mL of medium into the lid of the $\varnothing$ 100 mm "hanging drop" culture dish. Hold the lid at approximately a 45° angle to rinse the embryoid bodies (EBs) down to the bottom (Fig. 1). Using a sterile 2.5-mL pipettor (to avoid damage to the EBs), gently transfer the total suspension to a 100-mm $\varnothing$ bacterial petri dish filled with fresh CDM. Using the same technique, transfer the content of a second hanging drop dish into the same bacterial dish.

3.2. Osteogenic Differentiation

For osteogenic differentiation of hESCs and primate ESCs, form EBs as described in Chapter 4 using CDM, then follow the protocol below, which is the same for ESCs of all species.

1. On day 5 of the differentiation, fill a fresh bacteriological petri dish ($\varnothing$ 100 mm) with approximately 5 mL of prewarmed (37°C) trypsin/EDTA (see Note 7).

2. Gently rotate the dish containing the EBs in one direction to collect EBs at the center of the dish. Collect the aggregated EBs from the center of the dish using a 1-mL pipette.

3. Transfer EBs into the petri dish filled with trypsin/EDTA. Try to transfer as little medium as possible. Incubate for 8 min at 37°C.

4. Mechanically disperse the EBs in the petri dish by repeatedly pipetting up and down. Check by microscope if the suspension consists of single cells (see Note 8).

5. Transfer the cell suspension into a Falcon tube with 5 mL of CDM and centrifuge at $200 \times g$ for 5 min.

6. Take off the supernatant and resuspend the pellet in ODM1 for those cells that should be differentiated towards the osteogenic lineage and CDM for the control cells. Count the cells and plate them at a density of $5 \times 10^4/cm^2$ onto cell culture plates coated with 0.1% gelatin (see Notes 2 and 9).

7. After 24 h, change medium in the osteogenic cultures to ODM2. Be careful not to disrupt the cells. Do not change medium in the control cultures at this point.

8. Cultivate the cells for 17 days in humidified atmosphere with 5% CO_2 at 37°C, and change the medium every second day using CDM for control cultures and ODM2 for osteogenic cultures (see Note 10).

9. On day 23 of the differentiation, change medium in the osteogenic cultures to ODM3. Keep changing the medium in the control cultures (CDM). Again, change medium to fresh ODM3 every second day.

10. Look for black deposit in phase contrast microscopy throughout the culture period, as this is the osteoblast specific matrix (see Note 11). Examples of respective appearance of osteogenic cultures for murine, human, and primate ESCs are shown in Fig. 2. Cultures are fully mature around day 28 when osteocalcin expression is highest.

3.3. Chondrogenic Differentiation

1. Prepare a solution of ESCs as described in Subheading 3.1, steps 1–8 in CCM. Prepare the cell suspension in a 50-mL Falcon tube and adjust the cell concentration to 4×10^4 cells/mL.

2. Keep the cells in suspension by frequent gentle agitation during the following steps and leave at room temperature only for the shortest time period necessary.

3. Using a Repeator® pipettor, prepare hanging drops as described in Subheading 3.1, steps 10 and 11.

4. On day 3 of the differentiation, follow Subheading 3.1, step 12 using CCM for the untreated control cultures and CCDM1 for the chondrogenic cultures.

5. Cultivate this EB suspension culture for 2 days in a humidified atmosphere with 5% CO_2 at 37°C.

6. On day 5 of the differentiation, pipette 1 mL of medium into each well of a 24-well tissue culture plate. Use CCDM2 for the cells that you want to differentiate into chondrocytes and use CCM for the control culture.

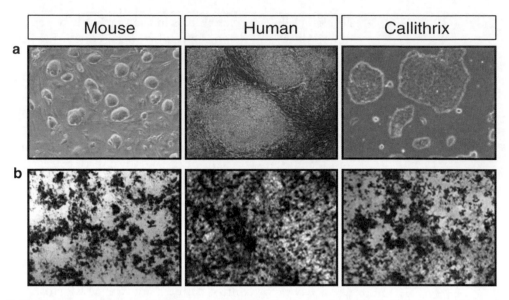

Fig. 2. Morphological appearance of embryonic stem cells and mineralized cultures in three species. (a) Images of undifferentiated ESCs colonies. Note the difference in 3D structure. Human ESC colonies are flatter than those of *Callithrix* and mouse. (b) Images of ESCs differentiated into osteoblasts with the basic VD_3 protocol. The images were acquired on day 30 using the settings described in Subheading 3.5.

7. Add one EB per well in a small volume (≤40 µL) using a 1-mL blue tip or shortened 100 µL yellow tip. Incubate 24-well plates until day 32 of culture in a humidified atmosphere with 5% CO_2 at 37°C.

8. Alternatively, up to ten EBs may be seeded together in one T25 cm² flask in 8 mL of medium.

9. Change medium from day 10 of culture onward, three times per week (typically Mondays, Wednesdays, and Fridays) (see Note 12).

3.4. RT-PCR and Quantitative PCR

Differentiating cells can be analyzed for presence of osteoblast or chondrocyte markers such as mineral and proteoglycan deposition, calcium content, or alkaline phosphatase activity as described in Chapter 17 and for protein expression as described in Chapter 15. Alternatively, RNA may be isolated and gene expression profiles studied as follows.

3.4.1. RNA Isolation

1. In order to harvest the cells for RNA isolation, aspirate off the medium and wash cells/EBs in 1× PBS before trypsinizing them.

2. Add 1 mL of trypsin/EDTA to each well of a 6-well plate of cells or each T25-cm² flask. If using 24-well plates, apply 500 µL of trypsin/EDTA per well. Incubate for 5 min at 37°C (see Note 13).

3. Collect the content of each 6-well plate well/T25-cm² flask into a 2-mL microfuge tube. Pool content of minimum

12 wells of a 24-well plate into a 15-mL Falcon tube. Stop the trypsinization with equal amounts of CDM or CCM.

4. Centrifuge for 5 min at $300 \times g$ with refrigeration. Discard the supernatant. If you are decanting, take care not to pour the cell pellet away.

5. Add 750 μL of RNA lysis buffer to each tube and incubate for 5 min. If you were using 15-mL Falcon tubes, transfer to 2-mL microfuge tubes at this point (see Note 14).

6. Transfer your lysates to individual Qiashredder columns and centrifuge at $16,000 \times g$ twice.

7. Add an equal amount of 70% ethanol (750 μL) to the homogenized flow-through.

8. Using a 21-G needle and syringe, shear the nucleic acids by aspirating the content and pressing it back into the microfuge tube several times. Avoid foaming.

9. Transfer the tube content to the Qiagen RNA isolation columns and follow the manufacturer's instructions.

10. Elute the RNA with RNase-free TE buffer at the end of the procedure.

3.4.2. cDNA Synthesis

1. Measure the concentration of RNA in your sample using the NanoDrop spectrophotometer or the RiboGreen RNA quantitation reagent and kit (Invitrogen).

2. Prepare a cDNA master mix containing 1× reaction buffer, 0.5 mM of dNTPs, 12 ng/μL of random hexamer primers, 2 U/μL of RNase Inhibitor, and 3.2 U/μL of Reverse Transcriptase. Add 625 ng of RNA template and DEPC-H_2O to a final volume of 25 μL.

3. Collect the liquid at the bottom of the tube by quickly centrifuging the samples at $2,000 \times g$ in a multifuge with a plate holder adapter.

4. Place your reaction tubes into the thermocycler and run the following program for cDNA synthesis: 25°C for 10 min, 42°C for 50 min, and 70°C for 15 min with the cycler lid temperature set to 80°C.

5. Add 2 Units of RNase H and incubate at 37°C for another 20 min.

3.4.3. RT-PCR

1. Assuming that the RNA is transcribed in a linear manner, use 25–50 ng of cDNA for your PCR (1–2 μL).

2. Prepare a PCR master mix containing 1× PCR buffer, 0.1 mM of dNTPs, 0.8 μM of both forward and reverse primers, 2 mM of $MgCl_2$, and 0.08 U/μL of Taq Polymerase. Add your cDNA template and DEPC-H_2O to a final volume of 25 μL.

3. Centrifuge the samples at $2,000 \times g$ in a multifuge with a plate holder adapter for 3–5 min.

4. Use a two-step PCR program when the primers are optimized for such a procedure as noted in Table 1. Program the thermocycler to run the following program: denaturation at 94°C for 5 min, then cycle for 32–35 times between 94°C for 45 s and 60°C (annealing and elongation) for 60 s. Insert a final step at 16°C to cool the samples, if desired. For a three-step PCR insert a separate elongation step at 72°C (60 s).

3.4.4. Quantitative PCR

1. Analyze each cDNA template in technical triplicates. Include those technical triplicates of three individual biological replicates in your statistic analysis (total of nine wells per sample).

2. Add 1 µL of 2.5 mM forward and reverse primers each and 5.5 µL of DEPC-H_2O to 12.5 µL of ABGene qPCR mix. Aliquot 20 µL of this master mix into the well of a PCR plate. Add 5 µL of cDNA template, previously diluted to 10 ng/µL with TE buffer or DEPC-H_2O.

3. Centrifuge down the liquid in the plate in a multifuge with a plate adapter for 5 min at $2,000 \times g$ at 4°C.

4. Insert the plate into the qPCR cycler. Follow the software in setting up and programming your run. Use the following cycle protocol: 94°C (5 min), 40× (94°C for 30 s followed T_a for 30 s), 16°C for 10 min with T_a being the annealing temperature of the primer pair. Run a melting curve after each run to control the presence of side-products.

5. Calculate the n-fold upregulation for each gene of interest over the housekeeper gene according to the delta-delta-Ct method using the following formula:

$$2^{-(C_T \text{ gene of interest} - C_T \text{ housekeeper})_{\text{sample}} - (C_T \text{ gene of interest} - C_T \text{ housekeeper})_{\text{control}}}$$

A typical example of quantitative PCR results for chondrogenic cultures is shown in Fig. 3 along with the morphological appearance of chondrogenic clusters.

3.5. IMAGE Analysis

The degree of mineralization in the course of osteogenic differentiation can be quantified with morphometric image analysis. In the following paragraphs we will explain this technique using a typical experiment as an example. ESCs were treated with VD_3 and a given compound X; the degree of mineralization was subsequently evaluated. It is necessary to include a culture of spontaneously differentiating ESCs as a control, which is not cultured with osteogenic supplements and will therefore not mineralize.

1. Capture an image of one EB or a section of the monolayer culture using a 4× magnification objective.

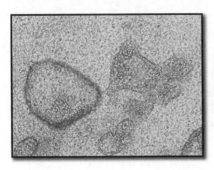
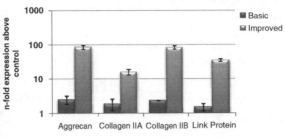

Fig. 3. Chondrocytes after 32 days of differentiation. (**a**) Morphology. (**b**) Gene expression profiles of cells cultured with the basic chondrogenic protocol and the improved chondrogenic medium (21). Quantitative PCR was performed as described in Subheading 3.5. Gene expression in chondrogenic cultures was normalized to that of GAPDH and standardized to non-treated spontaneously differentiating controls, $n=3\pm SD$.

2. Set the navigation software of your microscope to an exposure of 8 ms with a brightness of 1.6 and a gamma adjustment to a value of 1.8. Set the pixel bit depth to a monochromatic value of 8 bpp. Reducing the contrast of the pictures is important as it makes the black mineralized cells easier to distinguish against the lighter background of the other cells.

3. Take a minimum of three gray-scale pictures for each culture (treatment), preferably from different wells. Be particularly objective regarding the choice of the area to be photographed; make sure this area represents what you see in the entire culture.

4. This typically results in an image size of 300×240 pixels (number of columns $\times$ number of rows of pixels). To each pixel, an intensity value in the range between 0 (white) and 255 (black) is assigned automatically (see Note 15).

5. Generate a histogram of each picture with the IMAGE J 1.33u software. Open the picture file and choose Edit>Selection>Select all. Use the Analyze>Histogram command. A new window with a histogram will open. A mean black pixel value of the image is assigned by the software (Fig. 4).

6. Calculate the mean black value of each picture according to the following formula:

$$100-\left(\left(x\,/\,256\right)\times100\right)$$

7. Then, take the difference between the control cultures and all osteogenic cultures (in our example the VD_3 culture and compound X). The mean mineralization of the osteogenic cultures can then be expressed as n-fold upregulation over the control cultures. The difference between the control and the VD_3 culture is set to 100%.

8. Calculate the average of all three pictures and the standard deviation and plot them (Fig. 4).

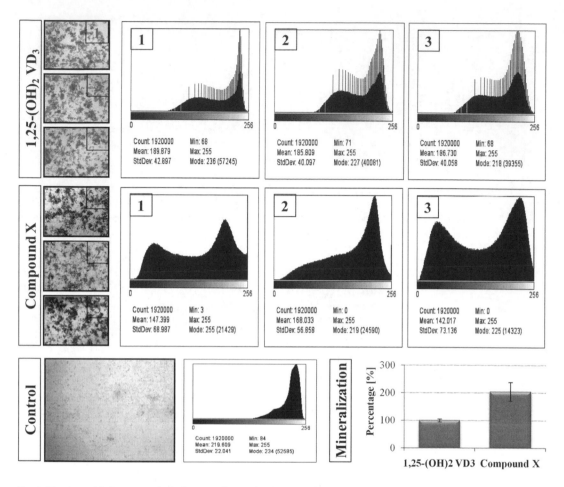

Fig. 4. Morphometric image analysis. Images of control cultures and osteogenic cultures [$1\alpha,25$-(OH)$_2$ VD$_3$ and compound X] were acquired. Histograms were generated with IMAGE J, and the degree of mineralization in compound X cultures was expressed as a percentage of VD$_3$ cultures.

4. Notes

1. The capability of ESCs to differentiate into chondrocytes and osteoblasts varies among different murine ESC lines; the most efficiently differentiating line is D3. This osteogenic differentiation protocol has, however, been successfully used to differentiate human ESCs (CA-1, derived by Dr. Andras Nagy, University of Toronto) and ESCs from the common marmoset monkey *Callithrix jacchus* (cESC-6, Dr. Erika Sasaki, Center for Experimental Animals, Japan).

2. Nonprecoated plasticware coated with gelatin may be used as described in Chapters 3 and 5. Alternatively, precoated Primaria culture plastic may be used (BD Biosciences).

3. If using human ESCs, do not break colonies into single cells. Rather use dispase, collagenase IV, or TrypLE (Invitrogen)

to loosen the cells from the plastic. Follow Chapter 4, Subheading 3.3, steps 1–8 (omitting steps 6 and 7) for EB formation from human ESCs using the medium described here at the indicated differentiation days.

4. Against common belief, we have observed that vigorous pipetting does not hurt the differentiation potential of the cells. However, keep foaming of medium to a minimum.

5. From here on, work quickly to expose the cells to room temperature as shortly as possible. To avoid cells adhering to the plastic, every now and then gently flip the closed Falcon tube upside down while you are pipetting the droplets onto the lids.

6. A yellow 100-μL pipette tip should be used as an extension of the dispenser combitip. This allows an easy change of the pipette tip in case you touch the outside of the dish, without having to discard the entire dispenser combitip.

7. Do not perform the trypsinization step in Falcon tubes, since excessive clumping may occur. Do not trypsinize more than the content of three 100-mm dishes of EBs together in one trypsin/EDTA dish. Adjust the number of trypsin/EDTA dishes according to your experiment.

8. This step is critical for the success of the differentiation. If the EBs have not been digested into single cells completely after the set time, place the dish back into the incubator only for a very short period of time. Longer incubation will result in the generation of strings of DNA, which will lead to complete clumping of your cell suspension. If you are uncertain about your technique, rather increase your initial incubation before you disaggregate the EBs with the pipette.

9. The number of cells that you need on day five of your experiment dictates the number of hanging drop dishes that you need to prepare on day 0. We typically assume that by day 5, one EB has grown to contain approximately 15,000 cells based on the fact that murine ESCs typically undergo four to five population doublings in this period. By dividing the number of cells that you need on day 5 by 15,000, you may calculate the number of total EBs that you need. Further assuming that you can fit 75–100 hanging drops on one dish, you can calculate the number of dishes needed on day 0.

10. Use independent preparations of the reagents in the second and third experiment.

11. Avoid prolonged exposure to light (e.g. under the microscope) when using VD_3. Check your cells with a microscope before changing the medium. Use light-tight tubes or wrap the tubes with aluminum foil to store the VD_3 solution.

12. Toward the end of the culture, the EBs may start to lift off the plates. This is by and large a sign of collagenous matrix secretion.

13. The goal is not to obtain a single cell suspension, but rather lift all cells off the plastic.

14. At this point you may interrupt the procedure and store your lysates at −80°C.

15. Depending on the settings of your microscope, these values may be different. The histogram will always show the maximum value assigned. Use this value for the calculations if it differs from 255.

Acknowledgment

NzN acknowledges the support of the German Ministry for Science, Education and Research (grant 0312314). NzN was funded through a postdoctoral fellowship from the Alberta Heritage Foundation for Medical Research while developing the improved osteogenic protocol.

References

1. Duplomb, L., Dagouassat, M., Jourdon, P., and Heymann, D. (2007) Concise review: embryonic stem cells: a new tool to study osteoblast and osteoclast differentiation. *Stem Cells*. **25(3)**, 544–552.

2. Buttery, L.D.K., Bourne, S., Xynos, J.D., Wood, H., Hughes, F.J., Hughes, S.P.F., et al. (2001) Differentiation of osteoblasts and in vitro bone formation from murine embryonic stem cells. *Tissue Eng.* **7**, 89–99.

3. zur Nieden, N. I., Kempka, G., and Ahr, H. J. (2003) In vitro differentiation of embryonic stem cells into mineralized osteoblasts. *Differentiation*. **71**, 18–27.

4. Phillips, B.W., Belmonte, N., Vernochet, C., Ailhaud, G., and Dani, C. (2001) Compactin enhances osteogenesis in murine embryonic stem cells. *Biochem. Biophys. Res. Commun.* **284(2)**, 478–484.

5. Sottile, V., Thomson, A., and McWhir, J. (2003) In vitro osteogenic differentiation of human ES cells. *Cloning Stem Cells*. **5**, 149–155.

6. Tremoleda, J.L., Forsyth, N.R., Khan, N.S., Wojtacha, D., Christodoulou, I., Tye, B.J., et al. (2008) Bone tissue formation from human embryonic stem cells in vivo. *Cloning Stem Cells*. **10(1)**, 119–132.

7. Arpornmaeklong, P., Brown, S.E., Wang, Z., and Krebsbach, P.H. (2009) Phenotypic characterization, osteoblastic differentiation, and bone regeneration capacity of human embryonic stem cell-derived mesenchymal stem cells. *Stem Cells Dev.* **18(7)**, 955–968.

8. Duester, G. (2008) Retinoic acid synthesis and signaling during early organogenesis. *Cell*. **134(6)**, 921–931.

9. zur Nieden, N.I., Price, F.D., Davis, L.A., Everitt, R.E., and Rancourt, D.E. (2007) Gene profiling on mixed embryonic stem cell populations reveals a biphasic role for beta-catenin in osteogenic differentiation. *Mol. Endocrinol.* **21(3)**, 674–685.

10. Doss, M.X., Chen, S., Winkler, J., Hippler-Altenburg, R., Odenthal, M., Wickenhauser, C., et al. (2007) Transcriptomic and phenotypic analysis of murine embryonic stem cell derived BMP2+ lineage cells: an insight into mesodermal patterning. *Genome Biol.* **8(9)**, R184.

11. Davis, L.A. and zur Nieden, N.I. (2008) Mesodermal fate decisions of a stem cell: the Wnt switch. *Cell. Mol. Life Sci.* **65(17)**, 2658–2674.

12. Bourne, S., Polak, J.M., Hughes, S.P., and Buttery, L.D. (2004) Osteogenic differentiation of mouse embryonic stem cells: differential gene expression analysis by cDNA microarray and purification of osteoblasts by cadherin-11 magnetically activated cell sorting. *Tissue Eng.* **10(5–6)**, 796–806.

13. Okazaki, M., Takeshita, S., Kawai, S., Kikuno, R., Tsujimura, A., Kudo, A., et al. (1994) Molecular cloning and characterization of OB-cadherin, a new member of cadherin family expressed in osteoblasts. *J. Biol. Chem.* **269(16)**, 12092–12098.

14. Sui, Y., Clarke, T., and Khillan, J.S. (2003) Limb bud progenitor cells induce differentiation of pluripotent embryonic stem cells into chondrogenic lineage. *Differentiation.* **71**, 578–585.

15. Vats, A., Bielby, R.C., Tolley, N., Dickinson, S.C., Boccaccini, A.R., Hollander, A.P., et al. (2006) Chondrogenic differentiation of human embryonic stem cells: the effect of the micro-environment. *Tissue Eng.* **12(6)**, 1687–1697.

16. Worster, A.A., Brower-Toland, B.D., Fortier, L.A., Bent, S.J., Williams, J., and Nixon, A.J. (2001) Chondrocytic differentiation of mesenchymal stem cells sequentially exposed to transforming growth factor-beta1 in monolayer and insulin-like growth factor-I in a threedimensional matrix. *J. Orthop. Res.* **19**, 738–749.

17. Estes, B.T., Wu, A.W., and Guilak, F. (2006) Potent induction of chondrocytic differentiation of human adipose-derived adult stem cells by bone morphogenetic protein 6. *Arthritis Rheum.* **54**, 1222–1232.

18. Wang, W.-G., Loua, S.-Q., Ju, X.-D., Xia, K., and Xia, J.-H. (2003) In vitro chondrogenesis of human bone marrow-derived mesenchymal progenitor cells in monolayer culture: activation by transfection with TGF-β2. *Tissue Cell.* **35**, 69–77.

19. Schonherr, E. and Hausser, H.J. (2000) Extracellular matrix and cytokines: a functional unit. *Dev. Immunol.* **7**, 89–101.

20. Majumdar, M.K., Wang, E., and Morris, E.A. (2001) BMP-2 and BMP-9 promotes chondrogenic differentiation of human multipotential mesenchymal cells and overcomes the inhibitory effect of IL-1. *J. Cell. Physiol.* **189**, 275–284.

21. Hatakeyama, Y., Tuan, R.S., and Shum, L. (2004) Distinct functions of BMP4 and GDF5 in the regulation of chondrogenesis. *J. Cell. Biochem.* **91**, 1204–1217.

22. zur Nieden, N.I., Kempka, G., Rancourt, D.E., and Ahr, H.J. (2005) Induction of chondro-, osteo- and adipogenesis in embryonic stem cells by bone morphogenetic protein-2: effect of cofactors on differentiating lineages. *BMC Dev. Biol.* **5**, 1.

23. Ballock, R.T. and Reddi, A.H. (1994) Thyroxine is the serum factor that regulates morphogenesis of columnar cartilage from isolated chondrocytes in chemically defined medium. *J. Cell. Biol.* **126(5)**, 1311–1318.

24. Johnstone, B., Hering, T.M., Caplan, A.I., Goldberg, V.M., and Yoo, J.U. (1998) In vitro chondrogenesis of bone marrow-derived mesenchymal progenitor cells. *Exp. Cell. Res.* **238**, 265–272.

25. Mackay, A.M., Beck, S.C., Murphy, J.M., Barry, F.P., Chichester, C.O., and Pittenger, M.F. (1998) Chondrogenic differentiation of cultured human mesenchymal stem cells from marrow. *Tissue Eng.* **4(4)**, 415–428.

26. Denker, A.E., Nicoll, S.B., and Tuan, R.S. (1995) Formation of cartilage-like spheroids by micromass cultures of murine C3H10T1/2 cells upon treatment with transforming growth factor-beta 1. *Differentiation.* **59**, 25–34.

27. Tanaka, H., Murphy, C.L., Murphy, C., Kimura, M., Kawai, S., and Polak, J.M. (2004) Chondrogenic differentiation of murine embryonic stem cells: effects of culture conditions and dexamethasone. *J. Cell. Biochem.* **93(3)**, 454–462.

28. Toh, W.S., Yang, Z., Liu, H., Heng, B.C., Lee, E.H., and Cao, T. (2007) Effects of culture conditions and bone morphogenetic protein 2 on extent of chondrogenesis from human embryonic stem cells. *Stem Cells.* **25(4)**, 950–960.

29. Hwang, N.S., Kim, M.S., Sampattavanich, S., Baek, J.H., Zhang, Z., and Elisseeff, J. (2006) Effects of three-dimensional culture and growth factors on the chondrogenic differentiation of murine embryonic stem cells. *Stem Cells.* **24(2)**, 284–291.

30. Hwang, N.S., Varghese, S., Zhang, Z., and Elisseeff, J. (2006) Chondrogenic differentiation of human embryonic stem cell-derived cells in arginine-glycine-aspartate-modified hydrogels. *Tissue Eng.* **12(9)**, 2695–2706.

Chapter 15

Differentiation of Mouse Embryonic Stem Cells in Self-Assembling Peptide Scaffolds

Núria Marí-Buyé and Carlos E. Semino

Abstract

Here, we describe the capacity of mouse embryonic stem cells (mESCs) to differentiate into osteoblast-like cells in a three-dimensional (3D) self-assembling peptide scaffold, a synthetic nanofiber biomaterial with future applications in regenerative medicine. We have previously demonstrated that classical tissue cultures (two-dimensional) as well as 3D-systems promoted differentiation of mESCs into cells with an osteoblast-like phenotype expressing osteopontin (OPN) and collagen type I (Col I), as well as high alkaline phosphatase (Alk Phos) activity and calcium phosphate mineralization. Interestingly, in 3D self-assembling peptide scaffold cultures, the frequency of appearance of embryonic stem-cell-like colonies was substantially enhanced, suggesting that this particular 3D microenvironment promoted the generation of a stem-cell-like niche that allows the maintenance of a small pool of undifferentiated cells. We propose that the 3D system provides a unique microenvironment permissive to promote differentiation of mESCs into osteoblast-like cells while maintaining its regenerative capacity.

Key words: mESCs, 3D-cultures, Self-assembling peptides, Differentiation, Osteogenesis

1. Introduction

Murine embryonic stem cells (mESCs) are derived from the inner cell mass of blastocysts and present unique characteristics such as almost unlimited expansion capacity and the potential to differentiate into diverse cell lineages, a feature called pluripotency (1, 2). When culturing them in presence of leukemia inhibitory factor (LIF), these cells can be expanded in vitro while maintaining their pluripotent phenotype (3, 4). This property is normally assessed in vitro by preparing embryoid bodies (EB), as described in the previous chapters, where mESCs can grow and differentiate into the three main embryonic germ lines: endoderm, mesoderm, and ectoderm (5, 6). Using this methodology, mESCs have been suc-

Nicole I. zur Nieden (ed.), *Embryonic Stem Cell Therapy for Osteo-Degenerative Diseases*, Methods in Molecular Biology, vol. 690, DOI:10.1007/978-1-60761-962-8_15, © Springer Science+Business Media, LLC 2011

cessfully differentiated into neural cells (7–9), smooth muscle cells (10), endothelial cells (11), hematopoietic cells (12, 13), adipocytes (14), chondrocytes (15), and osteoblasts (16). Their vast pluripotent capacity together with emerging bioengineering technologies has promoted an increasing interest in combining multiple disciplines to develop functional tissue constructs in vitro. Although mESC differentiation represents a complex process for experimental tissue engineering approaches, the incorporation of other disciplines into the field, such as matrix biology and biomechanics, will improve the development of appropriate and reliable functional cells. For instance, during endochondral ossification process, the cartilage tissue serves as a model for further bone formation. In the last stage of the process, hypertrophic chondrocytes secrete numerous small membrane-bound vesicles into the extracellular matrix. These vesicles contain enzymes that are active in the generation of calcium and phosphate ions and initiate the mineralization process within the cartilaginous matrix (17). Most of the osteogenic differentiation studies found in the literature have focused on exploring the experimental conditions to induce osteoblast-specific differentiation using two-dimensional (2D) cultures, which do not really mimic the in vivo environment. Although these systems are successful in promoting osteogenic commitment, it is known that cell–cell and cell–matrix interactions are crucial during embryonic development and also important during bone remodeling and healing processes in vivo (18–21). These interactions regulate a variety of cell signaling pathways to efficiently promote the development of tissues (22). For this reason, three-dimensional (3D) systems or scaffolds could substantially improve the differentiation process by creating an environment that can better simulate the in vivo milieu. That is to say, bioengineered 3D-culture systems could give the necessary structural pattern to cells, thereby allowing the correct organization of their extracellular matrix as well as enhancing their proliferative and differentiation capacity (23, 24). The main focus of this chapter is to describe the osteogenic differentiation of mESCs in a synthetic matrix, such as the self-assembling peptide scaffold PuraMatrix, a defined nanofiber scaffold that structurally mimics extracellular matrices and offers real possibilities for future applications in regenerative medicine (see Fig. 1) (25–27).

We used a transgenic derivative of the R1mESC line expressing the green fluorescent protein (GFP) under the transcriptional control of the Oct-4 promoter (28) in a 3D-culture technique that uses a synthetic nanofiber scaffold as extracellular matrix analog (25–27). Parallel experiments were carried out on classical culture dishes (2D) to compare the differentiation capacity between both culture systems under osteogenic conditions. We have demonstrated previously that both 2D- and 3D-culture systems promoted differ-

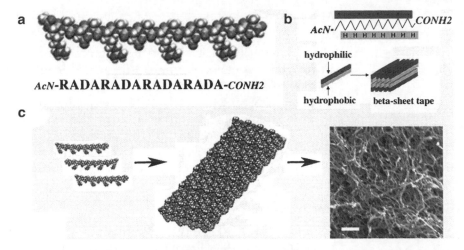

Fig. 1. Peptide RAD16-I self-assembles into a nanofiber network. This peptide is an example of a class of self-assembling peptide scaffolds of nanometric fiber with injectable properties. The scaffold is biocompatible, biodegradable, and will allow cell seeding. The material per se has no instructive capacity for cells, only structural features (nanofiber network). (a) Molecular model of peptide RAD16-I. (b) Molecular model of the nanofiber developed by self-assembly of RAD16-I molecules. Note: The nanofiber is formed by a double tape of assembled RAD16-I molecules in an antiparallel β-sheet configuration. (c) RAD16-I nanofiber network as seen in SEM. The *white bar* represents 200 nm.

entiation of mESCs into cells with osteoblast-like phenotype expressing bone markers including osteopontin (OPN), collagen type I (Col I), alkaline phosphatase (Alk Phos), and calcium mineralization. We propose that the 3D-culture system can be used to explore the potential of mESCs to differentiate into other mesenchymal tissues including cartilage, muscle, and fat.

2. Materials

2.1. Cell Culture, Differentiation, and Isolation

1. Murine embryonic stem cell line R1, transgenic for GFP expression under the control of the Oct-4 promoter. These ES R1 Oct4-GFP cells were generously obtained from Dr. Ali Khademhosseini at the Massachusetts Institute of Technology (28).

2. Murine embryonic stem cell medium (mESCM): Dulbecco's Modified Eagle's Medium (DMEM), high-glucose 4,500 mg/mL (Invitrogen) containing 1,000 U/mL recombinant mouse LIF (Millipore), 15% fetal bovine serum (FBS; Hyclone), 1 mM sodium pyruvate (Invitrogen), 0.1 mM nonessential amino acids (Invitrogen), 4 mM L-glutamine (Invitrogen), 1% (v/v) penicillin-streptomycin (Invitrogen), and 0.1 mM 2-mercaptoethanol.

3. Murine embryonic stem cell medium without LIF (mESCM/-LIF): Same formulation as mESCM except LIF.

4. Osteogenic medium: DMEM (high-glucose 4,500 mg/mL, Invitrogen) containing the Osteogenic SingleQuot kit (Cambrex), which is composed of mesenchymal cell growth supplement (MCGS), 1% (v/v) penicillin-streptomycin, 4 mM L-glutamine, 0.05 mM ascorbate, 10 mM beta-glycerophosphate, and 0.1 µM dexamethasone. Moreover, 50 nM 1α,25-OH$_2$ vitamin D$_3$ (VD$_3$) was added to the osteogenic medium (16) (see Note 1).

5. 0.05% trypsin/ethylenediaminetetraacetic acid (EDTA; Invitrogen).

6. 0.1% (w/v) gelatin solution: Dissolve 0.1 g gelatin in 100-mL sterile water. Sterilize by autoclaving.

7. Bacteriological nonadherent petri dishes (100-mm diameter).

8. 12-well cell culture plates.

9. Regular culture dishes, sterile pipettes (2, 5, 10, and 25 mL), usual and wide-bore pipette tips, and 15-mL conical tubes.

10. Fluorescence microscope with long-distance objectives.

2.2. Embryonic Stem Cell Encapsulation

1. 10% (w/v) sucrose solution: dissolve 10 g sucrose per 100 mL of distilled water. Sterilize by filtering through a 0.22-µm filter.

2. RAD16-I peptide solution at 0.5% (w/v): Dilute RAD16-I peptide [BD™ PuraMatrix™ peptide hydrogel, 1% (w/v), BD Biosciences, see Note 2] at 1:2 using 10% (w/v) sucrose solution (final pH 3.5). Thoroughly mix the components by placing the solution tube into an ultrasonic bath for 20 min. The resulting solution may be stored at 4°C for several months.

3. Transwell inserts, 10-mm diameter, 0.78 cm^2 area, pore size = 0.2 µm.

4. 6-well cell culture plates.

2.3. Cell and Construct Fixation

1. 4% (w/v) paraformaldehyde (PFA) solution: weigh 4 g PFA in a flask with narrow neck, suspend the powder in approximately 80 mL of PBS, and stir. Add 1 M sodium hydroxide (NaOH) until complete dissolution of the powder and use 1 M hydrochloric acid (HCl) and distilled water to finally adjust the pH at 7.4 and the volume to 100 mL. Aliquot and store the resulting 4% PFA solution at –20°C. When needed, thaw the solution and dilute 1:4 with PBS to obtain a 1% PFA solution (see Note 3).

2. Phosphate-buffered saline (PBS): 1 mM KH$_2$PO$_4$, 10 mM Na$_2$HPO$_4$, 137 mM NaCl, 2.7 mM KCl, pH 7.4.

3. 1 M NaOH.

4. 1 M HCl.

2.4. von Kossa Staining

1. 1% (w/v) PFA solution (see Subheading 2.3).

2. 5% (w/v) silver nitrate ($AgNO_3$): Dissolve 0.5 g silver nitrate in 10 mL of distilled water. Store in the dark at 4°C.

3. 2% (w/v) agarose solution: Weight 2 g agarose in a flask and add 100 mL of distilled water. Use the microwave to heat up to approximately 100°C to dissolve the powder.

4. Distilled water.

5. Dissecting microscope.

2.5. Immuno-fluorescence

1. Culture media: mESCM/-LIF and osteogenic medium (see Subheading 2.1).

2. 1% (w/v) PFA solution (see Subheading 2.3).

3. Blocking buffer: PBS (see above) containing 20% (v/v) FBS (Hyclone), 0.1% (v/v) Triton X-100, and 1% (v/v) dimethyl sulfoxide.

4. Primary antibody for osteopontin (OPN): OPN (AKm2A1) mouse monoclonal immunoglobulin (IgG), 200 µg/mL (Santa Cruz Biotechnologies). Primary antibody for Oct-4: Oct-3/4 (H-134) rabbit polyclonal immunoglobulin (IgG), 200 µg/mL (Santa Cruz Biotechnologies). Primary antibodies are diluted 1:200 to a final concentration of 1 mg/mL in blocking buffer prior to use.

5. Secondary antibody goat anti-mouse IgG-R (rhodamine conjugated, 200 µg/0.5 mL, Santa Cruz Biotechnologies). Secondary antibody donkey anti-rabbit IgG-R (rhodamine conjugated, 200 µg/0.5 mL, Santa Cruz Biotechnologies). Secondary antibodies are used at a final concentration of 1 mg/mL and diluted 1:400 in blocking buffer just prior to use.

6. 4′,6-Diamidino-2-phenylindole dihydrochloride (DAPI) solution: Dissolve DAPI in methanol andwater (50:50) at 10 mg/mL. This stock solution must be stored at –20°C. This stock solution is diluted 1:10,000 in PBS to yield a 1 µg/mL DAPI working solution.

7. Fluorescence microscope with long-distance objectives.

2.6. Western Blotting

1. Lysis buffer: PBS (see above), containing 0.1% (v/v) Triton X-100 and one tablet (per 10 mL of buffer) of protease inhibitor cocktail (complete mini protease inhibitor cocktail, Roche Diagnostics).

2. RIPA buffer (commercial, i.e. Sigma).

3. DC Protein Assay (Bio-Rad Laboratories), which contains an alkaline copper tartrate solution (Reagent A), a dilute Folin reagent (Reagent B), a surfactant solution (Reagent S), and bovine serum albumin standard (BSA). Reagent A must be prepared (20 µL of Reagent S to each mL of Reagent A needed), and it is stable for only 1 week. If a precipitate forms, warm and vortex before use. BSA standard solutions (3–5 dilutions) must be also prepared with concentrations between 0.2 mg/mL and 1.5 mg/mL.

4. Test tubes (13 mm × 100 mm).

5. Spectrophotometer.

6. Sample preparation buffer (NuPAGE® LDS Sample Buffer 4×, Invitrogen).

7. Running buffer (NuPAGE® MOPS SDS Running buffer 20×, Invitrogen).

8. Transfer buffer: Dissolve 3.03 g Tris(hydroxymethyl)aminomethane (Tris base) and 14.4 glycine in 800 mL of distilled water and add 200 mL of methanol.

9. Blocking buffer: PBS containing 4% (w/v) nonfat powdered milk and 0.1% (v/v) Triton X-100.

10. Tris buffered saline solution (TBS): 50 mM Tris base, 150 mM NaCl, pH 7.4. Dissolve 6.1 g Tris(hydroxymethyl)aminomethane (Tris base) and 9.0 g sodium chloride in 100 mL of distilled water.

11. TBS/Tween: TBS containing 0.05% Tween-20.

12. Polyacrylamide gel electrophoresis (PAGE) system: NuPAGE® Novex 10% Bis-Tris gel 1.0 mm, 10-well.

13. Polyvinylidene fluoride membranes (PVDF, 0.2-mm pore size).

14. SeeBlue® Plus2 prestained protein standards (Invitrogen) or similar.

15. Collagen type I, from rat tail.

16. Primary antibody for collagen type I: Anti-collagen I rabbit polyclonal (Novocastra). Primary antibody for Runx2: anti-Runx2 (PEBP2aA) rabbit polyclonal, 200 µg/mL (Santa Cruz Biotechnologies). Dilute both 1:200 in blocking buffer just prior to use.

17. Secondary antibody goat anti-rabbit IgG-HRP-conjugated (200 µg/0.5 mL, Santa Cruz Biotechnologies). Use at a final concentration of 1 mg/mL (1:400 dilution in blocking buffer).

18. Western blotting luminol reagent (Santa Cruz Biotechnologies), which consists of two solutions (solutions A and B). Mix equal volumes of luminol reagent solutions A and B immediately before use.

19. Plastic wrap, i.e. Saran wrap.

20. Chemiluminescent chamber with camera, i.e. Kodak or Bio-Rad.

2.7. Kinetic Studies

1. mESCM and mESCM/-LIF cell culture media (see Subheading 2.1).

2. 0.05% trypsin/EDTA (Invitrogen).

3. 24-well cell culture plates.

2.8. Lysis of 3D Cultures for RNA Isolation

1. TRIzol® Reagent (Invitrogen) or similar RNA lysis buffer.

2. 10-mL BD Luer-Lok™ syringe with 20G×1½ in. BD PrecisionGlide™ needle.

3. Methods

3.1. Osteogenic Differentiation of mESCs in 2D- and 3D-Cultures

The transgenic murine ESC line ES R1 Oct4-GFP is used to obtain an embryonic cell lineage with osteogenic potential (mesoderm), when it is encapsulated in a self-assembling peptide scaffold to study the osteogenic differentiation in this 3D system. The specific stages followed are summarized in Fig. 2. Briefly, embryoid bodies (EBs) are formed from expanded mESCs by following a classic differentiation protocol (29) (Fig. 2, stages 1 and 2). When cells lose their pluripotency, EBs are dissociated and the resulting EB-derived cells (EB-dcs) are encapsulated in the peptide scaffold (3D system) or cultured on classical culture dishes (2D-system), as control (Fig. 2, stages 3 and 4). Then, cells are maintained for several days until medium is replaced for osteogenic medium to induce differentiation into osteoblast-like cells (Fig. 2, stage 5).

3.1.1. Differentiation Initiation

1. Maintain ES R1 Oct4-GFP cells at 37°C in humidified air with 5% carbon dioxide (CO_2) in gelatin-coated flasks with mESCM as described in Chapter 2.

2. Remove medium of the flask and rinse cells with 1 mL of 0.05% trypsin/EDTA. Aspirate the liquid and add 2 mL of 0.05% trypsin/EDTA. Incubate at 37°C up to a maximum of 2 min until cells are detached and quickly add 5 mL of medium to inactivate the trypsin. Mix well and break clusters by pipetting up and down and transfer the suspension into a 15-mL conical tube.

3. Centrifuge at $75 \times g$ for 5 min and suspend the pellet in 10 mL of mESCM/-LIF at a concentration of 1.5×10^5 cells/mL. Culture the suspension in a nonadherent petri dish (100 mm diameter) to form the embryoid bodies (EBs).

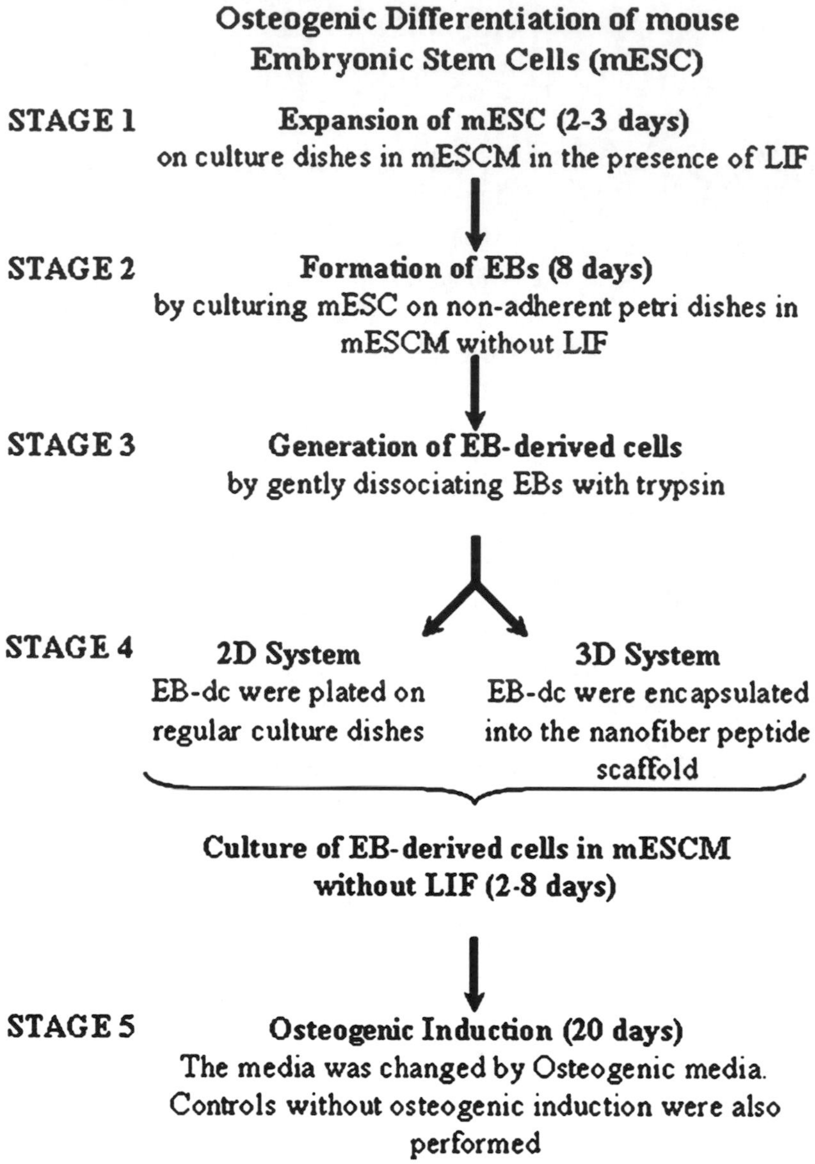

Osteogenic Differentiation of mouse Embryonic Stem Cells (mESC)

STAGE 1   **Expansion of mESC (2-3 days)**
on culture dishes in mESCM in the presence of LIF

STAGE 2   **Formation of EBs (8 days)**
by culturing mESC on non-adherent petri dishes in
mESCM without LIF

STAGE 3   **Generation of EB-derived cells**
by gently dissociating EBs with trypsin

STAGE 4   **2D System**   **3D System**
EB-dc were plated on   EB-dc were encapsulated
regular culture dishes   into the nanofiber peptide
                                        scaffold

**Culture of EB-derived cells in mESCM
without LIF (2-8 days)**

STAGE 5   **Osteogenic Induction (20 days)**
The media was changed by Osteogenic media.
Controls without osteogenic induction were also
performed

Fig. 2. Flow diagram of the differentiation protocols. Schematic representation of the protocol used for the osteogenic induction of the mouse embryonic stem cell line ES R1 Oct4-GFP.

4. Maintain the EB cultures at 37°C in a humidified atmosphere with 5% CO_2 and feed them every 3 days by transferring them into a 15-mL conical tube and allowing EBs to settle down. Then, remove 5 mL of the old medium and replace it with 5 mL of fresh medium. Gently pipette EBs with a wide-bore pipette and transfer them to a new petri dish.

5. Fluorescence microscopy may be used to monitor GFP expression during EB formation. When EBs present reduced GFP expression (approximately after 8 days), they are ready to

be encapsulated in the synthetic peptide scaffold (3D system) or cultured in classical culture dishes (2D-system) (see Note 4).

3.1.2. Three-Dimensional Culture of mESCs

1. Harvest the EBs and allow them to settle in a conical tube. Carefully remove the medium and dissociate EBs by gently treating with trypsin/EDTA to generate EB-dcs.

2. Add medium to inhibit trypsin and count the EB-dcs (see Note 5).

3. Place the transwell inserts (one for each encapsulation) into a 6-well plate.

4. Centrifuge cells at $75 \times g$ for 5 min and suspend them in 10% sucrose at a final concentration of 4×10^6 cells/mL (see Note 6).

5. Mix the cell suspension with an equal volume of liquid RAD16-I peptide solution (0.5% in 10% sucrose, see Note 7) to obtain a final suspension of 2×10^6 cells/mL in 0.25% RAD16-I in 10% sucrose (see Note 8).

6. Load 100 μL of the resulting suspension to each insert and immediately add 200 μL of mESCM/-LIF to the bottom of the insert, allowing the membrane to wet (see Note 9). Let the peptide to gel for approximately 20 min (see Note 10).

7. Slowly add 50 μL of culture medium to the top of the previously formed hydrogel and let it drain. Repeat the process three times. Change the medium in the well with 2 mL of fresh medium and also add 0.5 mL of medium onto the hydrogel (see Note 11). An example is shown in Fig. 3.

8. Maintain the 3D-cultures in a humidified incubator equilibrated with 5% CO_2. Change medium every day (see Note 12).

3.1.3. Two-Dimensional Culture of EBs

1. Harvest the EBs and allow them to settle in a tube. Carefully remove the medium and dissociate EBs by gently treating with trypsin/EDTA to generate EB-dcs.

2. Add medium to inhibit trypsin and count EB-dcs.

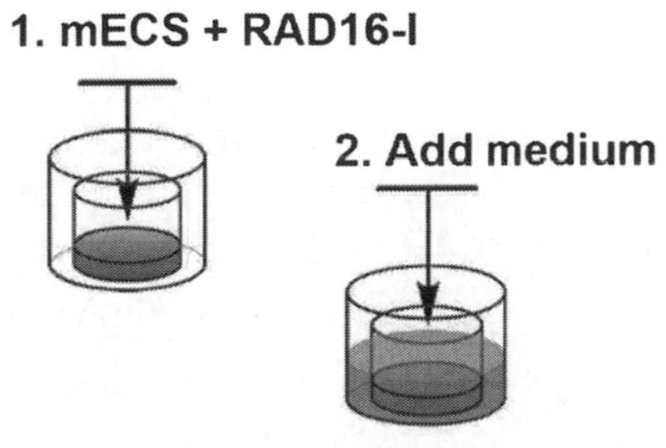

1. mECS + RAD16-I

2. Add medium

Fig. 3. Three-dimensional culture (3D-culture) system used. Murine ESCs are mixed with a solution of self-assembling peptide scaffolds in 10% sucrose (1), and medium is used to initiate scaffold gelation (2).

3. Centrifuge cells at $75 \times g$ for 5 min and resuspend them in mESCM/-LIF. Plate the EB-dcs into a 12-well plate at a density of 2×10^5 cells/cm^2 (7.6×10^5 cells/mL).

4. Maintain the 2D-cultures at 37°C with 5% CO_2. Change medium every 2 days (see Note 13).

3.1.4. Osteogenic Differentiation in 2D- and 3D-Cultures

1. Allow 2D- and 3D-cultures to grow in mESCM/-LIF for 2 days before osteogenic induction (see Note 14).

2. Change the medium to osteogenic medium, but depending on your experimental setup, also run controls in mESCM/-LIF.

3. Maintain the cultures for 20–22 days at 37°C in a humidified atmosphere with 5% CO_2. Change medium every day in 3D-cultures, as indicated in Note 12, and every 2 days in 2D-cultures.

3.2. Cell Isolation and Culture of Isolated Cells

Some characterization techniques, such as kinetics studies and immunofluorescence for some markers, require the postculture of the cells isolated from the 2D- and 3D-systems.

1. Treat cells from 2D- and 3D-cultures with 0.05% trypsin/EDTA. In the case of 3D-constructs, also use the micropipette to mechanically disrupt the peptide scaffold cultures.

2. Add complete medium (the appropriate in each case: mESCM, mESCM/-LIF, or osteogenic medium) and make sure by checking culture wells under a phase contrast microscope that single cells are obtained.

3. Count cells and centrifuge the cells at $75 \times g$ for 5 min.

4. Suspend the cells and culture them as needed (refer to the protocol of the final analysis for specific culture conditions). You may also analyze your cultures using the Ca^{2+} assay and the quantitative alkaline phosphatase assay described in Chapter 17 or with the techniques described below.

3.3. von Kossa Staining for Mineralization

Calcium phosphate deposits are detected by the widely used von Kossa technique (30), where silver cations substitute calcium ions mainly in carbonate and phosphate salts and they subsequently reduce forming black nodules. Therefore, this technique allows estimating when the 2D- and 3D-matrices start to mineralize, which is an indication of osteogenic commitment.

1. Wash 2D- and 3D-cultures twice with PBS.

2. Fix the constructs with 1% PFA solution for 1 h at room temperature.

3. Strictly rinse the fixed cultures with distilled water until PFA and PBS are completely removed (see Notes 15 and 16).

Control   **Osteogenic**

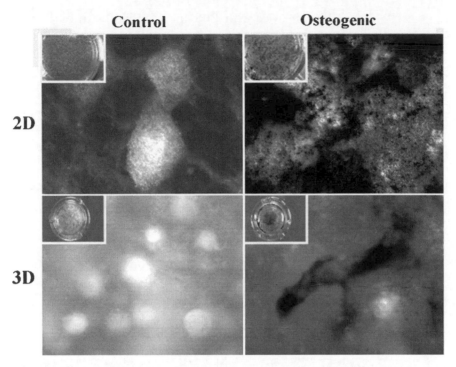

Fig. 4. von Kossa staining for mineralized calcium. von Kossa staining was performed after the osteogenic induction of 2D- and 3D-cultures of EB-dc after 22 days in mESCM/-LIF (control, negative) or in osteogenic medium (osteogenic, positive). For better visualization of the mineralized nodules, a lower magnification of each well is shown in each *top-left corner.*

4. Cover the samples with a solution of 5% (w/v) silver nitrate. Leave it to react for 1 h in the dark.

5. Gently wash the samples with distilled water to remove the excess silver nitrate solution (see Note 17). Place the sample under a strong light source for 10 min.

6. Inspect the samples visually or through the microscope in reflection mode. Calcium in mineralized nodules stains dark as depicted in Fig. 4.

3.4. Immunofluo-rescence for OPN and Oct-4

Immunostaining may be used to detect the presence of osteo-pontin (OPN), a noncollagenous protein present in natural bone matrix and considered as an early osteogenic marker, in the iso-lated cells from 2D- and 3D-cultures after the osteogenic induc-tion of mESCs. In addition, Oct-4 (an indicator of pluripotency) may be assayed by immunofluorescence on mESC cultures before and after the differentiation process.

1. Isolate cells as indicated in Subheading 3.2 (see Note 18).

2. Suspend cells in their culture medium (mESCM/-LIF or osteogenic medium) and culture them in regular multiwell plates for 4–6 days.

3. Fix the cells with 1% PFA solution for 1 h.

4. Wash twice with PBS and incubate with blocking buffer for 4 h at room temperature by placing the plate onto an orbital shaker.

5. Incubate cells with the primary antibody overnight at room temperature.

6. The next day, remove the primary antibody and wash three times with blocking buffer.

7. Incubate cells with the corresponding secondary antibody for 2 h in the dark at room temperature.

8. Remove the secondary antibody by washing three times with blocking buffer.

9. To identify the cell nuclei, cover cells with DAPI working solution and incubate for 5 min in the dark.

10. The samples are observed through a fluorescent microscope (equipped with DAPI, Rhodamine, and FITC filters) (see Figs. 5 and 8).

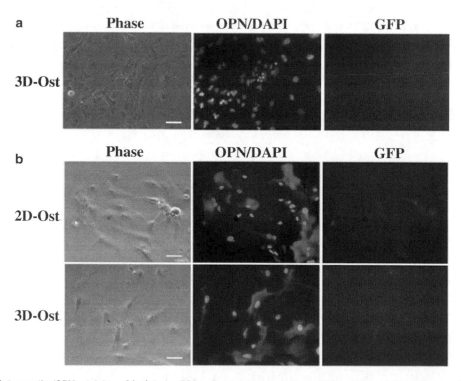

Fig. 5. Osteopontin (OPN) staining of isolated mESCs after osteogenic induction from 2D- and 3D-cultures. (a) EB-dcs cultured for 2 days before osteogenic induction. Cells isolated from 3D-cultures after 22 days in osteogenic medium show only slightly detectable OPN expression. OPN was not detected in 2D osteogenic cultures (not shown). (b) EB-dcs cultured for 8 days before osteogenic induction. Cells isolated both from 2D- and 3D-cultures after 20 days in osteogenic medium present high OPN expression. In both cases, no residual GFP signal was seen, indicating no remaining Oct-4 expression. *Bar* 50 μm.

3.5. Western Blotting for Collagen Type I and Runx2

Expression of the transcription factor Runx2 and the extracellular matrix protein collagen type I (Col I) is indicative of an osteogenic commitment. Both markers may be studied in cell lysates from samples before and after differentiation.

1. Wash the cell cultures twice with PBS and suspend them in lysis buffer. To help cell disruption, sonicate the cell suspension for 5 min.

2. Transfer to a microcentrifuge tube and centrifuge at $16,000 \times g$ for 5 min in a tabletop centrifuge set to 4°C.

3. Determine the total amount of protein in the supernatant fraction using the protein detection kit (DC Protein Assay), which is based on the Lowry method. Pipette 100 μL of the samples and standard (BSA) solutions into test tubes, add 500 μL Reagent A and vortex. Add 4 mL of Reagent B into each tube and vortex immediately. After 15 min, read the absorbance at 750 nm (solutions are stable for 1 h, see Note 19).

4. Suspend each pellet in sample preparation buffer containing SDS and 2 mercaptoethanol and heat the suspensions at 80°C for 10 min.

5. Equilibrate the 10% PAGE system with MOPS SDS running buffer and load the samples into the different wells. Also load a protein standard ladder to identify the molecular weight (for example SeeBlue® Plus2 or as desired) and 5 μg/well from a solution of Col I (3.5 mg/mL) from rat tail as a positive control.

6. Run the gel at 110–115 mA through the stacking gel and 60–70 mA through the separating gel.

7. Wash a PVDF transfer membrane before use, once with methanol and twice with transfer buffer.

8. Transfer the separated proteins to a PVDF membrane for 2 h using transfer buffer.

9. Incubate the membrane with blocking buffer at room temperature for 2 h in a small container using an orbital shaker.

10. Incubate the membrane for 1 h with the corresponding primary antibody at working dilution at room temperature.

11. Remove the primary antibody by washing the membrane three times with blocking buffer (30 min each).

12. Incubate the membrane with the secondary antibody for 1 h at room temperature.

13. Remove the excess antibody by washing the membrane once with blocking buffer for 30 min and then three times with TBS/Tween, 5 min each time. Finish the washing steps by rinsing in TBS only for 5 min.

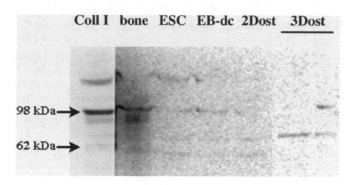

Fig. 6. Western blot analysis of collagen type I during osteogenic differentiation. *Col I* collagen type I standard (rat), *bone* mouse bone, *ESC* ES R10ct4-GFP ESCs, *EB-dc* EB-derived cells, *2Dost* cells from 2D-cultures after 22 days in osteogenic medium, *3Dost* two different samples of cells from 3D cultures in osteogenic medium: EB-dcs cultured for 2 or 8 days before osteogenic induction. In general, 3D-osteogenic cultures presented higher collagen type I expression than 2D-osteogenic culture.

	mESC	EB-dc	2D-Ost.	3D-Ost.
Duplication Time /h	12.6	16.5	19.6	32.2

Fig. 7. Average duplication times for mESCs, EB-dcs, and cells isolated after osteogenic induction in 2D- and 3D-cultures. Murine ESCs show the characteristic exponential growth known in mouse embryonic stem cells. The proliferation rate is lower in EB-dcs and even lower for the osteogenic systems, suggesting that cells undergo differentiation at a higher frequency.

14. Reveal the membrane staining using a chemiluminescent substrate reaction kit (luminol reagent). In order to do so, cover the membrane with the previously mixed reagent for 1 min at room temperature.

15. Pour off the excess of reagent and wrap the membrane with plastic wrap (avoid bubbles).

16. Monitor the membrane staining with a chemiluminescent chamber with a video camera (see Fig. 6).

3.6. Kinetic Growth Rate Determination

1. Isolate cells as indicated above (see Subheading 3.2).

2. Seed cells at a known density into 24-well plates, one plate for each cell type. Use mESCM/-LIF for EB-dcs and differentiated cells from 2D- and 3D-cultures, including control and osteogenic, and mESCM for the mESCs (see Note 20).

3. Harvest cells with 0.05% trypsin/EDTA from three wells every other day and count them to obtain a growth curve. The first day after seeding is taken as the initial growth point (0 h). Calculate standard deviation for each point.

4. From the growth curve, duplication times (the time in that cell population becomes twice) may be calculated (Fig. 7).

3.7. Lysis of 3D Cultures for RNA Extraction

RT-PCR is a very useful technique to monitor the expression of various osteogenic markers during ESC differentiation. Here, the experimental protocol to lyse the 3D-constructs is reported, which may be followed by the usual RNA extraction, cDNA synthesis, and the actual RT-PCR analysis (for details see Chapter 5).

1. Aliquot 0.5-mL TRIzol® or alternative RNA lysis buffer into RNase-free microcentrifuge tubes (1.5 mL, as many as you have samples).

2. Carefully extract the culture medium from the inserts and wells of the samples.

3. Transfer the 3D-cultures to the tubes with the lysis reagent. Use a spatula to help handling the constructs.

4. Disrupt the sample by pipetting up and down with a micropipette (see Note 21).

5. The homogenized samples may be stored at −80°C for at least 1 year or can sit at room temperature for several hours to proceed with the RNA extraction according to the manufacturer's instructions of the respective kits used.

4. Notes

1. For an alternative composition of osteogenic medium for mESCs, see Chapters 9 and 14.

2. The peptide sequence of RAD16-I is AcN-RADARAD-ARADARADA-$CONH_2$.

3. Due to the toxicity of PFA, take extreme care when handling it. Wear protective gloves and a mask and perform the preparation in a fume hood.

4. EB-dcs are extracted from EBs when these show reduced Oct-4 expression, since this indicates the loss of cellular pluripotency and subsequent differentiation into embryonic tissues, including ectoderm, mesoderm, and endoderm, as previously described (5, 6). Oct-4 expression is easily monitored in different stages of the process by visual inspection of GFP expression through the fluorescent microscope (Fig. 8a). In order to confirm the fluorescent signal (from GFP) from mESCs and EB-dc with Oct-4 expression, colonies from these stages were immunostained with an anti-Oct-4 antibody resulting in colocalization of GFP with Oct-4, as expected (Fig. 8b). In addition, Western blot analysis was performed to follow the Oct-4 expression during the entire differentiation process (Fig. 8c), confirming the results from the immunostain.

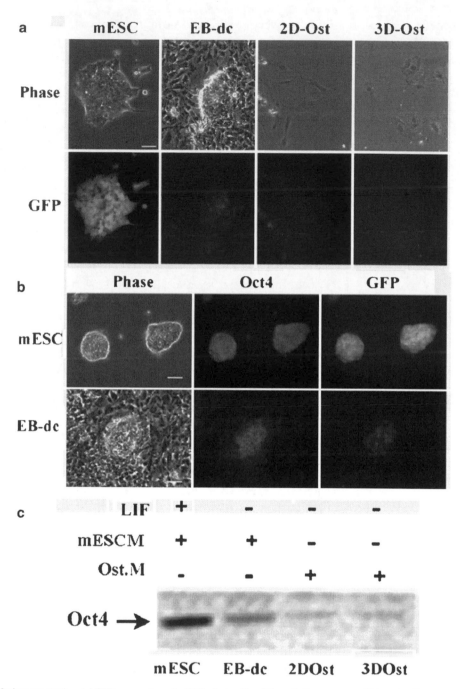

Fig. 8. Marker expression. (a) GFP expression of mESCs during the differentiation process. (b) Immunofluorescent staining of mESCs and EB-dcs with Oct-4, compared to GFP expression. (c) Western blot analysis of Oct-4 during the differentiation process. All three analyses show a strong decrease of Oct-4 expression after EB formation in EB-derived cells (EB-dcs), although EB-dc cultures still present a remaining population of cells expressing Oct-4. At the end of the osteogenic induction, cells from both 2D- and 3D-cultures (2D-Ost and 3D-Ost) do not show any GFP expression, suggesting that cells are fully differentiated. *Bar* 50 μm.

5. Consider that 2×10^5 cells are needed for each encapsulation, but the cell number obtained may decrease after the next centrifugation. For better accuracy, count cells again after the next centrifugation once they will be resuspended in sucrose solution.

6. Sucrose solution is isotonic and nonionic at the same time. Therefore, it allows viable cells while avoiding peptide gelling during the mixing process.

7. It is recommended to sonicate the peptide solution for approximately 5 min before use. Do not sonicate much longer, since the material tends to become stiffer.

8. Cells may be also encapsulated in a composite of the self-assembling peptide and hydroxyapatite particles (or other desired calcium phosphate), which has been shown to promote enhancement of osteogenic differentiation (31). To do so, mix hydroxyapatite particles with the commercial liquid peptide at a ratio of 1:1 and homogenize by pipetting and subsequent sonication. Use this mixture to prepare the 0.5% (w/v) peptide, which is then mixed with the cells.

9. Since peptide solution has a pH value as low as 3.5, try to minimize the time that cells are in this hostile condition. Quickly mix the cell suspension with the liquid peptide by pipetting approximately ten times, load the inserts, and equilibrate with the medium as soon as possible. Be careful not to create bubbles.

10. During this time, the higher ionic strength and the neutral pH of the medium induce the spontaneous self-assembly of the peptide as the medium diffuses through the peptide from the bottom up (32). Do not skip this waiting time, as the addition of medium on top of the suspension when the gel is not properly formed increases the chances of getting a broken scaffold.

11. The addition of medium in consecutive small portions favors the leaching of the sucrose. The remaining medium in the well, which is rich in sucrose, is then aspirated and replaced with fresh medium.

12. Medium change is performed every day by removing 0.5 mL of medium from the well and adding 0.5 mL of fresh mESCM/-LIF into the insert.

13. The difference in cell numbers seeded in 2D- versus 3D-cultures maintains more similar cell densities between the systems. As such, cells at a density of 2×10^6 cells in 3D present a similar cell-to-cell distance as 2×10^5 cells in 2D.

14. The culture of these EB-dcs cells for longer than 2 days before the addition of the osteogenic medium seemed to lead to a

delay in osteogenic commitment (32). Our observations support the notion that osteogenic specification is time-dependently regulated by Wnt/beta-catenin signaling as put forward in a recent review (33), as both inducers in our osteogenic medium affect nuclear activity of beta-catenin (34, 35).

15. Since the peptide construct is quite fragile and difficult to transfer into another well without breaking, the washing might be performed in the following way. Attach a silastic tube (10-cm length, 9-mm diameter) to the top of the insert and fill it up with water. Let the water drain through the construct overnight.

16. In order to easily manipulate the 3D-cultures, a few drops of 2% (w/v) agarose solution may be added to the top of the constructs. After cooking agarose in the microwave, let the solution cool down to 50–60°C before you pour it onto the samples. Then, leave the constructs at room temperature and wait until gelation.

17. Do not attempt to use PBS for washing steps as precipitation will occur.

18. Isolation of cells after osteogenic induction from either 2D- or 3D-cultures is necessary to avoid interferences from mineralized calcium in the detection of the mentioned markers.

19. For a variation of this protocol with smaller volumes in microwell plates, see Chapter 9.

20. A small fraction of the cells, from either EB or osteogenic differentiation cultures in 2D and 3D, appear to develop into GFP[+] colonies with an ESC-like phenotype. To study the frequency of appearance of these GFP[+]/ESC-like colonies, total cells from each culture condition can be isolated, counted, and subcultured in mESCM or mESCM/-LIF at 5×10^3 cells/well in regular 24-well culture plates. After several days in culture, GFP[+]/ESC-like colonies are identified and counted using the fluorescence microscope. Data can be expressed as a percentage of the number of ESC-like colonies per 5,000 initial cultured cells as shown in Fig. 9.

21. If this is not enough to homogenize the sample, use a syringe (20G × 1½ needle).

Acknowledgments

NMB acknowledges financial support from DURSI (Generalitat de Catalunya) and the European Social Foundation (2006FI 00447). CES was supported by the Translational Centre for Regenerative Medicine (TRM), University of Leipzig, Germany, Award 1098 SF.

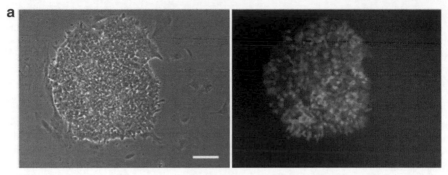

	Frequency of mES-like colonies (%)	
	-LIF	**+LIF**
2D Control	0.003 (±0.008)	0.030 (±0.024)
2D Osteogenic	0.007 (±0.016)	0.003 (±0.008)
3D Control	0.023 (±0.025)	0.715 (±0.072)
3D Osteogenic	0.005 (±0.009)	0.185 (±0.061)

Fig. 9. The 3D-culture system promotes the development of a stem cell niche. (a) Example of an ESC-like GFP+ colony observed after osteogenic induction. Bar is 100 μm. (b) Frequencies of ESC-like GFP+ colonies found in 2D- and 3D-cultures of EB-dc maintained in mESCM/-LIF (control) or in osteogenic medium (osteogenic) for 20 days. Standard deviations are given in parentheses. Frequencies of these ESC-like GFP+ colonies are very low. However, the 3D-culture system seems to enhance their presence in comparison with the classical 2D-culture system when cells are maintained in the presence of LIF.

References

1. Evans, M. J. and Kaufman, M. H. (1981) Establishment in culture of pluripotential cells from mouse embryos. *Nature* **292**, 154–156.

2. Martin, G. R. (1981) Isolation of a pluripotent cell line from early mouse embryos cultured in medium conditioned by teratocarcinoma stem cells. *Proc. Natl. Acad. Sci. USA* **78**, 7634–7638.

3. Smith, A. G., Heath, J. K., Donaldson, D. D., Wong, G. G., Moreau, J., Stahl, M., et al. (1988) Inhibition of pluripotential embryonic stem cell differentiation by purified polypeptides. *Nature* **336**, 688–690.

4. Williams, R. L., Hilton, D. J., Pease, S., Willson, T. A., Stewart, C. L., Gearing, D. P., et al. (1988) Myeloid leukaemia inhibitory factor maintains the developmental potential of embryonic stem cells. *Nature* **336**, 684–687.

5. Bradley, A., Evans, M., Kaufman, M. H., and Robertson, E. (1984) Formation of germ-line chimaeras from embryo-derived teratocarcinoma cell lines. *Nature* **309**, 255–256.

6. Wobus, A. M. (2001) Potential of embryonic stem cells. *Mol. Aspects Med.* **22**, 149–164.

7. Brustle, O., Jones, K. N., Learish, R. D., Karram, K., Choudhary, K., Wiestler, O. D., et al. (1999) Embryonic stem cell-derived glial precursors: a source of myelinating transplants. *Science* **285**, 754–756.

8. Lee, S. H., Lumelsky, N., Studer, L., Auerbach, J. M., and McKay, R. D. (2000) Efficient generation of midbrain and hindbrain neurons from mouse embryonic stem cells. *Nat. Biotechnol.* **18**, 675–679.

9. Wichterle, H., Lieberam, I., Porter, J. A., and Jessell, T. M. (2002) Directed differentiation of embryonic stem cells into motor neurons. *Cell* **110**, 385–397.

10. Drab, M., Haller, H., Bychkov, R., Erdmann, B., Lindschau, C., Haase, H., et al. (1997) From totipotent embryonic stem cells to

spontaneously contracting smooth muscle cells: a retinoic acid and db-cAMP in vitro differentiation model. *FASEB J.* **11**, 905–915.

11. Yamashita, J., Itoh, H., Hirashima, M., Ogawa, M., Nishikawa, S., Yurugi, T., et al. (2000) Flk1-positive cells derived from embryonic stem cells serve as vascular progenitors. *Nature* **408**, 92–96.

12. Schmitt, R. M., Bruyns, E., and Snodgrass, H. R. (1991) Hematopoietic development of embryonic stem cells in vitro: cytokine and receptor gene expression. *Genes Dev.* **5**, 728–740.

13. Wiles, M. V. and Keller, G. (1991) Multiple hematopoietic lineages develop from embryonic stem (ES) cells in culture. *Development* **111**, 259–267.

14. Dani, C., Smith, A. G., Dessolin, S., Leroy, P., Staccini, L., Villageois, P., et al. (1997) Differentiation of embryonic stem cells into adipocytes in vitro. *J. Cell. Sci.* **110**, 1279–1285.

15. Kramer, J., Hegert, C., Guan, K., Wobus, A. M., Muller, P. K., and Rohwedel, J. (2000) Embryonic stem cell-derived chondrogenic differentiation in vitro: activation by BMP-2 and BMP-4. *Mech. Dev.* **92**, 193–205.

16. zur Nieden, N. I., Kempka, G., and Ahr, H. J. (2003) In vitro differentiation of embryonic stem cells into mineralized osteoblasts. *Differentiation* **71**, 18–27.

17. Wu, L. N., Genge, B. R., Dunkelberger, D. G., LeGeros, R. Z., Concannon, B., and Wuthier, R. E. (1997) Physicochemical characterization of the nucleational core of matrix vesicles. *J. Biol. Chem.* **272**, 4404–4411.

18. Armant, D. R. (2005) Blastocysts don't go it alone. Extrinsic signals fine-tune the intrinsic developmental program of trophoblast cells. *Dev. Biol.* **280**, 260–280.

19. Imai, S., Kaksonen, M., Raulo, E., Kinnunen, T., Fages, C., Meng, X., et al. (1998) Osteoblast recruitment and bone formation enhanced by cell matrix-associated heparin-binding growth-associated molecule (HB-GAM). *J. Cell. Biol.* **143**, 1113–1128.

20. Maroto, M., Dale, J. K., Dequeant, M. L., Petit, A. C., and Pourquie, O. (2005) Synchronised cycling gene oscillations in presomitic mesoderm cells require cell–cell contact. *Int. J. Dev. Biol.* **49**, 309–315.

21. Stains, J. P. and Civitelli, R. (2005) Cell-to-cell interactions in bone. *Biochem. Biophys. Res. Commun.* **328**, 721–727.

22. Geiger, B., Bershadsky, A., Pankov, R., and Yamada, K. M. (2001) Transmembrane crosstalk between the extracellular matrix–cytoskeleton crosstalk. *Nat. Rev. Mol. Cell. Biol.* **2**, 793–805.

23. Ferrera, D., Poggi, S., Biassoni, C., Dickson, G. R., Astigiano, S., Barbieri, O., et al. (2002) Three-dimensional cultures of normal human osteoblasts: proliferation and differentiation potential in vitro and upon ectopic implantation in nude mice. *Bone* **30**, 718–725.

24. Kale, S., Biermann, S., Edwards, C., Tarnowski, C., Morris, M., and Long, M. W. (2000) Three-dimensional cellular development is essential for ex vivo formation of human bone. *Nat. Biotechnol.* **18**, 954–958.

25. Genove, E., Shen, C., Zhang, S., and Semino, C. E. (2005) The effect of functionalized self-assembling peptide scaffolds on human aortic endothelial cell function. *Biomaterials* **26**, 3341–3351.

26. Semino, C. E., Kasahara, J., Hayashi, Y., and Zhang, S. (2004) Entrapment of migrating hippocampal neural cells in three-dimensional peptide nanofiber scaffold. *Tissue Eng.* **10**, 643–655.

27. Semino, C. E., Merok, J. R., Crane, G. G., Panagiotakos, G., and Zhang, S. (2003) Functional differentiation of hepatocyte-like spheroid structures from putative liver progenitor cells in three-dimensional peptide scaffolds. *Differentiation* **71**, 262–270.

28. Viswanathan, S., Benatar, T., Mileikovsky, M., Lauffenburger, D. A., Nagy, A., and Zandstra, P. W. (2003) Supplementation-dependent differences in the rates of embryonic stem cell self-renewal, differentiation, and apoptosis. *Biotechnol. Bioeng.* **84**, 505–517.

29. Lumelsky, N., Blondel, O., Laeng, P., Velasco, I., Ravin, R., and McKay, R. (2001) Differentiation of embryonic stem cells to insulin-secreting structures similar to pancreatic islets. *Science* **292**, 1389–1394.

30. McGee-Russell, S. M. (1958) Histochemical methods for calcium. *J. Histochem. Cytochem.* **6**, 22–42.

31. Garreta, E., Gasset, D., Semino, C., and Borros, S. (2007) Fabrication of a three-dimensional nanostructured biomaterial for tissue engineering of bone. *Biomol. Eng.* **24**, 75–80.

32. Garreta, E., Genove, E., Borros, S., and Semino, C. E. (2006) Osteogenic differentiation of mouse embryonic stem cells and mouse embryonic fibroblasts in a three-dimensional self-assembling peptide scaffold. *Tissue Eng.* **12**, 2215–2227.

33. Davis, L. A. and zur Nieden, N. I. (2008) Mesodermal fate decisions of a stem cell: the Wnt switch. *Cell. Mol. Life Sci.* **65**, 2658–2674.

34. zur Nieden, N. I., Price, F. D., Davis, L. A., Everitt, R., and Rancourt, D. E. (2007) Gene array analysis on mixed ES cell populations: a biphasic role for beta-catenin in osteogenesis. *Mol. Endocrinol.* **21**, 674–685.

35. Smith, E. and Frenkel, B. (2005) Glucocorticoids inhibit the transcriptional activity of LEF/TCF in differentiating osteoblasts in a glycogen synthase kinase-3β-dependent and -independent manner. *J. Biol. Chem.* **280**, 2388–2394.

Chapter 16

Methods for Investigation of Osteoclastogenesis Using Mouse Embryonic Stem Cells

Motokazu Tsuneto, Toshiyuki Yamane, and Shin-Ichi Hayashi

Abstract

Investigation of osteoclastogenesis in vivo, especially in early development, has proven difficult because of the accessibility of these early embryonic stages. Our ability to culture embryonic stem cells (ESCs) in vitro has overcome this difficulty as these versatile cells can be expanded endlessly. Thus, the whole process of osteoclastogenesis can be monitored in these cultures through the microscope and with the help of molecular biology techniques. We have developed two methods to induce osteoclasts, the bone matrix remodeling cells, from murine ESCs. Surprisingly, one of these induction methods produces osteoclasts, osteoblasts, and also endothelial cells in the same culture dish. Hence, it is likely that ESCs in culture mimic the in vivo development of osteoclasts.

Key words: Embryonic stem cell, Osteoclast, Osteoblast, Endothelial cell

1. Introduction

Homeostasis of bone is maintained with a balance between production and resorption of bone matrix by two different cell types, the osteoblasts and the osteoclasts, respectively, being so diverse that they stem from different lineages. Osteoclasts are derived from hematopoietic stem cells (1) and are indispensable for bone remodeling, bone marrow formation, and tooth eruption (2–4). In case space in the bone marrow cavity is limited due to osteoclast dysfunction, the site of hematopoiesis shifts to extramedullary organs. B lymphopoiesis in the marrow is most sensitively and severely affected by the limitation of marrow space (5). Thus, osteoclasts are unique to an appropriate hematopoietic microenvironment in bone marrow.

The hematopoietic origin of osteoclasts is also verified by the observation that no osteoclasts are generated from embryonic

Nicole I. zur Nieden (ed.), *Embryonic Stem Cell Therapy for Osteo-Degenerative Diseases*, Methods in Molecular Biology, vol. 690, DOI 10.1007/978-1-60761-962-8_16, © Springer Science+Business Media, LLC 2011

stem cells (ESCs) lacking SCL/tal-1, a transcription factor essential for the development of all blood cell lineages (6–8). Furthermore, addition of VEGFR-1/Fc in early days of ESC culture, which is expected to disrupt Flk-1 function at the stage of mesoderm or hemangioblast commitment, a postulated common progenitor for hematopoietic cells and endothelial cells, impairs the development of osteoclasts and endothelial cells (9, 10).

Mice deficient for the transcription factor PU.1 specifically lack osteoclasts and B lymphocytes, while other blood cell lineages are not severely affected (11, 12). The fact that exogenous PU.1 expression in SCL/tal-1-deficient ESCs rescues osteoclastogenesis (13) suggests that PU.1 expression is a key event in osteoclast development after hematopoietic specification regulated by the SCL/tal-1 transcription factor has occurred.

Osteoclasts exist only in bone surfaces, while osteoclast precursors can be detected also in spleen and peritoneal cavity, as well as in bone marrow (14, 15). It is still unknown whether these extramedullary precursors go to the bone surfaces to differentiate into osteoclasts in the bone.

Osteoclasts are also differentiated from colony-forming unit granulocyte macrophage colony-stimulating factor (CFU-GM), and CFU-M in bone marrow as well as hematopoietic stem cells (16). Cell sorting experiments indicated that cells positive for the stem cell factor c-kit and macrophage colony-stimulating factor (M-CSF) receptor (c-Kit$^+$c-Fms$^+$) or c-Kit$^+$c-Fms$^-$ cells are the main population of osteoclast progenitors in bone marrow, although osteoclast precursors in peritoneal cavity are predominantly found in c-Fms$^+$ cells (16, 17). Mature macrophages and dendritic (18) cells are also reported to have the potential to differentiate into osteoclasts. Therefore, many kinds of hematopoietic cells at different stages might possess a potential to differentiate into osteoclasts.

Two cytokines, M-CSF and RANK ligand (RANKL), are essential for osteoclast formation (19, 20). $Csf1^{op}/Csf1^{op}$ mice carrying point mutations in the M-CSF gene and therefore lacking functional M-CSF have severe defects in osteoclastogenesis, which results in failure of tooth eruption and bone marrow cavity formation. M-CSF seems to be important especially for the survival of the osteoclast lineage, as osteoclastogenesis in $Csf1^{op}/Csf1^{op}$ mouse can partially be rescued by enforced expression of $Bcl-2$, an antiapoptotic gene (21). Mice deficient in RANKL, which is normally produced by osteoblasts and stromal cells, and its receptor RANK completely lack ostcoclasts (19, 20).

RANKL can be substituted by tumor necrosis factor alpha (TNFα) (22, 23) in culture. Furthermore, lipopolysaccharide (LPS) can substitute RANKL if cells from mouse carrying a deficiency of protein tyrosine phosphatase Src homology 2-domain phosphatase-1 ($Ptpn6^{me-v}/Ptpn6^{me-v}$) are used as osteoclast precursors (24). RANKL, TNFα, and LPS share nuclear factor κB (NFκB) and TNF receptor-associated factor 6 (TRAF6) in their downstream signaling cascade. Consistent with these observations, TNFα and LPS are known to induce osteolysis in vivo (25, 26). Recently, it has been reported that the master regulator of osteoclastogenesis is thought to be nuclear factor of activated T-cells, cytoplasmic, calcineurin-dependent 1 (*Nfatc1*), which is induced following RANKL stimulation (27). RANKL also induces and activates NFATc1 through calcium signaling, and calcineurin inhibitors such as FK506 and cyclosporin A strongly inhibit osteoclastogenesis. Therefore, these downstream molecules might become targets to cure patients with osteolysis (26).

We have established three alternative culture systems for osteoclastogenesis from undifferentiated ESCs utilizing cocultures with stromal cell lines (28). The system is useful in investigating the function of critical factors and genes involved in the process of osteoclastogenesis. In one system, the hematopoietic cell lineage is induced from ESCs by means of coculture with OP9 cells, which is a stromal cell line derived from fetal calvaria of the M-CSF-deficient *Csf1*op/*Csf1*op mouse (29). After induction of hematopoietic cells including osteoclast precursors for 5 days on OP9, the cells are transferred onto a bone marrow-derived ST2 stromal cell line to induce the formation of mature osteoclasts in the presence of 1α,25-(OH)$_2$ vitamin D$_3$ (VD$_3$) and dexamethasone, a synthetic glucocorticoid (30, 31). In this maturation step, M-CSF is constitutively supplied from ST2, and RANKL expression is induced by VD$_3$. Expression of a decoy receptor for RANK, osteoprotegerin (OPG), on the other hand is blocked by dexamethasone. Osteoclasts can be detected in cultures after 6 days of culture on ST2. This culture is referred to "2-step culture" (Figs. 1 and 2). Insertion of one more step on OP9 without

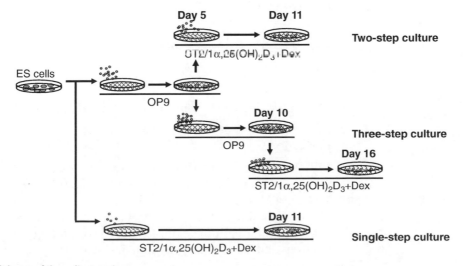

Fig. 1. Scheme of the culture systems.

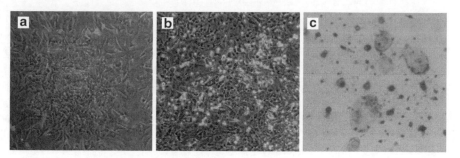

Fig. 2. Appearance of ESC cultures. (**a**) A single ESC colony at day 5. (**b**) Hematopoietic cell clusters at day 10. (**c**) Osteoclasts at day 16 identified with TRAP staining.

addition of VD_3 and dexamethasone between the induction step on OP9 and the maturation step on ST2 allows for a 20-fold expansion of osteoclast progenitors during this period. This culture system is referred to as "3-step culture."

In the third culture system, undifferentiated ESCs are directly seeded on ST2 cells and cultivated in the presence of VD_3 and dexamethasone for 11 days. This is referred to as the "single-step culture" (Fig. 1), and this culture system enables us to investigate the complete osteoclastogenic process without any manipulations except for regular medium changes. In this culture system, a single ESC forms a colony consisting of several cell lineages including osteoblasts and endothelial cells. Osteoclasts, which express tartrate-resistant acid phosphatase (TRAP), appear at the edge of colonies, and osteoblasts, which express alkaline phosphatase (Alk Phos), are formed close to the osteoclasts (Fig. 3a, c, c′) (32). Endothelial cells are formed radially within colonies (Fig. 3b). Therefore, this third culture system allows the development of some microenvironmental element resembling the bone marrow cavity.

2. Materials

2.1. Maintenance and Preparation of OP9 Cells

1. 100-mm ø dishes, i.e. Greiner Bio-One or similar.
2. Fetal calvaria-derived OP9 cells (American Type Culture Collection).
3. OP9 medium: Alpha Minimum Essential Medium (αMEM) without ribonucleosides and deoxyribonucleosides (Invitrogen) with 2 mM L-glutamine and 1.5 g/L sodium bicarbonate also containing 20% fetal bovine serum (FBS) (JRH Biosciences). Medium is supplemented with 50 µg/mL streptomycin and 50 U/mL penicillin and stored at 4°C.
4. 1× PBS (Invitrogen).

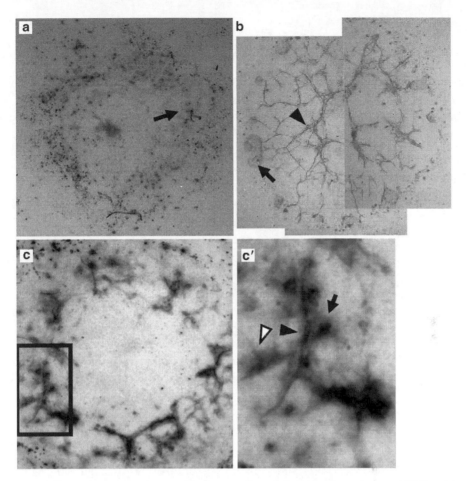

Fig. 3. Appearance of single colonies with various types of staining in the single-step cultures. (**a**) TRAP staining: *arrow*: TRAP positive cells. (**b**) CD31 positive and TRAP positive cells. *Arrow*: TRAP positive cells, *arrowhead*: CD31 positive cells. (**c, c′**) ALP, TRAP and CD31 staining in a single colony: CD31 positive cells (*arrowhead*) are *round-shaped* and TRAP positive cells (*arrow*) are present around them. (**c′**) is a higher magnification of (**c**). *White arrowhead* shows ALP positive cells.

5. 0.1% trypsin/Ethylenediaminetetraacetic acid (EDTA, Invitrogen). Store at 4°C.

6. OP9 freezing medium: 90% OP9 medium plus 10% dimethyl sulfoxide.

7. 2-mL cryovials, i.e. Nunc Nalgene.

8. Falcon tubes (BD Biosciences) or similar, 15 mL.

2.2. Maintenance and Preparation of ST2 Cells

1. ST2 cells (bone marrow stroma cell-derived), available from RIKEN Cell Bank.

2. ST2 medium: RPMI medium 1640 (RPMI-1640) containing 5×10^{-5} M 2-mercaptoethanol and 5% FBS. ST2 medium is supplemented with 50 µg/mL streptomycin and 50 U/mL penicillin.

3. 1× PBS.

4. 0.05% trypsin/EDTA (Invitrogen). Store at 4°C.

5. ST2 freezing medium: 90% FBS plus 10% dimethyl sulfoxide.

2.3. ESC Differentiation into Osteoclasts

1. Mouse leukemia inhibitory factor (mLIF): mouse LIF is derived from medium conditioned by Chinese hamster ovary (CHO) cells transfected with a LIF expression vector. Commercial mLIF (ESGRO™, Millipore) can be substituted for this conditioned medium at more than 1,000 U/mL for maintaining ESCs.

2. ESC Medium: Dulbecco's Modified Eagle's Medium (DMEM) containing 15% heat-inactivated FBS (JRH Biosciences) (see Notes 1 and 2), 2 mM L-glutamine (200 mM), 0.1 mM 100× MEM nonessential amino acids (Invitrogen), 1,000 U/mL CHO-LIF, and 0.1 mM 2-mercaptoethanol.

3. 0.25% trypsin in PBS with 0.5 mM EDTA: commercial 2.5% trypsin (Invitrogen) is divided into small aliquots in 15-mL tubes and stored at –20°C. When an aliquot is used, it is thawed, diluted with 1× PBS, and supplemented with EDTA for 0.25% trypsin/0.5 mM EDTA. After dilution, it is stored at 4°C.

4. 1α,25-$(OH)_2$ vitamin D_3 (VD_3): as VD_3 is unstable, it is diluted to 10^{-4} M with ethanol and divided into small aliquots in cryostat tubes. The tubes are stored in the dark at –80°C.

5. Dexamethasone: dexamethasone is stored at 10^{-3} M in ethanol at 4°C.

6. Osteoclast Differentiation Medium A (OsDM-A): minimum essential medium alpha medium (αMEM) supplemented with 20% fetal bovine serum. The medium also contains streptomycin and penicillin at final concentrations of 50 μg/mL and 50 U/mL, respectively. The medium is stored at 4°C.

7. Osteoclast Differentiation Medium B (OsDM-B): same as OsDM-A, but 10% FBS instead of 20%. Additionally contains 10^{-8} M VD_3 and 10^{-7} M dexamethasone.

2.4. Identification of Osteoclasts

1. Sodium tartrate dehydrate.

2. Sodium acetate trihydrate.

3. Naphthol AS-MX phosphate, store at –20°C.

4. Fast red violet LB salt, store at –20°C.

5. 10% Formalin solution (3.7% formaldehyde): Add 2 mL of 37% formaldehyde to 18 mL of 1× PBS.

6. Ethanol/acetone solution: combine ethanol and acetone to equal parts.

7. TRAP staining buffer: to 450 mL distilled water, add 5.75 g sodium tartrate dehydrate (final concentration 59.3 M), 6.8 g sodium acetate trihydrate to a final concentration of 165.7 M, and 250 mg Naphthol AS-MX phosphate. Stir TRAP solution until the solutes have dissolved. Adjust the pH to 5.0 with acetic acid and the volume to 500 mL with distilled water. Store TRAP staining buffer at 4°C in the dark. Just before performing TRAP staining, dissolve fast red violet LB salt in TRAP staining solution at a final concentration of 0.5 mg/mL.

8. 1× PBS.

2.5. Immunocyto-chemistry for CD31

1. 4% PFA: dissociate 8 g of paraformaldehyde (PFA) in 200 mL of 1× PBS (Invitrogen), with heating at 70°C. Do not boil. Add 400 μL NaOH and then add 700 μL of 2N HCl to dissociate the solutes completely. Confirm that the pH is in the range of 7–8. If not, adjust pH with NaOH or HCl.

2. 0.3% H_2O_2 in methanol.

3. Ethanol.

4. DAB reagent sct (i.e. Kirkegaard & Perry Laboratories, Inc.).

5. Block Ace (i.e. Dainippon Seiyaku): It can be substituted for normal goat serum.

6. Primary antibody: monoclonal rat anti-mouse CD31 IgG (clone 390, Immunotech, Marseille, France). Dilute to 5 μg/mL with Block Ace, just prior to use.

7. Secondary antibody: horseradish peroxidase (HRP)-conjugated goat anti-rat IgG (ICN Pharmaceuticals, Inc.). Dilute to a 35 μg/mL working solution with Block Ace.

8. 1 M Tris–HCl (pH 8.5).

9. PBS-T: 1× PBS supplemented with 0.05% Tween 20 or polyoxyethylene (20) sorbitan monolaurate. PBS-T is stored at 4°C.

3. Methods

All cultures are maintained at 37°C in 95% air/5% CO_2 in a humidified incubator. Medium and PBS should be pre-warmed to 37°C.

3.1. Maintenance and Preparation of OP9 Cells

M-CSF-deficient fetal calvaria-derived OP9 cells are cultured in 100-mm ø dishes in OP9 medium.

3.1.1. Thawing of OP9 Cells

1. Quickly thaw frozen cells in a 37°C water bath until just a small bit of frozen solution remains in the tube.

2. Add 1 mL of medium gently to the cell suspension drop by drop, transfer the suspension to a 15-mL centrifuge tube, and gently add 7 mL of medium.

3. Centrifuge at $200 \times g$ for 5 min at 4°C.

4. Aspirate the supernatant and suspend the cell pellet in the appropriate medium.

5. Seed cells into culture dishes.

3.1.2. Maintenance Culture

1. Split OP9 cells every 3 days when they reach 70% confluency (Fig. 4c). Wash them with 1× PBS twice, then add 1 mL of 0.1% trypsin/EDTA and incubate at 37°C for 5 min.

2. Add 7 mL of OP9 medium and dissociate the cells by vigorous pipetting.

3. Transfer to a 15-mL Falcon tube and spin cells down by centrifugation at $200 \times g$ for 5 min.

4. Pour off or aspirate off the supernatant and resuspend the pelleted cells using fresh OP9 medium. Split cells 1:4 into new culture dishes (see Note 3).

5. When OP9 cells reach 70% confluency, you may use them either to initiate a differentiation experiment or to freeze them down.

6. For freezing, OP9 cells are recovered after centrifugation and resuspended in OP9 freezing medium. All cells from one 100-mm dish should be resuspended in 2 mL of freezing medium and divided into two cryovials.

7. For ESC differentiation, seed cells from one 100-mm dish to a six-well culture plate 2 days before starting differentiation of ESCs. After OP9 cells reach confluency, they are ready to use.

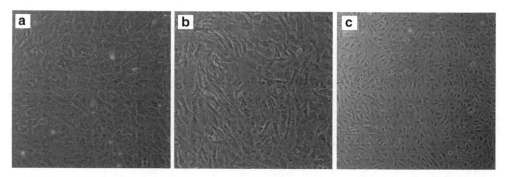

Fig. 4. Morphology of coculture cell lines. (**a**) Stromal cell line ST2. (**b, c**) Stromal cell line OP9. (**c**) a lower magnification of (**b**).

3.2. Maintenance and Preparation of ST2 Cells

Bone marrow-derived ST2 cells are cultured in 100-mm cell culture dishes in ST2 medium and are regularly passed every 3 days.

1. Thaw ST2 cells as described for OP9 cells in Subheading 3.1.1.

2. When ST2 cells reach 100% confluency, wash them with 1× PBS once, then add 1 mL of 0.05% trypsin/EDTA and incubate at 37°C for 2 min.

3. Add 7 mL of ST2 medium and dissociate the cells by vigorous pipetting. Transfer to a 15-mL Falcon tube, centrifuge at $200 \times g$ for 5 min, pour off or aspirate off the supernatant, and resuspend pellet in 4 mL of fresh ST2 medium.

4. Split ST2 cells at a 1:4 ratio into new culture dishes. In order to do so, add 1 mL of cell suspension to a new dish previously prepared with 9 mL of ST2 medium.

5. Confluent cells may be used to initiate a differentiation experiment with ESCs or to freeze ST2 stocks. After trypsinization, seed half of the confluent cells from one 100-mm dish to all wells of a 24-well culture plate the day before seeding of undifferentiated ESCs or differentiated ESCs. Cells are ready to be used when they reach confluency.

6. For freezing, ST2 cells are taken up in ST2 freezing medium after centrifugation. Cells from one 100-mm dish should be divided into two cryovials.

3.3. Differentiation of ESCs in 2- and 3-Step Cultures

Maintain ESCs on mouse embryonic fibroblasts as described in Chapter 2. This protocol will guide you through differentiation of undifferentiated ESCs into osteoclasts in 2- or 3-step cultures.

1. Prepare OP9 cells in a six-well culture plate as described under Subheading 3.1. Culture OP9 cells until the cells reach 100% confluency.

2. Grow ESCs to 70% confluency, aspirate ESC medium, and wash cells three times in 1× PBS.

3. Remove the PBS and add 1 mL 0.25% trypsin/EDTA. Incubate at 37°C for 5 min to loosen the cells from the plastic.

4. Stop the trypsinization by adding 7 mL of ESC medium. Pipette up and down to break up remaining cell clumps. Transfer the single cell suspension to a 15-mL Falcon tube and centrifuge at $200 \times g$ for 5 min.

5. After aspirating the supernatant, suspend the ESCs in OsDM-A.

6. Count viable cells by trypan blue exclusion as described in Chapter 8.

7. Seed 10^4 cells/well on OP9 cells (see Notes 4 and 5).

8. On day 3, replace half of the medium with fresh medium.

9. On day 5, wash the cells three times with 1× PBS and trypsinize them with 0.5 mL of 0.25% trypsin/EDTA per well by incubating at 37°C for 5 min (see Note 6).

10. After incubation, add 3.5 mL/well of OsDM-A to the wells and dissociate the cells into single cells by vigorous pipetting.

11. Transfer the suspension to a 15-mL Falcon tube and centrifuge at 250×g for 5 min at 4°C. When 2-step culturing is performed, go to step 17.

12. Aspirate the supernatant and suspend the cells in OsDM-A.

13. Seed 10^5 cells/well on freshly prepared OP9 cells in six-well culture plates (see Note 7).

14. On day 8, replace half of the medium with fresh medium.

15. On day 10, add 2 mL of OsDM-A, dissociate the cells into single cells by vigorous pipetting and transfer the suspension to a 15-mL centrifuge tube (see Note 8).

16. Leave it stand, to remove debris, for 5 min at 25°C. After the debris has settled, transfer the supernatant to a fresh 15-mL tube. Centrifuge at 250×g for 5 min at 4°C.

17. Aspirate the supernatant and suspend the cells in OsDM-B.

18. Prepare a suspension of 10^3 cells/mL in OsDM-B. Seed 1 mL/well of this cell suspension on ST2 cells previously prepared in a 24-well culture plate.

19. Three days later, change the medium to fresh OsDM-B.

20. After 6 days of the culture on ST2, perform TRAP staining to identify osteoclasts.

3.4. Induction of Osteoclasts, Osteoblasts, and Endothelial Cells from Undifferentiated ESCs in Single-Step Culture

At the surface of bone matrix, osteoclasts and osteoblasts reside in close proximity to each other. Similarly, colonies derived from single ESCs contain both osteoclasts and osteoblasts in this culture system. Osteoclasts and osteoblasts appear at the periphery of the colony (Fig. 3). Hematopoietic c-Kit-positive cells first appear at day 4 and TRAP-positive cells first appear at day 8 of the differentiation. This timing is similar to that observed during in vivo development, suggesting that the ESC differentiation system spatiotemporally mimics in vivo development. Specifically, in this culture system, αMEM/10%FBS (OsDM-B) is used to induce ESCs to differentiate into osteoclasts (see Note 9).

1. Prepare ST2 cells in a 24-well plate. Culture ST2 cells until the cells reach 100% confluency in the plate.

2. Grow ESCs to 70% confluency and dissociate them into single cells using 0.25% trypsin/EDTA as described in

Subheading 3.3. After centrifugation, aspirate the supernatant and suspend the ESCs in OsDM-B.

3. Count the cell number and seed 50–100 cells/well in OsDM-B.

4. Change the medium every 2 or 3 days.

5. On day 11, perform the desired staining.

3.5. Detection of Osteoclasts

Osteoclasts will be detected by staining with tartrate-resistant acid phosphatase (TRAP).

1. Aspirate the medium and add 1 mL/well of 10% formalin to cover the cultured cells.

2. Incubate cells at 25°C for 10 min (see Note 10).

3. Aspirate the 10% formalin solution off the cells and wash once with 1× PBS.

4. After aspiration of 1× PBS, cover the cells with 0.5 mL of ethanol/acetone solution and incubate them at 25°C for just 1 min, then take off ethanol/acetone immediately.

5. Fill each well with 1× PBS. Aspirate PBS off the wells and wash one more time with 1× PBS.

6. Cover the cells with 250 μL/well of TRAP solution and let the plates stand for 5 min at 25°C.

7. Aspirate off the staining solution and soak the culture dish in tap water for more than 30 min to remove unspecific staining.

8. Osteoblasts may be detected with staining for alkaline phosphatase as described in Chapter 4 or by measuring the activity of this enzyme (see Chapter 17).

3.6. Immunohistochemical Staining of CD31

1. To fix cultured cells, add 4% PFA to each well and keep the cells on ice for 15 min.

2. After having aspirated PFA, add ice-cold 1× PBS to each well and leave the dishes on ice for 5 min. Perform this washing step three times.

3. Remove PBS and add 500 μL of 0.3% H_2O_2/methanol and leave it stand on ice for 10 min.

4. After aspirating the 0.3% H_2O_2/methanol, add ice-cold 100% ethanol and leave the dishes on ice for just 1 min.

5. Wash the cells twice with 1× PBS, add Block Ace, and leave the plates on ice for 20 min.

6. Add the primary antibody diluted in Block Ace and let stand at 4°C overnight.

7. Aspirate the primary antibody solution and wash cells twice with ice-cold PBS-T on ice for 5 min.

8. Overlay cells with HRP-conjugated secondary antibody working solution and leave the plates on ice for 1 h.

9. Aspirate the secondary antibody solution and wash cells twice with ice-cold PBS-T on ice for 5 min.

10. Aspirate PBS-T and cover well with freshly prepared DAB solution. DAB solution is included in the DAB Reagent Set. Leave at 25°C for 10 min and then add distilled water to stop the reaction. Observe cells through an inverted microscope.

4. Notes

1. FBS should be used that supports the growth of ESCs well and prevents them from differentiating. Prepare ESC Medium containing several lots of FBS and check the growth rate of ESCs cultivated in these preparations of medium. To check whether ESCs are maintained in an undifferentiated state, cultivate ESCs in ESC Medium without mLIF and perform ALP staining as described in Chapter 4.

2. To inactivate complements, FBS is heated in a water bath at 56°C for 30 min.

3. Take special care to split at the right ration. If the density of the subcultures is too low, the culture will not reach confluence. However, do not overgrow either. If very large cells appear in your culture, the cultures are overgrown and will not support the maintenance of hematopoietic cells.

4. Similar results are obtained by using three ESC lines: D3 (33), J1 (34), and CCE (35), in our coculture system with stromal cell lines.

5. Seeding mixtures of ESCs and mouse embryonic fibroblast feeder cells on OP9 or ST2 cells does not cause a problem.

6. On day 5, differentiated (Fig. 2a) and undifferentiated colonies are observed (36). The appearance of undifferentiated colonies is smooth, and the boundary of the cells is not clear. These colonies may comprise, at most, one-third of the total colonies.

7. During the culture, OP9 cells sometimes differentiate into adipocytes, but this does not influence the experimental outcomes.

8. Trypsinization is not necessary at this point.

9. The efficiency of the formation of colonies by ESCs varies from 1 to 40% depending on the lot of serum. We recommend seeding ESCs at a density that allows the formation of less than 20 colonies/well in 24-well plates. High colony

density prevents ESCs from differentiating into osteoclasts. A preliminary experiment should be performed to check the plating efficiency by seeding 10–1,000 cells/well.

10. While the cells are being fixed, you may add the fast red violet LB salt to TRAP staining solution.

Acknowledgments

We thank Fritz Melchers (Max Planck Institute, Berlin) for critical reading of this manuscript. M. Tsuneto is a fellow of the Alexander von Humboldt Foundation in Germany. This work was supported by grants of a Grant-in-Aid for Scientific Research (C) (20590400) from JSPS, twenty-first Century COE Program from MEXT, and from JST.

References

1. Mundy, G.R., and Roodman, G.D. (1987) Osteoclast ontogeny and function. *J. Bone Miner. Res.* **5**, 209–279.

2. Yoshida, H., Hayashi, S., Kunisada, T., Ogawa, M., Nishikawa, S., Okamura, H., et al. (1990) The murine mutation osteopetrosis is in the coding region of the macrophage colony stimulating factor gene. *Nature* **345**, 442–444.

3. Niida, S., Abe, M., Suemune, S., Yoshiko, Y., Maeda, N., and Yamasaki, A. (1997) Restoration of disturbed tooth eruption in osteopetrotic (op/op) mice by injection of macrophage colony-stimulating factor. *Exp. Anim.* **46**, 95–101.

4. Yoshino, M., Yamazaki, H., Yoshida, H., Niida, S., Nishikawa, S., Ryoke, K., et al. (2003) Reduction of osteoclasts in a critical embryonic period is essential for inhibition of mouse tooth eruption. *J. Bone Miner. Res.* **18**, 108–116.

5. Tagaya, H., Kunisada, T., Yamazaki, H., Yamane, T., Tokuhisa, T., Wagner, E.F., et al. (2000) Intramedullary and extramedullary B lymphopoiesis in osteopetrotic mice. *Blood* **95**, 3363–3370.

6. Shivdasani, R.A., Mayer, E.L., and Orkin S.H. (1995) Absence of blood formation in mice lacking the T-cell leukaemia oncoprotein tal-1/SCL. *Nature* **373**, 432–434.

7. Porcher, C., Swat, W., Rockwell, K., Fujiwara, Y., Alt, F.W., and Orkin, S.H. (1996) The T cell leukemia oncoprotein SCL/tal-1 is essential for development of all hematopoietic lineages. *Cell* **86**, 47–57.

8. Yamane, T., Kunisada, T., Yamazaki, H., Nakano, T., Orkin, S.H., and Hayashi, S.I. (2000) Sequential requirements for SCL/tal-1, GATA-2, macrophage colony-stimulating factor, and osteoclast differentiation factor/osteoprotegerin ligand in osteoclast development. *Exp. Hematol.* **28**, 833–840.

9. Okuyama, H., Tsuneto, M., Yamane, T., Yamazaki, H., and Hayashi, S. (2003) Discrete types of osteoclast precursors can be generated from embryonic stem cells. *Stem Cells* **21**, 670–680.

10. Choi, K., Kennedy, M., Kazarov, A., Papadimitriou, J.C., and Keller, G. (1998) A common precursor for hematopoietic and endothelial cells. *Development* **125**, 725–732.

11. Scott, E.W., Simon, M.C., Anastasi, J., and Singh, H. (1994) Requirement of transcription factor PU.1 in the development of multiple hematopoietic lineages. *Science* **265**, 1573–1577.

12. Tondravi, M.M., McKercher, S.R., Anderson, K., Erdmann, J.M., Quiroz, M., Maki, R., et al. (1997) Osteopetrosis in mice lacking haematopoietic transcription factor PU.1. *Nature* **386**, 81–84.

13. Tsuneto, M., Tominaga, A., Yamazaki, H., Yoshino, M., Orkin, S.H., and Hayashi, S. (2005) Enforced expression of PU.1 rescues osteoclastogenesis from embryonic stem cells lacking Tal-1. *Stem Cells* **23**, 134–143.

14. Abe, E., Miyaura, C., Tanaka, H., Shiina, Y., Kuribayashi, T., Suda, S., et al. (1983) 1

alpha,25-dihydroxyvitamin D3 promotes fusion of mouse alveolar macrophages both by a direct mechanism and by a spleen cell-mediated indirect mechanism. *Proc. Natl. Acad. Sci. USA* **80**, 5583–5587.

15. Hayashi, S., Miyamoto, A., Yamane, T., Kataoka, H., Ogawa, M., Sugawara, S., et al. (1997) Osteoclast precursors in bone marrow and peritoneal cavity. *J. Cell. Physiol.* **170**, 241–247.

16. Yamazaki, H., Kunisada, T., Yamane, T., and Hayashi, S.I. (2001) Presence of osteoclast precursors in colonies cloned in the presence of hematopoietic colony-stimulating factors. *Exp. Hematol.* **29**, 68–76.

17. Arai, F., Miyamoto, T., Ohneda, O., Inada, T., Sudo, T., Brasel, K., et al. (1999) Commitment and differentiation of osteoclast precursor cells by the sequential expression of c-Fms and receptor activator of nuclear factor kappaB (RANK) receptors. *J. Exp. Med.* **190**, 1741–1754.

18. Rivollier, A., Mazzorana, M., Tebib, J., Piperno, M., Aitsiselmi, T., Rabourdin-Combe, C., et al. (2004) Immature dendritic cell transdifferentiation into osteoclasts: a novel pathway sustained by the rheumatoid arthritis microenvironment. *Blood* **104**, 4029–4037.

19. Yasuda, H., Shima, N., Nakagawa, N., Yamaguchi, K., Kinosaki, M., Mochizuki, S., et al. (1998) Osteoclast differentiation factor is a ligand for osteoprotegerin/osteoclastogenesis-inhibitory factor and is identical to TRANCE/RANKL. *Proc. Natl. Acad. Sci. USA* **95**, 3597–3602.

20. Kong, Y.Y., Yoshida, H., Sarosi, I., Tan, H.L., Timms, E., Capparelli, C., et al. (1999) OPGL is a key regulator of osteoclastogenesis, lymphocyte development and lymph-node organogenesis. *Nature* **397**, 315–323.

21. Lagasse, E., and Weissman, I.L. (1997) Enforced expression of Bcl-2 in monocytes rescues macrophages and partially reverses osteopetrosis in op/op mice. *Cell* **89**, 1021–1031.

22. Li, J., Sarosi, I., Yan, X.Q., Morony, S., Capparelli, C., Tan, H.L., et al. (2000) RANK is the intrinsic hematopoietic cell surface receptor that controls osteoclastogenesis and regulation of bone mass and calcium metabolism. *Proc. Natl. Acad. Sci. USA* **97**, 1566–1571.

23. Kobayashi, K., Takahashi, N., Jimi, E., Udagawa, N., Takami, M., Kotake, S., et al. (2000) Tumor necrosis factor alpha stimulates osteoclast differentiation by a mechanism independent of the ODF/RANKL-RANK interaction. *J. Exp. Med.* **191**, 275–286.

24. Hayashi, S., Tsuneto, M., Yamada, T., Nose, M., Yoshino, M., Shultz, L.D., et al. (2004) Lipopolysaccharide-induced osteoclastogenesis in Src homology 2-domain phosphatase-1-deficient viable motheaten mice. *Endocrinology* **145**, 2721–2729

25. Merkel, K.D., Erdmann, J.M., McHugh, K.P., Abu-Amer, Y., Ross, F.P., and Teitelbaum, S.L. (1999) Tumor necrosis factor-alpha mediates orthopedic implant osteolysis. *Am. J. Pathol.* **154**, 203–210.

26. Takayanagi, H., Ogasawara, K., Hida, S., Chiba, T., Murata, S., Sato, K., et al. (2000) T-cell-mediated regulation of osteoclastogenesis by signalling cross-talk between RANKL and IFN-gamma. *Nature* **408**, 600–605.

27. Takayanagi, H., Kim, S., Koga, T., Nishina, H., Isshiki, M., Yoshida, H., et al. (2002) Induction and activation of the transcription factor NFATc1 (NFAT2) integrate RANKL signaling for terminal differentiation of osteoclasts. *Dev. Cell* **3**, 889–901.

28. Yamane, T., Kunisada, T., Yamazaki, H., Era, T., Nakano, T., and Hayashi, S.I. (1997) Development of osteoclasts from embryonic stem cells through a pathway that is c-fms but not c- kit dependent. *Blood* **90**, 3516–3523.

29. Kodama, H., Nose, M., Niida, S., and Yamasaki, A. (1991) Essential role of macrophage colony-stimulating factor in the osteoclast differentiation supported by stromal cells. *J. Exp. Med.* **173**, 1291–1294.

30. Nishikawa, S., Ogawa, M., Nishikawa, S., Kunisada, T., Kodama, H. (1988) B lymphopoiesis on stromal cell clone: stromal cell clones acting on different stages of B cell differentiation. *Eur. J. Immunol.* **18**, 1767–1771.

31. Udagawa, N., Takahashi, N., Akatsu, T., Sasaki, T., Yamaguchi, A., Kodama, H., et al. (1989) The bone marrow-derived stromal cell lines MC3T3-G2/PA6 and ST2 support osteoclast-like cell differentiation in cocultures with mouse spleen cells. *Endocrinology* **125**, 1805–1813.

32. Hemmi, H., Okuyama, H., Yamane, T., Nishikawa, S., Nakano, T., Yamazaki, H., et al. (2001) Temporal and spatial localization of osteoclasts in colonies from embryonic stem cells. *Biochem. Biophys. Res. Commun.* **280**, 526–534.

33. Doetschman, T.C., Eistetter, H., Katz, M., Schmidt, W., Kemler, R. (1985) The in vitro

development of blastocyst-derived embryonic stem cell lines: formation of visceral yolk sac, blood islands and myocardium. *J. Embryol. Exp. Morphol.* **87**, 27–45.

34. Li, E., Bestor, T.H., and Jaenisch, R. (1992). Targeted mutation of the DNA methyltransferase gene results in embryonic lethality. *Cell* **69**, 915–926.

35. Robertson, E., Bradley, A., Robertson, E.J., and Evans, M.J. (1986) Germ-line transmission of genes introduced into cultured pluripotential cells by retroviral vector. *Nature* **323**, 445–448.

36. Nakano, T., Kodama, H., and Honjo, T. (1994) Generation of lymphohematopoietic cells from embryonic stem cells in culture. *Science* **265**, 1098–1101.

Chapter 17

Absorption-Based Assays for the Analysis of Osteogenic and Chondrogenic Yield

Lesley A. Davis, Anke Dienelt, and Nicole I. zur Nieden

Abstract

The typical characteristics of cartilage and bone tissue are their unique extracellular matrices on which our body relies for structural support. In the respective tissue, the cells that create these matrices are the chondrocyte and the osteoblast.

During in vitro differentiation from an embryonic or any other stem cell, specific cell types must be unequivocally identifiable to be able to draw the conclusion that a specific cell type has indeed been generated. Here, gene expression profiling can be helpful, but examining functional properties of cells is a lot more conclusive. As proteoglycans are found in and are part of the function of cartilage tissue, their detection and quantification becomes an important diagnostic tool in tissue engineering. Likewise, in bone regeneration therapy and in research, alkaline phosphatase is a known marker to detect the degree of development and function of differentiating osteoblasts. Calcification of the maturing osteoblast is the last stage in its development, and thus, the quantification of deposited calcium can aid in determining how many cells in a given culture have successfully matured into fully functioning osteoblasts. This chapter describes methods ideal for testing of proteoglycan content, alkaline phosphatase activity, and calcium deposit during in vitro chondro- and osteogenesis.

Key words: Proteoglycan, Alkaline phosphatase, Calcium, Chondrocyte, Osteoblast, Extracellular matrix

1. Introduction

In the bony skeleton there are two main cell types, the osteoclast, which absorbs excess bone, and the osteoblast, which secretes hormones and factors for growth and repair of the tissue. At the end of the long bones reside the chondrocytes, which form the hyaline cartilage, providing a smooth surface for frictionless joint movement.

Nicole I. zur Nieden (ed.), *Embryonic Stem Cell Therapy for Osteo-Degenerative Diseases*, Methods in Molecular Biology, vol. 690, DOI 10.1007/978-1-60761-962-8_17, © Springer Science+Business Media, LLC 2011

The development of bone is linked to cartilage formation in that bone is formed from a cartilage template or anlage. During bone growth as well as repair, the chondrocyte and the osteoblast are responsible for the laying down of the extracellular matrix (ECM). On a cellular level, the ECM provides support and anchorage for cells and acts as a local depot for growth factors. Although both the cartilage and the bone ECM are predominantly composed of triple helical collagen fibers, they are unique with regard to the presence of additional molecules, which reflects the distinct functional requirements of both tissues.

The framework of bone is a composite material comprising noncollagenous proteins and lipids beside the already mentioned collagen (1). It is particularly the mineral phase (hydroxyapatite) that enables us to walk, jump, run, and simply stand, by providing us with a hard scaffolding throughout our body, and thus giving us strength. Therefore, molecules associated with the composition and maintenance of the mineralized matrix, which are discussed below, may serve as unique markers of bone tissue.

In contrast, cartilage contains a higher ratio of proteoglycans, the function of which is to distribute load. By virtue of their net negative charge, proteoglycans attract water molecules, keeping the ECM and resident cells hydrated and providing it with its typical viscoelasticity by which every move is cushioned.

1.1. Proteoglycans in Cartilage Matrix

Most proteoglycans are composed of a protein core, to which chondroitin, heparan, and keratan sulphate-rich glycosaminoglycan side chains are covalently attached. The most common proteoglycan in cartilage is aggrecan, a large molecule with a core protein of over 2,000 amino acids (2). Aggrecan contains chondroitin and keratan sulfate glycosaminoglycan side chains as well as asparagine-linked oligosaccharides and O-linked oligosaccharides and is expressed in embryonic stem cell (ESC)-derived chondrocytes (3).

Beside aggrecan, a variety of other mostly smaller proteoglycans are responsible for cartilage function, such as decorin, biglycan, and lumican, which all belong to the growing family of small leucine-rich repeat proteoglycans (SLRPs). SLRPs are not structurally related to aggrecan, but are characterized by multiple adjacent domains bearing a common leucine-rich motif (4).

The function of the SLRPs depends on both their core protein and their glycosaminoglycan chains. Decorin and biglycan may be classified as dermatan sulphate proteoglycans, whereas lumican is a keratin sulphate proteoglycan. However, in bone, decorin and biglycan are conjugated to chondroitin sulphate side chains (5–7). Typically, in auricular cartilage, the nonglycosylated forms of the proteins are found (8).

The function of the SLRPs is to help regulate fibril diameter during its formation and possibly fibril–fibril interaction in the extracellular matrix, which is achieved via their core proteins, which allow the SLRPs to interact with the fibrillar collagen that forms the framework of the tissue (9). Because of their molecular localization, they also appear to limit access of the collagenases to their unique cleavage site, protecting the collagen fibrils from proteolytic damage. For a detailed review, the reader is referred to an article by Roughley (10). Recent evidence suggests a role for the two dermatan sulphate proteoglycans of the SLRP family, decorin and biglycan, in influencing bone cell differentiation and proliferation, possibly also in regulating mineral deposition (11).

1.2. The Mineralized Matrix: Alkaline Phosphatase and Calcium

In vitro, osteogenesis from a mouse ESC requires the expression and presence of both general and bone-specific factors (12). The coordinated expression of these factors directs the cell toward an osteoblast fate. The three key factors added to the media during osteoblast differentiation include: ascorbic acid, beta-glycerophosphate, and bone active factors, such as $1\alpha,25$-$(OH)_2$ vitamin D_3 or dexamethasone (13–15). All three factors are needed for differentiation and mineralization of the mature osteoblast; beta-glycerophosphate specifically provides the cells with the source of inorganic phosphate, thus is the substrate for alkaline phosphatase (Alk Phos). Alk Phos is considered the major enzyme in mineralization and is expressed on the cell surface of the osteoblast. The removal of a phosphate by this enzyme increases the amount of inorganic phosphate that can either be ingested by the cell or react with molecules in the environment, such as calcium, in the case of bone growth and repair, to form one of the components of the mineralized extracellular matrix.

As the name suggests, Alk Phos is an enzyme that needs an alkaline environment to hydrolyze phosphate monoesters (16). Its catalytic sites require the presence of two zinc ions and one magnesium ion per monomer (17). The range of organisms that utilize Alk Phos as part of their cellular function extends from *E. coli* (18) to humans (17, 19), underlining the importance of this enzyme in the animal kingdom. In bacteria, Alk Phos is located in the periplasmic space and is thus considered secreted even though it is not released to the outside of the cell (20). Mammalian Alk Phos is located within the plasma membrane via a glycosyl phosphatidyl inositol (GPI) anchor and consists of two homodimers with each monomer having a GPI anchor on its C-terminal group (16, 17). In addition, it differs from bacterial Alk Phos in that it has higher activity and K_m values, needs a more alkaline environment, and is less heat-stable (17). The isozymes of Alk Phos are found in most tissues of the body with the forms found in kidney, liver, and bone labeled as tissue nonspecific (TN) (17, 19). TN Alk Phos is an established marker of ESC-derived preosteoblasts (13, 21, 22). In the embryo, its activity is high throughout the organism, specifically in the primordial germ cells (23).

This combined with evidence of Alk Phos activity in cells taken from mouse teratomas and ESCs has made this enzyme also a marker for pluripotent cells (23–27). A method for staining undifferentiated ESCs for Alk Phos is described in Chapter 4.

During ESC-osteogenesis, Alk Phos activity is decreased in differentiating cells and can be detected early when mineralization is initiated under influence of $1\alpha,25$-$(OH)_2$ vitamin D_3 (differentiation days 10–15) (13). A second wave of Alk Phos is seen later in differentiation directly before the expression of the mature osteoblast markers osteocalcin and bone sialoprotein (21). Therefore, the detection and quantification of the enzyme in the body and the differentiating cell become important in determining the stages of differentiation, development, and disease.

The inorganic phosphate provided by Alk Phos readily binds to ionic calcium, which is a necessity for the mammalian body for its role in building and maintaining strong bones (28). The processes put in place by the body to control and regulate the flow of calcium are vital for the function of that cell and the tissue in which it is associated. There are a number of diseases in which their cause or symptoms include a disruption of calcium production and collection within the body. Among these is osteoporosis, a common disease in which the sufferer has more osteoclasts than osteoblasts, thus preventing the body from utilizing dietary calcium, for the purpose of bone repair and cell turnover (29).

The regulation and maintenance of Ca^{2+} within the body and specifically the osteoblast has been linked to a number of factors such as: parathyroid hormone, $TGFbeta_1$, Integrin, vitamin D_3, and Calmodulin (28). These factors all contribute to or act on one or both of the following ways: expression of calcium-binding proteins and receptors or directly bind calcium for cell signaling (28, 30). Calmodulin isoforms in particular need Ca^{2+} to be activated, and those isoforms in osteoblasts may control both the differentiation and proliferation of the cell (28).

The release of Ca^{2+} to the outside of the cell is a possible method for the osteoblast to increase the concentration of calcium in its environment, thus enabling it to start laying down a mineralized matrix. The removal of calcium from the cell is usually done by active transport at the plasma membrane through Ca^{2+}ATPase channels or Na^+Ca^{2+} exchangers (31, 32). The transport of Ca^{2+} via these Na^+Ca^{2+} exchangers in the plasma membrane will enable intracellular calcium found in the cytoplasm to be utilized as part of the extracellular matrix during bone mineralization (32). However, most of the calcium is found outside the cell and is brought into the body through diet. Once Ca^{2+} has been released from the cell and enters the blood stream, calcium-sensing receptors found in the kidney and parathyroid will sense extracellular Ca^{2+} levels, thus enabling the body to release the appropriate hormone to either keep Ca^{2+} or release it. Two such hormones, vitamin D_3 and parathyroid hormone, stimulate the expression of calcium-binding proteins and receptors that will collect calcium from the intestine and plasma, for use in the body and by the developing osteoblast (28).

During in vitro osteogenesis, a number of factors are added to the medium to push the differentiating ESC toward an osteoblast fate. One factor added to the medium is $1\alpha,25\text{-}(OH)_2$ vitamin D_3, which is brought into the differentiating cell, thus signaling the start of osteoblast differentiation (13). As the stem cell slowly differentiates toward an osteoblast fate, it will undergo a period of mineralization where it will deposit a collagenous matrix, and if there is enough Ca^{2+} and inorganic phosphate present, hydroxyapatite crystals will form and thus the ECM mineralizes (13, 31). As collagen type I and calcium are the major components of the bony ECM, the quantification of calcium deposition is an accurate method to determine the number of cells in a culture which have developed into mature osteoblasts.

2. Material

2.1. DMMB Assay

It is absolutely important that you prepare your stock and working solutions in deionized water only.

1. Phenylmethylsulfonyl fluoride (PMSF), 98.5% (GC): store at room temperature; inhibits serine proteases such as trypsin and chymotrypsin, also inhibits cysteine proteases (reversible by reduced thiols) and mammalian acetylcholinesterase; half-life = 1 h at pH 7.5; soluble in dry solvents (ethanol, methanol, and 2-propanol); contact with liquid liberates toxic gas; handle in chemical hood only; make a 200 mM stock solution, which is stable for minimally 9 months at 4°C; toxic (see Note 1).

2. Proteoglycan extraction buffer (PEB, see Note 2): 4 M of guanidine hydrochloride/0.05 M sodium acetate, including 100 mM 6-amino caproic acid, 10 mM ethylenediaminetetraacetic acid (EDTA), 5 mM benzamidine hydrochloride, 10 mM N-ethylmaleimid, 0.4 mM Pepstatin A (see Note 3), 1 mM PMSF, and 1 μg/mL soybean trypsin inhibitor (dissolved in aqua dest). Set pH to 5.8 and store at 4°C.

3. Heidolph Polymax Wave Shaker (Brinkmann).

4. Refrigerated microcentrifuge, i.e. Eppendorf 5415R, and microcentrifuge tubes, 1.5 mL.

5. Flat 24-well assay plates (i.e. BD Biosciences).

6. Chondroitin sulfate C (sodium salt from shark cartilage): Make a 10 mg/mL stock solution in ddH$_2$O and store at 4°C.

7. Dispenser pipettor, e.g. Eppendorf Repeater® Plus or similar, with blue adapter and Eppendorf Combitips® Plus (25 mL), Biopur individually packed.

8. DMMB assay reagent: First, dissolve 1,9-dimethyl methylene blue (DMMB) in ethanol to make a 3.2% (w/v) solution (see Note 4). To make 1 L of DMMB assay reagent, use 5 mL of the 3.2% DMMB dye solution and add 2 g of sodium formate and 2 mL of formic acid (98%). Fill volume with autoclaved sterile aqua dest. Adjust pH to 3.5 and store at room temperature (see Note 5).

9. Temperature-controlled ELISA plate reader set to 535 nm, e.g. Tecan Safire 2™.

2.2. Lysing Cells

1. Phosphate-buffered saline (PBS): Combine 137 mM NaCl, 2.7 mM KCl, and 10 mM sodium phosphate dibasic (Na_2HPO_4). Adjust with aqua dest to final volume. Adjust pH to 7.4 and autoclave at 121°C for 60 min. Buffer is stable at room temperature; however, keep at 4°C for long-term storage.

2. RIPA buffer: 1× PBS containing 0.1% dodecyl sodium sulfate (SDS, see Note 6), 1% NP-40, and 0.5% sodium deoxycholate (see Note 7). Add sterile aqua dest to final volume and adjust pH to 7.2. This RIPA base may be stored at 4°C. Before each use, protease inhibitors should be added freshly to final concentrations of 1 mM PMSF (see Note 1), 10 mM benzamidine hydrochloride, and 4.2 μM leupeptin hemisulfate salt. At working concentrations, this solution is stable for only a few hours and should therefore be stored on ice for intermittent use over several hours.

3. Heidolph Polymax Wave Shaker (Brinkmann).

4. VWR Tube Rotator, with 36×1.5/2.0 mL rotisserie assembly.

5. Refrigerated microcentrifuge, e.g. Eppendorf 5415R and 1.5-mL microcentrifuge tubes.

6. Flat 96-well general assay plates (e.g. Corning Costar).

7. Hydrochloric acid, 1.0 N (HCl): used as HCl lysis buffer. When diluting the stock solution (i.e. when preparing standard curves), add product carefully to watery solutions, since this is an exothermic process. Stable at room temperature if kept sealed and away from bases and metals.

2.3. Alkaline Phosphatase Assay

1. Alkaline phosphatase yellow liquid substrate system for ELISA (pNPP): ready-to-use reagent; stable for 1 year when stored at 4°C (see Note 8).

2. Flat 96-well general assay plates (e.g. Corning Costar).

3. Dispenser pipettor, e.g. Eppendorf Repeater® Plus or similar with Eppendorf Combitips® Plus (5 mL), Biopur individually packed.

4. Temperature-controlled ELISA plate reader with a wavelength range between 350 and 750 nm, e.g. Tecan Safire 2™.

2.4. Ca²⁺ Assay

1. Calcium chloride dehydrate, ≥99%: Kept at a stock of 27.69 mg/mL ($\equiv$10 mg/mL Ca^{2+}) in deionized water at room temperature.

2. Flat 96-well general assay plates (e.g. Corning Costar).

3. Dispenser pipettor, e.g. Eppendorf Repeater® Plus or similar and 5-mL individually packed Combitips® Plus (Biopur).

4. Ca^{2+} reagent, 0.15 mM Arsenzo III (2,2′-bisbenzene-arsonic acid, DCL Toronto): ready-to-use reagent; kept at room temperature away from light.

5. Temperature-controlled ELISA plate reader with a wavelength range between 350 and 750 nm, e.g. Tecan Safire 2™.

2.5. Lowry Assay

1. Biorad DC protein reagent kit.

2. Bovine serum albumin: store powder at 4°C. Prepare a solution of 50 mg/mL, aliquot, and store at –20°C.

3. Temperature-controlled ELISA plate reader set to a wavelength of 750 nm, e.g. Tecan Safire 2™.

3. Methods

3.1. DMMB Assay

The overall proteoglycan content of chondrogenic ESC cultures may be determined with the metachromatic DMMB assay (3).

1. For each treatment group that you would like to assay, seed five wells to be assayed with this method. This will allow you to calculate adequate standard deviations. Extract the proteoglycans deposited by the cells in 300 µL of PEB (see Note 9). To do this, let the plates incubate under shaking for 48 h at 4°C (see Note 10).

2. Remove remaining nonlysed cellular debris by collecting the lysate into a microcentrifuge tube and centrifuging in a microcentrifuge at full speed ($16,000 \times g$, 20 min, 4°C). Keep your samples on ice during all following steps.

3. Prepare a standard concentration series with chondroitin sulphate C with PEB in the range between 0.3 and 300 µg/mL.

4. On a 24-well assay plate, mix two-thirds of the lysate or standard with ten times the volume of DMMB assay reagent (see Note 11), using a dispenser to add the reagent. Assay every

sample in triplicate. Include a blank with PEB only. If the lysate contains proteoglycans, the color should change from a dark blue to an intensive purple, the absorbance of which can be read at 535 nm.

5. Calculate the amount of proteoglycan in the samples from the concentration curve.

3.2. Lysing Cells for Alk Phos, Ca²⁺, and Lowry Assays

Differentiate mouse or human ESCs toward mature osteoblasts, as described in Chapters 9, 10, 14, or 15; for each treatment group, seed a minimum of five wells (see Note 12).

1. When starting your cell culture experiment, plan to seed a minimum of five wells of cells for each treatment group that you would like to assay (five biological replicates). Remove media from cells – be careful not to disrupt the cell surface. The use of an aspirator is preferred to dispose of all trace amounts of media. Wash cells twice with 1× PBS – this is an important step as you want to ensure that there is no serum containing media on the cells, which may contain Alk Phos or any calcium traces.

2. Overlay with 300 μL of RIPA buffer with proteinase inhibitors (per well of a 24-well plate). The cells may loosen and come off the plate, or the cell will lyse leaving behind its calcified matrix.

3. Scrape cells and collect into a microcentrifuge tube and rotate for 1 h at 4°C to lyse. Subsequently, spin down for 30 min at $16,000 \times g$ at 4°C.

4. Transfer the supernatant to clean microcentrifuge tubes. Aliquot and store at –80°C. Use for subsequent protein assay, Alk Phos, and Ca²⁺ measurements.

5. Wash the remaining pellet in ddH₂O and store it at –80°C.

6. To the rest of the matrix on your plate, add 1 N HCl (300 μL for 1/24 well). Incubate overnight at 4°C with shaking (see Note 10).

7. The next day, collect HCl and matrix from the plate and add to the tube, which already contains the pellet, and rotate overnight at 4°C to mix. The next day, centrifuge for 5 min at $16,000 \times g$.

8. Collect the supernatant into a fresh tube, aliquot, and store at –80°C.

3.3. Alk Phos Assay

There are a number of ways by which a researcher can detect and quantify Alk Phos activity during osteoblast differentiation: antibodies, naphthiol/fast red violet, and *p*-nitrophenyl phosphate substrate. The use of antibodies and fast red violet requires the cells to be fixed, while *p*-nitrophenyl phosphate allow the cells to be tested while alive or directly after cell lysis, thereby allowing the activity to be compared to the total protein present per given experimental sample. Moreover, the described method of using *p*-nitrophenyl phosphate enables the quantification of Alk Phos activity in comparison with other treatments.

3.3.1. Measurement

The principle of this assay lies with the conversion of the *p*-nitrophenyl phosphate by Alk Phos. Prior to reaction initiation, the substrate should appear as a colorless to pale yellow solution. Following the reaction with Alk Phos, the absorbance of the dark yellow reaction product *p*-nitrophenol may be read at 405 nm. This hydrolysis only occurs at an alkaline pH.

1. It is important for the purpose of quantification that each experimental sample (biological replicate) is tested in triplicate and that a blank (pNPP + RIPA buffer) is included in your test set.

2. Before you start mixing the reagent and the sample lysate, set your ELISA reader to a temperature of 37°C.

3. Pipette 10 μL of the RIPA cell lysate into a 96-well plate.

4. Add 90 μL of pNPP to each lysate using the dispenser.

5. Set the mixing function of the plate reader to 5 s and 5 flashes. Take an absorbance reading at 405 nm instantly (t_0). This absorbance is designated $A_{initial}$ (see Note 14). Then, incubate at 37°C for up to 30 min, preferably inside the plate reader chamber. The time may vary with different companies and lot numbers of the substrate (see Note 15). Take a second measurement at $t_x = 30$ min at 405 nm. Reading is designated as A_{final} (see Notes 16–20).

3.3.2. Calculations

1. Subtract the blank from each reading and calculate the units of enzyme from the following formula for each reading:

$$\text{ALP activity (Units/mL)} = ((A_{final} - A_{initial})/t_x) \times R/18.45$$

with 18.45 being the extinction coefficient ε. *R* is the dilution factor divided by the path length. For a conventional 96-well plate and a reaction volume of 100 μL, the path length is approximately 0.31 cm.

2. Next, you need to take into account the dilution factor of your sample, which is calculated by dividing the reaction volume by the used volume of the RIPA lysate; here, the calculation is 0.1/0.01 mL. This value gives the total enzyme activity per mL of your RIPA lysate. The total enzyme activity in your sample (0.3 mL volume) is then calculated by multiplying by 0.3. The result of your calculation is the enzyme activity in Units in your entire sample.

3. Next, you need to normalize this to the total amount of protein of the cell lysate, which is measured as described in Subheading 3.5. At the end of this calculation, you get the specific enzyme activity in Units/mg. How Alk Phos activity develops over time in ESC cultures is shown in Fig. 1.

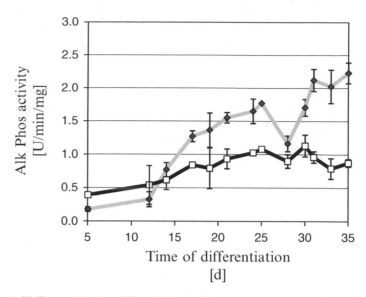

Fig. 1. Alk Phos activity in mESCs. Alk Phos activity was measured in control cultures (*black line*), which spontaneously differentiated into all lineages upon deprivation of LIF. When osteogenesis was induced with $1\alpha,25$-(OH)$_2$ vitamin D$_3$ (*grey line*), Alk Phos activity increased around differentiation day 11. Alk Phos activity is shown as mean ± standard deviation for three independent experiments.

3.4. Ca²⁺ Assay

This method is a reliable method for approximating the number of cells that have differentiated to osteoblasts during your study. There are a number of methods for detecting calcium production and deposition in vitro. There are two general ways to detect calcium: one is through fluorescence and the other absorbance. The fluorescent dyes can then be separated into those that increase in fluorescence in the presence of calcium and those that change their excitation wavelengths. The fluorescent dyes include: calcein, coelenterazine, dehydrocalcein fluo-3, fura-2, indo-1, and rhod-2. For absorbance, there is Arsenzo III. The fluorescent dyes are better for the detection of the distribution and quantity of intracellular calcium. However, for the purposes of detecting and quantitating the amount of calcium deposited by an osteoblast, the absorbant dye is the best. Arsenzo III or 2,2′-bisbenzene-arsonic acid forms a blue or purple complex when it comes in contact with calcium ions (Fig. 2) which can be read at 650 nm.

3.4.1. Standard Curves for RIPA and HCl Lysates

1. For HCl standard, prepare a concentration range of 0–0.1 mg/mL Ca²⁺ in HCl lysis buffer. To measure the OD, use 250 µL of Arsenazo III and 10 µL of HCl standards.

2. For RIPA standard, prepare a lower concentration range from 0–0.03 mg/mL Ca²⁺ using calcium chloride stock solution. Use 75 µL of Arsenazo III and 25 µL of RIPA standard for the measurement.

3. A pipetting scheme can be found below. Standards in bold letters need to be prepared only for the HCl standard curve.

3.4.2. Measurement

This method can be performed at any time once calcification has begun during osteogenesis. This description assumes that the cells are grown in 96-well plates. It is important that you perform the Ca^{2+} measurement from your RIPA AND from the HCl lysate (see Subheading 3.2).

1. In a 96-well plate, dispense Arsenazo III into three wells for each sample that you will be assaying, and prepare a standard curve as described above.

2. Use 250 μL of Arsenazo III for the HCl lysates and 75 μL of Arsenazo III for the RIPA lysates in each well.

3. Prepare three wells with Arsenazo III only as a blank (see Note 21).

4. Aliquot each lysate (10 μL of HCl lysate and 25 μL of the RIPA lysates) into the triplicate wells containing Arsenazo III reagent.

5. Set the mixing function of the plate reader to 5 s and 5 flashes and read absorbance at 650 nm. The reaction product should be stable for at least an hour.

6. Check the absorbance of the samples at 650 nm. Similar to the Alk Phos assay, make sure that your measured OD is lower than 1. If it is not, dilute your sample with the respective lysis buffer and redo the assay (see Table 1).

Table 1
Pipetting schematic for calcium standard curve

Standard

Dilution no.	Concentration (mg/mL)	Ca^{2+} 10 mg/mL	RIPA or HCl lysis buffer
(1)	0.1	10 μL	990 μL
(2)	0.07	**70 μL of dilution (1)**	**30 μL**
(3)	0.05	**50 μL of dilution (1)**	**50 μL**
(4)	0.04	**40 μL of dilution (1)**	**60 μL**
(5)	0.03	30 μL of dilution (1)	70 μL
(6)	0.02	20 μL of dilution (1)	80 μL
(7)	0.01	10 μL of dilution (1)	90 μL
(8)	0		Buffer only

Dilutions in bold are needed for HCl lysates only

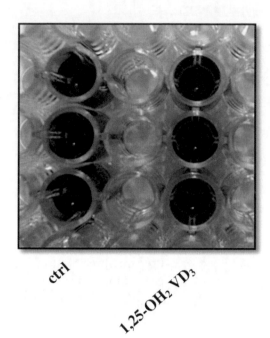

ctrl

1,25-OH₂ VD₃

Fig. 2. Calcium assay. The change in absorption (development of *purple* to *dark blue* color) is shown for 1α,25-(OH)₂ vitamin D₃ (VD₃)-treated murine ESCs. Lysates from control cultures (ctrl) without osteogenic supplements do not react with Arsenazo III.

3.4.3. Calculations

1. Draw a calibration curve of the standards (mean value–blank) ($y = m \times x$). An example is given in Fig. 3.

2. Calcium concentration: Mean value–blank divided by increase of the calibration curve (m; from $y = m \times x$).

3. Total calcium amount: Calcium concentration multiplied by volume of your sample (0.3 mL if you have followed what was described above).

4. Sum up your total calcium amount from the RIPA lysate and the HCl sample.

5. Relate you total amount of calcium to your total protein amount as measured by Lowry assay (see below).

3.5. Lowry Assay

3.5.1. Measurement

1. Add 20 μL of reagent S to each mL of reagent A that will be needed for the run. All reagents are provided in the kit. This solution is then designated reagent A'.

2. Prepare a protein standard curve according to Table 2.

3. Pipette 5 μL of standards and samples into a clean, dry microtiter plate (each sample in triplicate, see Note 22).

4. Add 25 μL of reagent A' into each well and then add 200 μL of reagent B into each well.

5. Gently agitate the plate to mix the reagents. After 15 min, read the absorbance at 750 nm. Set the mixing function of the plate

Table 2
Pipetting schematic for protein standard curve

Standard		BSA (50 mg/ml)	RIPA buffer
Dilution no.	Concentration (mg/mL)		
(1)	1.5	3.0 μL	97.0 μL
(2)	1	33.3 μL of dilution (1)	16.7 μL
(3)	0.75	25.0 μL of dilution (1)	25.0 μL
(4)	0.5	25.0 μL of dilution (2)	25.0 μL
(5)	0.2	20.0 μL of dilution (4)	30.0 μL
(6)	0		Buffer only

reader to 5 s and 5 flashes (see Note 23). Again, be sure that your measured OD is lower than 1. If not, dilute and redo.

3.5.2. Calculations

1. Take the average value of the standards and the samples.

2. Draw a calibration curve of the standards (mean value–blank) ($y = m \times x$), see Fig. 3 for an example.

3. Protein concentration: Mean value–blank divided by increase of the calibration curve (m; from $y = m \times x$). The total protein amount is the concentration multiplied by volume of your sample (0.3 mL if you have done as described above).

4. With these established values for your samples, normalize the Alk Phos activity and the calcium content to the total protein content of your samples by dividing them.

5. As a very last step, take the average value of samples and calculate the standard error.

6. Statistically relevant *p*-values may be calculated by using SigmaStat or the web-based Student's *t* test calculation matrix from the Physics Department of the College of Saint Benedict/ Saint John's University (http://www.physics.csbsju.edu/ stats/t-test.html).

4. Notes

1. PMSF is soluble in isopropanol at a maximum concentration of 35 mg/mL with heating but is extremely unstable in the presence of water (half-life of aqueous PMSF at 25°C at pH 8.0 is 35 min only).

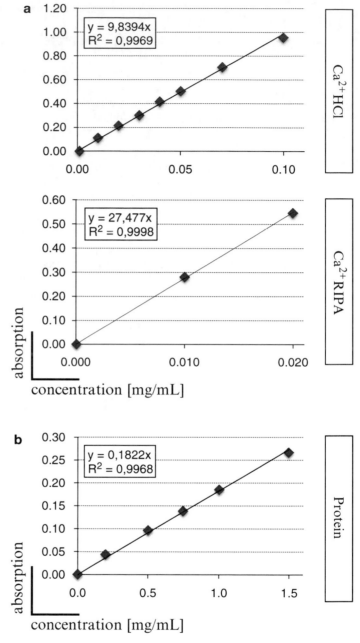

Fig. 3. Standard curves generated for Ca^{2+} and Lowry assays. (**a**) Calcium standard curve made with calcium chloride in HCl and RIPA lysis buffers and reacted with Arsenazo III. (**b**) Protein standard curve of a concentration range of BSA dilutions. Lowry assay measured at 750 nm.

2. PEB contains ingredients that are toxic or harmful. Always wear protective equipment when dissolving ingredients, and make the stock solutions as well as the buffer only in a fume hood.

3. Pepstatin A is a potent inhibitor of acid proteases, such as pepsin, rennin, and cathepsin D. It is stable for 24 h only and

needs to be made fresh every time. The solubility of Pepstatin A is related to the purity of the preparation. Purer forms of the chemical (minimum 90%) are insoluble at any concentration in methanol or DMSO. However, if you choose a purer form, you may dissolve it at 1 mg/mL in 10% (v/v) acetic acid in methanol (9:1 methanol:acetic acid). Stock solutions at 1 mg/mL should then be stable at least a week at 4°C. A 1 mM solution in methanol should be stable for months at −20°C. If solutions become more yellow, the reagent is hydrolyzing.

4. DMMB is also called Taylor's Blue. As it is irritating to the eyes, wear suitable protective clothing and gloves. Store in a cool dry place.

5. Keep container containing sodium formate tightly closed in a dry and well-ventilated place. The reagent is moisture-sensitive, and contact with strong oxidizing agents and strong acids should be prevented because of fire hazard. Suitable protective clothing, gloves, and eye/face protection must be worn when handling its acid form (formic acid).

6. SDS is highly flammable and harmful. A 10% (w/v) stock solution is stable at room temperature. In case, precipitation occurs after longer storage, solution may be warmed with caution. However, prolonged heating at 40°C or greater can cause decomposition of alkyl sulfates into fatty alcohols and sodium sulfate and should be prevented.

7. Sodium deoxycholate is useful for extraction of membrane receptors and other plasma membrane proteins and for nuclei isolation. Solubility is pH-sensitive, and the compound may precipitate when pH drops to lower than 5. It is a powder and should not be mistaken for sodium deoxycholic acid. Prepare a 10% working solution and freeze this stock at −20°C.

8. The reagent should be brought to room temperature before use. The reagent solution may be decanted to separate it from the insoluble stabilizer pellets. It is also light sensitive and should be protected from direct sunlight or UV sources.

9. This volume sufficiently overlays cells grown in 24-well plates. Adjust accordingly if using different size plates.

10. This can best be accomplished by using a belly-dancer type shaker (e.g. waver shaker) that has been placed in the cold room. If too much of your solution evaporates overnight, reduce the incubation time.

11. The ratios given are for 24-well plates. You may have to adjust accordingly if using different size plates.

12. We typically perform the Ca^{2+} assay from cells grown in 96-well plates and the Alk Phos assay from cells grown in 48-well plates.

It is also possible to take both measurements from the same plate. In this case, we recommend using 24-well plates. Volumes for all further steps are given for this type of plates.

13. Alternatively, when using 96-well plates, overlay with 150 μL of RIPA buffer.

14. The first reading for t_0 is vital and should be done immediately after the cell lysate is added to the substrate.

15. Ensure that the time is consistent between all experiments being compared.

16. Periodically, check your color development during the incubation time. The reagent may need less or more time to develop color depending on your Alk Phos content.

17. For end-point assays, the reaction can be stopped by the addition of a NaOH solution. Add 50 μL of 3 N NaOH for every 200 μL of substrate reaction. This also yields a yellow end product that can be read at 405 nm.

18. Alternatively, pNPP reagent may be added directly to the cells, since Alk Phos is a membrane-bound enzyme. In this case, wash cells twice with 1× PBS before you add the Alk Phos reagent and wait no longer than 10 min until second measurement. Lyse cells afterwards for protein measurement. Please note calcium testing can still be done on the extracellular matrix left behind after cell lysis.

19. Alk Phos enzyme activity will be at its highest directly after lysis, and it is not recommended that cell lysate be kept longer than 24 h for further Alk Phos tests.

20. Using p-nitrophenylphosphate as a substrate to test for Alk Phos activity is limited by the assumption that the spectrometer is able to detect minute color changes between different experimental conditions. In addition, this method assumes for comparative analysis that there is no absorbance at t_0 and that the second absorbance is read exactly after the same incubation time for each reading. If the substrate is left on the cells too long, all wells will be saturated, and the spectrometer will not be able to detect any difference between the control and the experimental wells. Therefore, the length of time the substrate is left on the cells should be tested before the actual experiment. Be certain that your measured OD is never above 1. If it is, redo the assay and incubate for a shorter time.

21. Calcium ions and mineral can be found on almost every surface and is found in most media and liquids used in the tissue culture room. Therefore, washing the cells with sterile PBS that does not contain calcium before lysing is vital for this experiment. Also, a blank well that does not contain any cell

tissue is also vital to normalize the wells to calcium found on the surface of the tissue culture plates. Be aware that there is an issue of quenching; if there is too much calcium in the plates, the spectrometer will not be able to detect the differences in experimental wells. Therefore, testing to ensure that you do not have too many cells in the well is suggested. Diluting your lysate can help in situations where there is a lot of calcification in your culture. If the cells you are testing are in suspension, it is suggested that the cells be washed and resuspended in PBS that does not contain calcium or magnesium.

22. It is critical to adhere to the described order of mixing reagents and sample.

23. Reaction product is only stable for 1 h.

References

1. Boskey, A.L. (2001) Bone mineralization. In: Cowin, S.C., editor. Bone Biomechanics. 3. CRC Press; Boca Raton, FL, pp. 5.1–5.34.

2. Doege, K.J., Sasaki, M., Kimura, T., and Yamada, Y. (1991) Complete coding sequence and deduced primary structure of the human cartilage large aggregating proteoglycan, aggrecan. Human-specific repeats, and additional alternatively spliced forms. *J. Biol. Chem.* **266**, 894–902.

3. zur Nieden, N.I., Kempka, G., Rancourt, D.E., and Ahr, H.J. (2005) Induction of chondro-, osteo- and adipogenesis in embryonic stem cells by bone morphogenetic protein-2: effect of cofactors on differentiating lineages. *BMC Dev. Biol.* **5**, 1.

4. Hocking, A.M., Shinomura, T., and McQuillan, D.J. (1998) Leucine-rich repeat glycoproteins of the extracellular matrix. *Matrix Biol.* **17**, 1–19.

5. Fisher, L.W., Termine, J.D., and Young, M.F. (1989) Deduced protein sequence of bone small proteoglycan I (biglycan) shows homology with proteoglycan II (decorin) and several nonconnective tissue proteins in a variety of species. *J. Biol. Chem.* **264**, 4571–4576.

6. Bianco, P., Fisher, L.W., Young, M.F., Termine, J.D., and Robey, P.G. (1990) Expression and localization of the two small proteoglycans biglycan and decorin in developing human skeletal and non-skeletal tissues. *J. Histochem. Cytochem.* **38**, 1549–1563.

7. Waddington, R.J., and Embery, G. (1991) Structural characterization of human alveolar bone proteoglycans. *Arch. Oral. Biol.* **36**, 859–866.

8. Roughley, P.J., White, R.J., Magny, M.C., Liu, J., Pearce, R.H., and Mort, J.S. (1993) Non proteoglycan forms of biglycan increase with age in human articular cartilage. *Biochem. J.* **295**, 421–426.

9. Vogel, K.G., Paulsson, M., and Heinegård, D. (1984) Specific inhibition of type I and type II collagen fibrillogenesis by the small proteoglycan of tendon. *Biochem. J.* **223**, 587–597.

10. Roughley, P.J. (2006) The structure and function of cartilage proteoglycans. *Eur. Cell Mater.* **12**, 92–101.

11. Waddington, R.J., Roberts, H.C., Sugars, R.V., and Schönherr, E. (2003) Differential roles for small leucine-rich proteoglycans in bone formation. *Eur. Cell Mater.* **6**, 12–21.

12. zur Nieden, N.I., Price, F.D., Davis, L.A., Everitt, R.E., and Rancourt, D.E. (2007) Gene profiling on mixed embryonic stem cell populations reveals a biphasic role for beta-catenin in osteogenic differentiation. *Mol. Endocrinol.* **21**, 674–685.

13. zur Nieden, N.I., Kempka, G., and Ahr, H.J. (2003) In vitro differentiation of embryonic stem cells into mineralized osteoblasts. *Differentiation* **71**, 18–27.

14. Buttery, L.D., Bourne, S., Xynos, J.D., Wood, H., Hughes, F.J., Hughes, S.P., et al. (2001) Differentiation of osteoblasts and in vitro bone formation from murine embryonic stem cells. *Tissue Eng.* **7**, 89–99.

15. Sottile, V., Thomson, A., and McWhir, J. (2003) In vitro osteogenic differentiation of human ES cells. *Cloning Stem Cells* **5**, 149–155.

16. Giocondi, M.C., Seantier, B., Dosset, P., Milhiet, P.E., and Le Grimellec, C. (2008)

Characterizing the interactions between GPI-anchored alkaline phosphatases and membrane domains by AFM. *Pflugers Arch.* **456**, 179–188.

17. Millan, J.L. (2006) Alkaline phosphatases: structure, substrate specificity and functional relatedness to other members of a large superfamily of enzymes. *Purinergic Signal.* **2**, 335–341.

18. Berg, P.E. (1981) Cloning and characterization of the *Escherichia coli* gene coding for alkaline phosphatase. *J. Bacteriol.* **146**, 660–667.

19. Kam, W., Clauser, E., Kim, Y.S., Kan, Y.W., and Rutter, W.J. (1985) Cloning, sequencing, and chromosomal localization of human term placental alkaline phosphatase cDNA. *Proc. Natl. Acad. Sci. U.S.A.* **82**, 8715–8719.

20. Beacham, I.R., Taylor, N.S., and Youell, M. (1976) Enzyme secretion in *Escherichia coli*: synthesis of alkaline phosphatase and acid hexose phosphatase in the absence of phospholipid synthesis. *J. Bacteriol.* **128**, 522–527.

21. Davis, L.A., and zur Nieden, N.I. (2008) Mesodermal fate decisions of a stem cell: the Wnt switch. *Cell. Mol. Life Sci.* **65**, 2658–2674.

22. Karp, J.M., Ferreira, L.S., Khademhosseini, A., Kwon, A.H., Yeh, J., and Langer, R.S. (2006) Cultivation of human embryonic stem cells without the embryoid body step enhances osteogenesis in vitro. *Stem Cells* **24**, 835–843.

23. Berstine, E.G., Hooper, M.L., Grandchamp, S., and Ephrussi, B. (1973) Alkaline phosphatase activity in mouse teratoma. *Proc. Natl. Acad. Sci. U.S.A.* **70**, 3899–3903.

24. Nicolas, J.F., Avner, P., Gaillard, J., Guenet, J.L., Jakob, H., and Jacob, F. (1976) Cell lines derived from teratocarcinomas. *Cancer Res.* **36**, 4224–4231.

25. Resnick, J.L., Bixler, L.S., Cheng, L., and Donovan, P.J. (1992) Long-term proliferation of mouse primordial germ cells in culture. *Nature* **359**, 550–551.

26. Iannaccone, P.M., Taborn, G.U., Garton, R.L., Caplice, M.D., and Brenin, D.R. (1994) Pluripotent embryonic stem cells from the rat are capable of producing chimeras. *Dev. Biol.* **163**, 288–292.

27. Pease, S., Braghetta, P., Gearing, D., Grail, D., and Williams, R.L. (1990) Isolation of embryonic stem (ES) cells in media supplemented with recombinant leukemia inhibitory factor (LIF). *Dev. Biol.* **141**, 344–352.

28. Zayzafoon, M. (2006) Calcium/calmodulin signaling controls osteoblast growth and differentiation. *J. Cell. Biochem.* **97**, 56–70.

29. Christenson, R.H. (1997) Biochemical markers of bone metabolism: an overview. *Clin. Biochem.* **30**, 573–593.

30. Brown, E.M., Chattopadhyay, N., and Yano, S. (2004) Calcium-sensing receptors in bone cells. *J. Musculoskelet. Neuronal. Interact.* **4**, 412–413.

31. Jung, S.Y., Park, Y.J., Cha, S.H., Lee, M.Z., and Suh, C.K. (2007) Na^+-Ca^{2+} exchanger modulates Ca^{2+} content in intracellular Ca^{2+} stores in rat osteoblasts. *Exp. Mol. Med.* **39**, 458–468.

32. Stains, J.P., Weber, J.A., and Gay, C.V. (2002) Expression of Na(+)/Ca(2+) exchanger isoforms (NCX1 and NCX3) and plasma membrane Ca(2+) ATPase during osteoblast differentiation. *J. Cell. Biochem.* **84**, 625–635.

Chapter 18

Identification of Osteoclasts in Culture

Nobuyuki Udagawa, Teruhito Yamashita, Yasuhiro Kobayashi, and Naoyuki Takahashi

Abstract

Osteoclasts are bone-resorbing multinucleated cells derived from the monocyte–macrophage lineage. Bone-forming osteoblasts play a role in the formation of osteoclasts. Osteoblasts/stromal cells express two cytokines essential for osteoclastogenesis: receptor activator of nuclear factor κB ligand (RANKL) and macrophage colony-stimulating factor (M-CSF). Using RANKL and M-CSF, osteoclasts can be induced from monocyte–macrophage lineage cells even in the absence of osteoblasts. We describe here methods for the identification of osteoclasts formed in vitro.

Key words: Osteoclast, RANKL, M-CSF, Bone marrow macrophage, RAW264.7 cell, Peripheral blood CD14-positive cell, TRAP, Vitronectin receptor, Pit assay

1. Introduction

Osteoclasts, the multinucleated giant cells that resorb bone, originate from hematopoetic cells of the monocyte–macrophage lineage (1, 2). We have developed a mouse coculture system comprised of primary osteoblasts and hematopoietic cells to examine the regulatory mechanism of osteoclastogenesis (1, 3). Bone resorbing factors such as $1\alpha,25$-dihydroxyvitamin D_3 [$1\alpha,25(OH)_2D_3$], parathyroid hormone (PTH), prostaglandin E_2 (PGE_2), and interleukin 11 (IL-11) stimulate the formation of osteoclasts in the coculture. A series of experiments using the coculture system have established the concept that osteoblasts/stromal cells or bone marrow-derived stromal cells have a key role in regulating osteoclast differentiation (1). Macrophage colony-stimulating factor (M-CSF, also called CSF-1) produced

Nicole I. zur Nieden (ed.), *Embryonic Stem Cell Therapy for Osteo-Degenerative Diseases*, Methods in Molecular Biology, vol. 690, DOI 10.1007/978-1-60761-962-8_18, © Springer Science+Business Media, LLC 2011

by osteoblasts/stromal cells (osteoblasts) was shown to be essential for the differentiation of osteoclasts from progenitor cells (4). Receptor activator of nuclear factor κB ligand (RANKL), a member of the tumor necrosis factor (TNF) family, was also identified as essential for osteoclastogenesis (5, 6). Osteoblasts express RANKL as a membrane-associated factor in response to various bone-resorbing factors (1, 5). Osteoclast precursors possess RANK, a TNF receptor family member, recognize RANKL through cell–cell interaction with osteoblasts/stromal cells, and differentiate into osteoclasts in the presence of M-CSF.

Bone marrow-derived macrophages differentiate into osteoclasts in the presence of M-CSF and RANKL (7, 8). Mouse macrophage-like RAW264.7 cells have been shown to differentiate into osteoclasts in response to RANKL even in the absence of M-CSF (9). Recent studies have shown that CD14-positive (CD14$^+$) monocytes prepared from human peripheral blood mononuclear cells efficiently differentiate into osteoclasts in cultures treated with M-CSF and RANKL (10, 11). We have also developed a method for obtaining a large number of functionally active osteoclasts from cocultures grown on collagen gel-coated dishes (12). Using osteoclasts recovered from the collagen gel culture, a reliable pit formation assay was established to investigate the regulatory mechanisms of osteoclast function (13, 14).

The most characteristic features of bone-resorbing osteoclasts are the presence of ruffled borders and sealing zones (15, 16). The sealing zone serves for the attachment of osteoclasts to the bone surface and isolation of the resorption area from the surroundings (17, 18). The sealing zone is defined as a unique large band of F-actin dots (actin ring). The formation of sealing zones in osteoclasts precedes the initiation of bone resorption, while the disruption of sealing zones results in the suppression of bone-resorbing activity of osteoclasts (19, 20). The resorbing area under the ruffled border is acidic, which favors the dissolution of bone minerals. In order to resorb bone matrix, osteoclasts highly express specific enzymes such as, tartrate resistant-acid phosphatase (TRAP), carbonic anhydrase II, matrix metalloproteinase 9, vacuolar type H$^+$-ATPase, and cathepsin K (15, 16). Osteoclasts also express abundant calcitonin receptors (21) and the vitronectin receptor αvβ3 integrin (22). Therefore, the detection of these enzymes and receptors is used to identify osteoclasts formed both in vivo and in vitro. When cultured on bone or dentine, osteoclasts form actin rings and resorb calcified substrates. Actin ring formation assays and pit formation assays are also used to identify osteoclasts formed in vitro. We describe here methods for the identification of osteoclasts formed in vitro.

2. Materials

2.1. TRAP Staining

1. Ca^{2+}- and Mg^{2+}-free phosphate-buffered saline (PBS, Sigma).
2. 3.7% Formaldehyde in PBS for fixation of cells.
3. Permeabilization solution: 0.1% Triton X-100 in PBS for permeabilization of cells.
4. Tartrate-resistant acid phosphatase (TRAP) staining solution: 5 mg of naphthol AS–MX phosphate (Sigma) is dissolved in 0.5 mL of N,N-dimethyl formamide in a glass container. Thirty milligrams of fast red violet LB salt and 50 mL of 0.1 M sodium acetate buffer, pH 5.0, containing 50 mM sodium tartrate are added to the mixture. TRAP-staining solution can be stored for 1 month at 4°C.

2.2. Immunostaining for Vitronectin Receptors

1. Ca^{2+}- and Mg^{2+}-free phosphate-buffered saline (PBS, Sigma).
2. Fixative: combine methanol and acetone to equal parts (50:50, vol/vol).
3. 3% Hydrogen peroxide (H_2O_2) to block endogenous peroxidase.
4. Histofine SAB-Po(M) kit (#426032, Nichirei Co., Tokyo, Japan): contains biotinylated anti-mouse IgG, avidin–biotin-conjugated peroxidase, and rabbit serum. Make a 10% dilution of the serum in PBS.
5. Primary antibody: monoclonal antibody against human vitronectin receptors (CD51/CD61), i.e. clone 23C6, BD Biosciences. Make a 1:500 working solution with PBS.
6. DAB substrate kit (e.g. Histofine, Nichirei Co., Tokyo, Japan).

2.3. Pit Formation Assay

2.3.1. Collagen-Coating of Dishes

1. 1 M N-2-hydroxyethylpiperazine-N-2-ethanesulfonic acid (HEPES) buffer, pH 7.4 containing 2.2% $NaHCO_3$ (7:2:1, by vol).
2. Type I collagen mixture: type I collagen solution (cell matrix type IA; Nitta Gelatin Co., see Note 1), 5× conc. α-Minimum Essential Medium (MEM), and 200 mM HEPES buffer with 2.2% $NaHCO_3$ are quickly mixed at a ratio of 7:2:1 by volume at 4°C just before use. Cell matrix type IA is suitable for this procedure.
3. 100-mm tissue culture treated dishes, i.e. Corning.

2.3.2. Crude Osteoclast Preparation

1. 0.2% Collagenase solution.
2. Shaking water bath (set to 60 cycles/min).
3. Aluminum foil.
4. Pit formation assay slice medium (PFAS medium): α-MEM with 10% fetal bovine serum.

2.3.3. Dentine Slice Culture	1. Dentine blocks (ivory): may be obtained from a local zoo.

2. Band saw (i.e. BS-3000, Exakt, Germany).

3. Cutting punch from Hardware store.

4. 70% Ethanol.

5. UV light source.

6. Ultrasonicator.

7. Mayer's hematoxylin (Wako Pure Chemical Industries).

8. Distilled water.

9. Light microscope.

10. Cotton buds.

11. Synthetic analogue of eel calcitonin (i.e. Elcatonin, Asahi Kasei Pharma, Tokyo).

2.4. Actin Ring Formation

1. Ca^{2+}- and Mg^{2+}-free phosphate-buffered saline (PBS, Sigma).

2. Rhodamine-conjugated phalloidin (Sigma): dissolve in a small volume of methanol, dilute with PBS to a final concentration of 0.3 mM, and store at 4°C in the dark.

3. 3.7% Formaldehyde in PBS for fixation of cells.

4. 0.1% Triton X-100 in PBS.

5. Distilled water.

6. Fluorescent microscope, for example, Carl Zeiss Axioplan 2.

3. Methods

3.1. TRAP Staining

Cytochemical staining for TRAP is widely used for identification of osteoclasts in vivo and in vitro.

1. Cells are fixed with 3.7% (v/v) formaldehyde in PBS for 10 min, treated with PBS containing 0.1% Triton X-100 for 1 min, and incubated with the TRAP-staining solution for 10 min at room temperature (see Note 2).

2. TRAP-positive osteoclasts appear as dark-red cells within 10 min (see Note 3).

3. After staining, cells are washed with distilled water, and TRAP-positive multinucleated cells having three or more nuclei as seen under a microscope are counted as osteoclasts (see Note 4) (Fig. 1).

3.2. Immunostaining for Vitronectin Receptors

Osteoclasts specifically express vitronectin receptors, cathepsin K, carbonic anhydrase II, and vacuolar type H^+-ATPase. Immunohistochemical staining of these markers is used for identification of osteoclasts formed in vitro (14, 15).

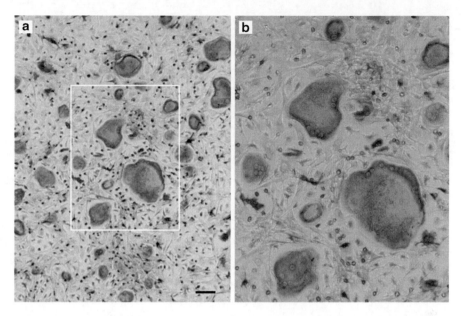

Fig. 1. Enzyme histochemistry for TRAP in mouse osteoclasts. The crude osteoclast preparation was prepared from the collagen-gel culture. The preparation was added at 0.1 mL per well to 48-well plates that contained 0.4 mL of α-MEM with 10% FBS per well. After 6 h of culture, the cells were fixed and stained for TRAP. Panel (**b**) shows a high-power view of the boxed region in (**a**). TRAP-positive cells appeared *red*. Scale bar = 100 μm.

1. Overlay cultures with cold fixative for 10 min on ice. Use 0.5 mL for each well of a 24-well plate and aspirate it off after incubation.

2. Treat cells for 10 min with 0.5 mL of 3% H_2O_2 to inactivate intrinsic peroxidase on ice. Remove the H_2O_2 solution by aspiration.

3. Add 1 mL of PBS to each well and incubate the plate for 5 min on ice. Aspirate supernatant and repeat this washing step three times.

4. Then, incubate cells for 10 min with 200 μL of 10% rabbit serum to block nonspecific binding. Remove the rabbit serum solution by aspiration and wash the cells with PBS three times, 5 min incubation for each wash.

5. Incubate cells with 200 μL/well primary antibody working solution or with a nonimmune mouse IgG control at room temperature. After incubation for 1 h, aspirate the serum solution and wash the cells with PBS three times for 5 min each.

6. Incubate cells for 10 min with biotinylated secondary antibody solution (200 μL) at room temperature and wash with PBS (5 min, three times).

7. Incubate cells for 10 min with avidin-conjugated peroxidase (200 μL) at room temperature and wash with PBS (5 min, three times).

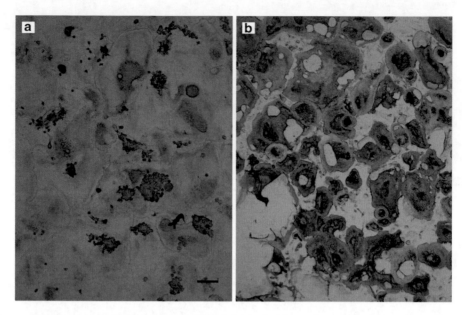

Fig. 2. Enzyme histochemistry for TRAP and immunohistochemistry for vitronectin receptors in human osteoclasts. Human CD14+ cells were prepared from human peripheral blood, and 5×10^5 cells/well were cultured with RANKL (100 ng/mL) and M-CSF (50 ng/mL) in 24-well plates. (**a**) After 6 days, the cells were fixed and stained for TRAP. TRAP-positive cells appeared red. (**b**) Cells were also stained with anti-vitronectin receptor antibody. Vitronectin receptor-positive cells appeared brown. Scale bar = 100 μm.

8. Incubate cells for 5 min with the DAB substrate (200 μL) at room temperature and wash with PBS (5 min, three times).

9. The bound antibodies are observed using a light microscope (Fig. 2).

Immunohistocytochemical staining of osteoclasts with antibodies against other specific markers such as carbonic anhydrase II and vacuolar proton ATPase is also utilized for the identification of osteoclasts (see Note 5). The pattern of immunohistochemical staining is essentially similar to that obtained with anti-vitronectin receptor antibodies.

3.3. Pit Formation Assay

3.3.1. Collagen-Coating of Dishes

Osteoclasts formed on plastic culture dishes are very difficult to detach by treatment with either trypsin/EDTA or bacterial collagenase. To obtain functionally active osteoclasts formed in cocultures with osteoblasts, a collagen gel culture is recommended.

1. Coat a 100-mm culture dish with 5 mL of the type I collagen mixture on ice.

2. Put the dish into a CO_2 incubator for 10 min to make the aqueous type I collagen gelatinous at 37°C.

3.3.2. Crude Osteoclast Preparation

1. Prepare a stem cell culture on collagen-coated plates as described above, and culture cells for 7 days with occasional medium changes when necessary (see Notes 6–8).

2. Next, prepare a crude osteoclast preparation from your stem cell differentiations.

3. Treat the dishes with 4 mL of a 0.2% collagenase solution for 20 min at 37°C in a shaking water bath. In order to do so, the culture dishes are carefully placed on a sheet of aluminum foil put on the water surface of the water bath to maintain the sterile condition of the dishes.

4. Collect the released cells into a 50-mL Falcon tube. Centrifuge at $250 \times g$ for 5 min, aspirate off or decant the supernatant, and suspend the pellet in 10 mL of PFAS medium (see Note 9). The crude osteoclast preparation is then used for biological and biochemical studies of osteoclasts.

3.3.3. Dentine Slice Culture

When osteoclasts are placed on dentine slices, they form resorption pits within 24 h in the presence of osteoblasts. A reliable pit formation assay was established using the crude osteoclast preparation recovered from the collagen gel culture and dentine slices (13, 14).

1. Prepare 200-μm thick dentine slices (ø 4 mm) from ivory blocks using a band saw and a cutting punch (see Note 10).

2. Clean dentine slices by ultrasonication in distilled water, sterilize using 70% ethanol, and dry under ultraviolet light. Store the slices at room temperature (see Note 11).

3. Place dentine slices in 96-well plates containing 0.1 mL/well of PFAS medium (one slice/well). Add a 0.1-mL aliquot of the crude osteoclast preparation onto the slices.

4. Let stand for 60 min at 37°C for setting down of cells on dentine slices.

5. Remove dentine slices and place them onto 24-well plates containing PFAS medium (0.5 mL/slice/well).

6. Culture cells on dentine slices for 24–48 h at 37°C.

7. Recover dentine slices from the culture. Strongly rub the surface of the slices with a cotton bud to remove all cells.

8. Place 10 μL of Mayer's hematoxylin on the surface of each dentine slice using surface tension for 1–2 min.

9. Wash dentine slices with distilled water. Rub the surface of the slices with a cotton bud to remove excess staining.

10. Resorption pits on dentine slices can be clearly visualized with Mayer's hematoxylin under transmitted light. Count the number of resorption pits formed on dentine slices under a light microscope. Alternatively, the resorbed area is measured using an image analysis system linked to a light microscope.

Resorption pits are first observed on dentine slices after culturing for 6–8 h, and the resorbed areas increase with time up to 72 h. Many resorption pits were observed on a dentine slice

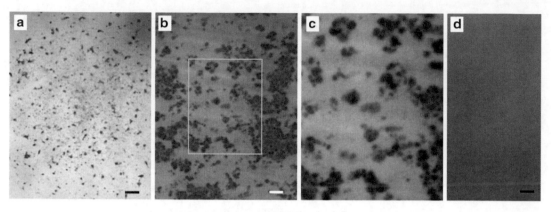

Fig. 3. Resorption pits formed by mouse osteoclasts on dentine slices. The osteoclast preparation recovered from a collagen-gel culture was placed on a dentine slice (4 mm ø) and cultured for 48 h in the presence (**d**) or absence (**a–c**) of eel calcitonin (10^{-9} M). The dentine slices were then treated with 3.7% (v/v) formaldehyde in PBS for 10 min, treated with PBS containing 0.1% Triton X-100 for 1 min, and incubated with the TRAP-staining solution for 10 min. The slices were washed with distilled water, and the TRAP-positive cells were observed through a microscope (**a**). After this observation, cells were removed from the dentine slices with a cotton bud. The slices were stained with Mayer's hematoxylin to visualize resorption pits (**b**). Panel (**c**) shows a high-power view of the boxed region in (**b**). Calcitonin strongly suppressed pit-forming activity of osteoclasts (**d**). Scale bar = 100 μm.

recovered from the culture for 48 h (Fig. 3) (see Note 12). When calcitonin at 10^{-9} M was added to the culture, the pit-forming activity of osteoclasts was completely inhibited (Fig. 3b) (19). Bisphosphonates also strongly inhibited the pit-forming activity of osteoclasts (20).

3.4. Actin Ring Formation

Osteoclasts adhere to bone through specialized discrete structures called "podosomes" in the clear zone, which consist mainly of dots containing F-actin (17, 18). Therefore, the ringed structure of podosomes (actin ring) is a characteristic of polarized osteoclasts (16–18). The actin rings are visualized by staining F-actin with rhodamine-conjugated phalloidin (see Note 13).

1. Fix cells cultured on dentine slices in 48-well plates with 0.4 mL of 3.7% formaldehyde in PBS for 10 min and wash three times with PBS.

2. Treat dentine slices with 0.4 mL of 0.1% Triton X-100 in PBS for 1 min.

3. Incubate dentine slices for 3 h with the rhodamine-conjugated phalloidin solution at 4°C. The rhodamine-conjugated phalloidin solution is recovered, and the slices are washed with water (see Note 14).

4. Actin rings formed by osteoclasts on dentine slices are detected with a fluorescence microscope.

Actin rings were observed on a dentine slice recovered from the culture for 24 h (Fig. 4a, c). One can confirm the presence of

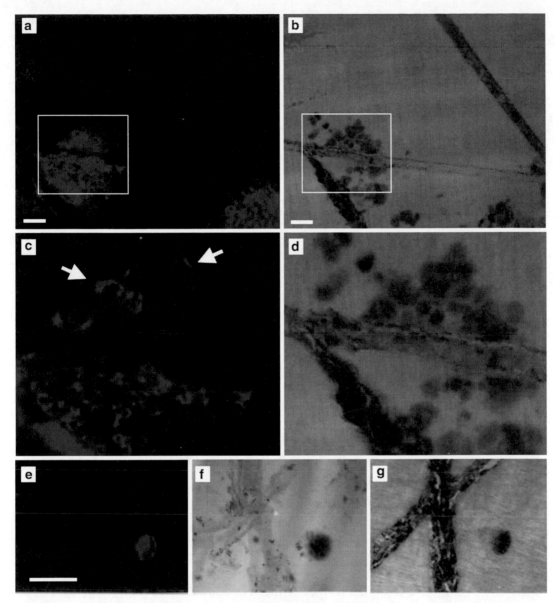

Fig. 4. Actin ring formation by mouse osteoclasts. The osteoclast preparation recovered from a collagen-gel culture was placed on a dentine slice (4 mm ø) and cultured for 24 h. Cells were then fixed with 3.7% formaldehyde. In this experiment, to identify both actin rings and resorption pits on the slices, several scratch lines were drawn on the slices with a blade. Then, slices were incubated with rhodamine-conjugated phalloidin to visualize the distribution of F-actin (**a, c**). After the observation of actin rings, the surface of dentine slices was rubbed with a cotton bud to remove all cells on the slices and subjected to staining with Mayer's hematoxylin (**b, d**). The two *scratch lines* indicate the same area of the slice shown in (**a**) and (**b**). Panels (**c**) and (**d**) show high-power views of the *boxed regions* in (**a**) and (**b**), respectively. One can perform triple staining of F-actin, TRAP, and resorption pits on the same dentine slice (**e–g**). Cells on the slice were fixed and then stained with rhodamine-conjugated phalloidin to visualize the distribution of F-actin (**e**). After the observation of actin rings, the slice was stained for TRAP (**f**). After this observation, cells were removed from the slices with a cotton bud. The slice was stained with Mayer's hematoxylin to visualize resorption pits (**g**). The two *scratch lines* indicate the same area of the slice shown in (**e–g**). Bars = 100 μm.

resorption pits on the slices after the observation of actin rings. The surface of dentine slices was rubbed strongly with a cotton bud to remove all cells and subjected to staining with Mayer's hematoxylin (Fig. 4b, d). Triple staining of actin rings, TRAP-positive cells, and resorption pits can be performed on the same dentine slices (Fig. 4e–g) (see Note 8).

4. Notes

1. Nitta Gelatin Co. supplies cell matrix type IA to scientists with an import permit in foreign countries.

2. TRAP is an intracellular enzyme. Treatment of cells with PBS containing 0.1% Triton X-100 is essential for staining TRAP. Instead of PBS containing 0.1% Triton X-100, ethanol–acetone (50:50, v/v) can be used in this step.

3. An incubation period longer than 10 min should be avoided, since cells other than osteoclasts become weakly positive for TRAP with time.

4. TRAP-positive mononuclear cells possess all characteristics of osteoclasts (23). Therefore, the total number of TRAP-positive cells can be counted.

5. Antibodies for carbonic anhydrase and vacuolar proton ATPase that have worked in our hands are #PA1-46408 and #PA1-10300, both Affinity BioReagents.

6. Initially, determine the number of cells that need to be seeded for your particular experiment.

7. Primary osteoblast–osteoclast cocultures (10, 11) may be used as positive controls for the stem cell differentiation assay. Primary osteoblasts (2×10^6 cells) and bone marrow cells (2×10^7 cells) are cocultured on the collagen gel-coated dish in 15 mL of α-MEM containing 10% FBS and 10^{-8} M $1\alpha,25$-$(OH)_2D_3$. The medium is changed every 2–3 days and the assay performed on day 7.

8. Incubate stem cell cultures as long as necessary in your stem cell differentiation medium of choice. It is recommended to perform a time course assay to determine the peak activity of the osteoclasts.

9. Usually, 4×10^4–1×10^5 osteoclasts are recovered from a 100-mm collagen gel-coated dish of control cultures containing primary osteoblasts and bone marrow cells, and the purity of osteoclasts is 2–3% in this crude preparation.

10. Bone slices prepared from bovine cortical bone can be used for the pit formation assay. However, we prefer dentine slices,

since dentine has a homogeneous structure and is free of vascular canals and osteocyte lacunae, which are present in bone slices. We obtained dentine blocks (ivory) through donation from a local zoo. Sperm whale dentine can be used as well.

11. A mark is put on one surface of each dentine slice with a pencil to check the orientation of the slice during the pit formation assay. The dentine slices are placed into wells of 96-well plates so that the surface without the mark faces up in the well.

12. To observe TRAP-positive cells on dentine slices, fix the slices with 3.7% (v/v) formaldehyde in PBS for 10 min and treat them with PBS containing 0.1% Triton X-100 for 1 min, and with the TRAP-staining solution for 10 min at room temperature. Once the TRAP-positive cells have been observed through a light microscope, rub the surface of the slices strongly with a cotton bud to remove all cells and stain the slices with Mayer's hematoxylin (Fig. 3).

13. Actin rings in osteoclasts cannot be observed after staining of TRAP. Therefore, the observation of the actin rings should precede the observation of TRAP-positive cells (Fig. 4).

14. The rhodamine-conjugated phalloidin solution can be used repeatedly until F-actin staining becomes weak.

References

1. Suda, T., Takahashi, N., and Martin, T. J. (1992) Modulation of osteoclast differentiation. *Endocr. Rev.* **13**, 66–80.

2. Boyle, W. J., Simonet, W. S., and Lacey, D. L. (2003) Osteoclast differentiation and activation. *Nature.* **423**, 337–342.

3. Takahashi, N., Akatsu, T., Udagawa, N., Sasaki, T., Yamaguchi, A., Moseley, J. M., et al. (1988) Osteoblastic cells are involved in osteoclast formation. *Endocrinology.* **123**, 2600–2602.

4. Yoshida, H., Hayashi, S., Kunisada, T., Ogawa, M., Nishikawa, S., Okamura, H., et al. (1990) The murine mutation osteopetrosis is in the coding region of the macrophage colony stimulating factor gene. *Nature.* **345**, 442–444.

5. Yasuda, H., Shima, N., Nakagawa, N., Yamaguchi, K., Kinosaki, M., Mochizuki, S., et al. (1998) Osteoclast differentiation factor is a ligand for osteoprotegerin/osteoclastogenesis-inhibitory factor and is identical to TRANCE/RANKL. *Proc. Natl. Acad. Sci. USA* **95**, 3597–3602.

6. Lacey, D. L., Timms, E., Tan, H. L., Kelley, M. J., Dunstan, C. R., Burgess, T., et al. (1998) Osteoprotegerin ligand is a cytokine that regulates osteoclast differentiation and activation. *Cell.* **93**, 165–176.

7. Kobayashi, K., Takahashi, N., Jimi, E., Udagawa, N., Takami, M., Kotake, S., et al. (2000) Tumor necrosis factor α stimulates osteoclast differentiation by a mechanism independent of the ODF/RANKL-RANK interaction. *J. Exp. Med.* **191**, 275–286.

8. McHugh, K. P., Hodivala-Dilke, K., Zheng, M. H., Namba, N., Lam, J., Novack, D., et al. (2000) Mice lacking beta3 integrins are osteosclerotic because of dysfunctional osteoclasts. *J. Clin. Invest.* **105**, 433–440.

9. Hsu, H., Lacey, D. L., Dunstan, C. R., Solovyev, I., Colombero, A., Timms, E., et al. (1999) Tumor necrosis factor receptor family member RANK mediates osteoclast differentiation and activation induced by osteoprotegerin ligand. *Proc. Natl. Acad. Sci. USA* **96**, 3540–3545.

10. Take, I., Kobayashi, Y., Nobuyuki, U., Tsuboi, H., Ochi, T., OKafuji, N., et al. (2005) Prostaglandin E_2 strongly inhibits human osteoclasts formation. *Endocrinology.* **146**, 5204–5214.

11. Tsuboi, H., Udagawa, N., Hashimoto, J., Yoshikawa, H., Takahashi, N., and Ochi, T. (2005) Nurse-like cells from patients with rheumatoid arthritis support survival of

osteoclast precursors via macrophage-colony stimulating factor production. *Arthritis Rheum.* **52**, 3819–3828.

12. Akatsu, T., Tamura, T., Takahashi, N., Udagawa, N., Tanaka, S., Sasaki, T., et al. (1992) Preparation and characterization of a mouse multinucleated cell population. *J. Bone Miner. Res.* **7**, 1297–1306.

13. Tamura, T., Takahashi, N., Akatsu, T., Sasaki, T., Udagawa, N., Tanaka, S., et al. (1993) A new resorption assay with mouse osteoclast-like multinucleated cells formed in vitro. *J. Bone Miner. Res.* **8**, 953–960.

14. Suda, T., Nakamura, I., Jimi, E., and Takahashi, N. (1997) Regulation of osteoclast function. *J. Bone Miner. Res.* **12**, 869–879.

15. Väänänen, H. K., Zhao, H., Mulari, M., and Halleen, J. M. (2000) The cell biology of osteoclast function. *J. Cell Sci.* **113**, 377–381.

16. Takahashi, N., Ejiri, S., Yanagisawa, S., and Ozawa, H. (2007) Regulation of osteoclast polarization. *Odontology.* **95**, 1–9.

17. Saltel, F., Destaing, O., Bard, F., Eichert, D., and Jurdic, P. (2004) Apatite-mediated actin dynamics in resorbing osteoclasts. *Mol. Biol. Cell.* **15**, 5231–541.

18. Luxenburg, C., Geblinger, D., Klein, E., Anderson, K., Hanein, D., Geiger, B., et al.

(2007) The architecture of the adhesive apparatus of cultured osteoclasts: from podosome formation to sealing zone assembly. *PLoS One* **2**, e179.

19. Suzuki, H., Nakamura, I., Takahashi, N., Ikuhara, T., Matsuzaki, K., Isogai, Y., et al. (1996) Calcitonin-induced changes in cytoskeleton are mediated by a signal pathway associated with protein kinase A in osteoclasts. *Endocrinology.* **137**, 4685–4690.

20. Takami, M., Suda, K., Sahara, T., Itoh, K., Nagai, K., Sasaki, T., et al. (2003) Involvement of vacuolar H^+-ATPase in specific incorporation of risedronate into osteoclasts. *Bone.* **32**, 341–349.

21. Sexton, P. M., Findlay, D. M., and Martin, T. J. (1999) Calcitonin. *Curr. Med. Chem.* **11**, 1067–1093.

22. Lakkakorpi, P. T., Horton, M. A., Helfrich, M. H., Karhukorpi, E. K., and Väänänen, H. K. (1991) Vitronectin receptor has a role in bone resorption but does not mediate tight sealing zone attachment of osteoclasts to the bone surface. *J. Cell Biol.* **115**, 1179–1186.

23. Yagi, M., Miyamoto, T., Sawatani, Y., Iwamoto, K., Hosogane, N., Fujita, N., et al. (2005) DC-STAMP is essential for cell–cell fusion in osteoclasts and foreign body giant cells. *J. Exp. Med.* **202**, 345–351.

Chapter 19

Analysis of Glycosaminoglycans in Stem Cell Glycomics

Boyangzi Li, Haiying Liu, Zhenqing Zhang, Hope E. Stansfield, Jonathan S. Dordick, and Robert J. Linhardt

Abstract

Glycosaminoglycans (GAGs) play a critical role in the binding and activation of growth factors in cell signal transduction required for biological development. A glycomics approach can be used to examine GAG content, composition, and structure in stem cells in order to characterize their general differentiation. Specifically, this method may be used to evaluate chondrogenic differentiations by profiling for the GAG content of the differentiated cells. Here, embryonic-like teratocarcinoma cells, NCCIT, a developmentally pluripotent cell line, were used as a model for establishing GAG glycomic methods, but will be easily transferrable to embryonic stem cell cultures.

Key words: Glycosaminoglycans, NCCIT cells, Chondroitin sulfate, Dermatan sulfate, Heparin, Heparan sulfate, Purification, Enzymatic digestion, Disaccharide analysis, LC-MS

1. Introduction

1.1. Glycomics

Glycomics is the study of the structure and function of glycans, glycoconjugates including glycosphingolipids, glycoproteins, such as proteoglycans (PGs), and glycan-binding proteins. An understanding of the cellular glycome should explain some mysteries associated with these frequent and important posttranslational modifications (1). The structural and functional glycomics of the glycosaminoglycan (GAG) chains (GAGome) of PGs from different tissues and cells are under intensive study in our laboratory (2–5). GAGs are linear, sulfated, heterogeneous polysaccharides consisting of various repeating disaccharide and are mainly located on both the external membrane of eukaryotic cells and within the extracellular matrix (6, 7). There are four distinct families of GAGs: chondroitin/dermatan sulfate (CS/DS), heparin/heparan sulfate (HS), keratan sulfate (KS), and hyaluronan (HA). GAGs are involved in

Nicole I. zur Nieden (ed.), *Embryonic Stem Cell Therapy for Osteo-Degenerative Diseases*, Methods in Molecular Biology, vol. 690, DOI 10.1007/978-1-60761-962-8_19, © Springer Science+Business Media, LLC 2011

numerous biological activities and are important as molecular coreceptors, in cell–cell interactions, cell adhesion, cell migration, cell signaling, cell growth, and cell differentiation (8–10).

1.2. GAGs in Embryonic Stem Cells

Embryonic stem cells (ESCs) have enormous potential as a source of cells for cell replacement therapy and have been used as in vitro models to study specific aspects of early embryonic development (11, 12). GAGs, particularly HS and CS, within stem cells play key roles in maintaining cell proliferation and differentiation (5). Understanding the glycomics of ESCs should shed light on development, including the differentiation of chondrocytes from mesenchymal cells (13–15). In our lab, we are using teratocarcinoma cells (NCCIT), a developmentally pluripotent cells, to offer a convenient model for ESCs. Methods for the elucidation of the GAGome within NCCIT cells, generally applicable for ESCs and their derivatives, such as chondrocytes, are described here for use in better understanding cell pluripotency and differentiation.

1.3. GAGs in the Extracellular Matrix of Chondrocytes

ESCs are able to differentiate into the mesenchymal cells that ultimately give rise to chondrocytes and endochondral ossification (13–15). Cartilage is a specialized connective tissue that provides support for other tissues or prevents friction of the joints. The cartilage is comprised of chondrocytes that sparsely distribute in extracellular matrix filled with collagen fibrils and PGs. The fibrous structure of collagen provides support and maintains tissue shape, while PGs form gels and act as filler to facilitate compressibility and prevent friction as well as perform other critical signaling functions. The PGs in cartilage, such as decorin, biglycan, and aggrecan, are glycosylated with one or multiple GAG chains. In cartilage, the GAG components are mainly CS and KS and HA as well as smaller amounts of DS and HS. CS, the most abundant GAG in the cartilage, is composed of 4-O-sulfo, 6-O-sulfo, and 4,6-di-O-sulfo sequences (16, 17). Research has shown that both the relative amounts of these sequences within CS and the length of CS chains change in aging and in diseases such as osteoarthritis (18–21). The GAG profile of differentiating ESCs may therefore help to elucidate whether or not chondrocytes have been formed, to what extent and may help to characterize the quality of the generated ESC-derived chondrocytes.

2. Materials

2.1. Cell Culture

1. 70% (v/v) ethanol.

2. NCCIT cells (ATCC: CRL-2073), frozen and preserved in 95% culture medium and 5% DMSO, were from American

Type Culture Collection. Cells are stored in liquid nitrogen until immediately prior to use (see Note 1).

3. NCCIT medium: 500 mL RPMI-1640 medium with L-glutamine, 50 mL fetal bovine serum (FBS, Invitrogen) and 5 mL 10,000 U/mL penicillin/streptomycin stock solution. FBS and penicillin/streptomycin stock solution are stored at –20°C before use and after mixing the culture medium is stored at 4°C before use.

4. Cell detachment process solution: 0.25% trypsin (Invitrogen) and 1 mM ethylenediamine tetraacetic acid (EDTA). Store at –20°C.

5. Trypan blue stain from Invitrogen for cell viability measurements.

6. Falcon® sterile polystyrene disposable aspirating pipettes (1, 5, 10, and 25 mL), sterile centrifuge tubes (15 and 50 mL) and sterile tissue culture flasks with vented cap (canted-neck; growth area: 25 cm²; total volume of the flask: 50 mL), for example from BD Biosciences.

7. Hemocytometer set (e.g. Hausser Scientific).

8. Microscope (e.g. CKX41, Olympus).

2.2. Recovery and Purification of GAGs

1. Defatting solution prepared from HPLC purity chloroform and HPLC purity methanol.

2. Proteolysis enzyme solution, actinase E (5 mg/mL in water) (see Note 2). Store at –20°C.

3. Protein and peptide denaturing solution: 8 M urea with 2% 3-[(3-cholamidopropyl)dimethylammonio]-1-propanesulfonate (CHAPS) adjusted to pH 8.3 using 1 M HCl.

4. Prewash solution: 200 mM sodium chloride (NaCl).

5. GAG collection solution: 16% (w/w) NaCl.

6. Methanol used as GAG precipitation solvent.

7. Millex™ 0.22-µm syringe driven filter unit from Millipore to remove particulates.

8. Vivapure MINI QH columns (Viva science) for GAG recovery.

9. Microcon® Centrifugal Filter Units-Microcon Ultracel YM-3 (3,000 MWCO), i.e. from Millipore for desalting.

10. 0.1 M sodium hydroxide (NaOH).

11. ColorpHast® pH strips (universal, pH ranging from 0 to 14), EMD Chemicals or similar.

2.3. Molecular Weight Analysis of GAGs by PAGE

1. Resolving gel buffer and lower chamber buffer: 100 mM boric acid, 100 mM Tris, and 1 mM disodium ethylenediaminetetraacetic acid (EDTA) at pH of 8.3. Store at room temperature.

2. Upper buffer: 1.24 M glycine and 200 mM Tris as written. Store at room temperature.

3. Front gel unpolymerized solution: 20.02% (w/v) acrylamide, 2% (w/v) *N*,*N*-methylenebisacrylamide, and 15% (w/v) sucrose in resolving gel buffer. Store at 4°C.

4. Stacking gel unpolymerized solution: 4.75% (w/v) acrylamide, 0.25% (w/v) *N*,*N*-methylenebisacrylamide in resolving gel buffer at pH 6.3 using 1 M HCl. Store at 4°C.

5. Polymerization reagents: *N*,*N*,*N*,*N*′-tetramethyl-ethylenediamine (TEMED), and 10% (w/v) aqueous ammonium persulfate (APS). Store separately at 4°C.

6. 50% (w/v) sucrose in water for density increase in GAGs. Store at 4°C.

7. Heparin (e.g. from Celsus Laboratory). Heparin oligosaccharides mixture, as the standard heparin ladder for molecular weight calculation can be obtained from mixing several oligosaccharides (e.g. tetrasaccharide, octasaccharide, decasaccharide, and dodecasaccharide) available from Iduron. Alternatively, heparin can be partially digested by heparinase I (Seikagaku) and used as substitute set of standards. In this protocol, we used partially digested heparin and a pure heparin-derived octasaccharide standard prepared in our laboratory (22). Phenol red solution can be added to aid in real-time visualization during electrophoresis.

8. Gel staining reagent: 0.5% (w/v) Alcian blue in 2% (v/v) aqueous acetic acid.

9. Mini-gel electrophoresis system PowerPac 1000 from Bio-Rad.

10. Gel-loading pipette tips (200 µL).

11. UN-SCAN-IT™ digitizing software, i.e. Silk Scientific or similar.

2.4. Enzymatic Lyase Depolymerization of GAGs and Recovery of GAG Disaccharides

1. 20 mM Tris(hydroxymethyl)-aminomethane (Tris)–HCl buffer, pH 7.2.

2. Chondroitin sulfate depolymerization enzymes: 10 mU of chondroitinase ABC and 5 mU of chondroitinase ACII (Seikagaku) prepared in 20 mM, pH 7.2 Tris–HCl buffer containing 0.1% BSA. Store at –20°C.

3. Heparin/heparan sulfate depolymerization enzymes: Heparinase I, II, and III (Seikagaku) prepared as a mixture 5 mU each in 20 mM pH 7.2 PBS buffer. Store at –20°C.

4. Desalting columns: Microcon® Centrifugal Filter Units-Microcon Ultracel YM-3 (3,000 MWCO) from Millipore.

2.5. Disaccharide Analysis by LC-MS

1. Unsaturated CS/DS disaccharides standards: ΔDi-0S, ΔUA-GalNAc; Δdi-4S, ΔUA-GalNAc4S; Δdi-6S, ΔUA-GalNAc6S; Δdi-UA2S, ΔUA2S-GalNAc; Δdi-diS$_B$, ΔUA2S-GalNAc4S; Δdi-diS$_D$, ΔUA2S-GalNAc6S; Δdi-diS$_E$, ΔUA-GalNAc4S6S; and Δdi-triS, ΔUA2S-GalNAc4S6S (Seikagaku Corporation).

2. Unsaturated heparin/HS disaccharides standards: Δdi-0S, ΔUA-GlcNAc; Δdi-NS, ΔUA-GlcNS; Δdi-6S, ΔUA-GlcNAc6S; Δdi-UA2S, ΔUA2S-GlcNAc; Δdi-UA2SNS, ΔUA2S-GlcNS; Δdi-NS6S, ΔUA-GlcNS6S; Δdi-UA2S6S, ΔUA2S-GlcNAc6S; and Δdi-triS, ΔUA2S-GlcNS6S (Iduron).

3. Disaccharide detection system: LC-MS system (Agilent, LC/MSD trap MS).

4. HPLC solution A for CS/DS disaccharide analysis: 0% (v/v) HPLC grade acetonitrile in HPLC grade water, 15 mM hexylamine (HXA) and 100 mM 1,1,1,3,3,3,-hexafluoro-2-propanol (HFIP).

5. HPLC solution B for CS/DS disaccharide analysis: 75% (v/v) HPLC grade acetonitrile in HPLC grade water, 15 mM HXA, and 100 mM HFIP.

6. HPLC solution C for heparin/HS disaccharide analysis: 15% (v/v) HPLC grade acetonitrile in HPLC grade water, 37.5 mM NH_4CH_3COO, and 11.25 mM tributylamine (TBA), pH 6.5 adjusted with glacial acetic acid.

7. HPLC solution D for heparin/HS disaccharide analysis: 65% (v/v) HPLC grade acetonitrile in HPLC grade water, 37.5 mM NH_4CH_3COO, and 11.25 mM TBA, pH 6.5 adjusted with glacial acetic acid.

8. ACQUITY UPLC™ BEH C18 column (Waters, 2.1 × 150 mm, 1.7 µm) for CS/DS disaccharide analysis and Zorbax SB-C18 column (Agilent, 0.5 × 250 mm, 5 µm) for heparin/HS disaccharide analysis.

3. Methods

Disaccharide analysis is useful for assessing the structure of the GAGome in pluripotent cells such as teratocarcinoma cells and embryonic stem cells. Changes in the GAGome can then be correlated to alteration in the transcription levels for enzymes involved in GAG biosynthesis, PG core proteins, GAG-binding proteins such as growth factors, growth factor receptors, chemokines, and adhesion proteins. An improved knowledge of structural glycomics of GAGs should result in a better understanding of the relationship of the GAGome to the functional glycomics associated with stem cell differentiation.

The following protocol will explain how to characterize the GAGome of embryonic stem cells and their differentiated progeny using teratocarcinoma cells as a model. In brief, teratocarcinoma cells are grown to confluency and 10^7 cells are collected. The washed cell pellet is defatted, proteolyzed and GAGs are extracted into CHAPS/Urea. Spin column-based ion chromatography is then used to recover the CS and HS GAGs that are washed and then released in salt. After membrane-based desalting, the molecular weights of purified CS and HS GAGs are analyzed by PAGE. The CS and HS GAGs are then individually depolymerized to disaccharide mixtures using either chondroitinases or heparinases. The resulting disaccharide mixtures are analyzed by reversed-phase ion-pairing high performance liquid chromatography and detected by UV and MS.

3.1. Preparation of Cells

Before performing the following steps, media should be taken out of the refrigerator and warmed to 37°C using a water bath. Make certain that the temperature never rises above 40°C. All steps performed in Subheading 3.1 were done in a laminar flow hood in a Biosafety Level 2 laboratory. The hood was sterilized with 70% (v/v) ethanol and exposed to UV light before use. All items taken into the hood were swabbed with 70% (v/v) ethanol.

3.1.1. NCCIT Cell Culture Inoculation and Maintenance

1. Remove a vial of cells from the liquid nitrogen tank and thaw quickly by swirling in the 37°C water bath. Make sure not to submerge the vial. Transfer the contents of the vial to a 15-mL centrifuge tube. With swirling add in a dropwise fashion 9 mL of warmed (37°C) NCCIT medium.

2. Centrifuge the cells at $250 \times g$ for 5 min. Discard the supernatant and resuspend the pellet in 10 mL of prewarmed (37°C) NCCIT medium. Transfer the cells to a T25-cm² flask and place in the 37°C incubator.

3. Change culture medium when the color indicator in the medium changes from rosy pink to yellow (approximately every 2 days, change the media at least three times per week).

4. Take the cell culture flask out of the 37°C incubator and remove the medium carefully using sterile glass pipette. Avoid touching the side where cells are growing and add 10 mL of fresh medium to the flask before returning the cell culture flask to the incubator.

5. NCCIT cells should be passaged when they reach 80% of the confluency on the wall of cell culture flask (in our experience about 4 days) (Fig. 1). For passaging, take the cell culture flask out of the 37°C incubator. Using a microscope (20-fold magnification), estimate the confluency of cells growth on the flask wall.

Fig. 1. View of cultured teratocarcinoma cells using phase-contrast microscopy.

6. Take the cell culture flask out of the incubator and remove the medium carefully using sterile glass pipette. Avoid touching the side where cells are growing and add 1 mL of the cell detachment solution.

7. Lay the flask down to let the solution completely contact with the layer of cells. Wait for about 5 min.

8. Agitate flask to make sure the cells are completely detached. Add 9 mL of NCCIT medium. Repeatedly rinse the flask wall with the medium in the flask to detach as many cells as possible from the wall of the flask. Make sure most of the cells are at the bottom of the flask suspended in the medium. Transfer the entire suspension into a 15-mL centrifuge tube and cetrifuge at 1,000 × g for 3 min. Remove the supernatant and resuspend the pellet in 10 mL of fresh NCCIT medium. Transfer an equal volume of 2.5 mL cell suspension into four new culture flasks. Add 7.5 mL of fresh and warmed (37°C) culture medium to each of those new cell culture flasks. Return the cell culture flasks to the 37°C incubator.

3.1.2. NCCIT Cell Harvest

1. Take the cell culture flask out of the incubator and remove the medium carefully using sterile glass pipette. Avoid touching the side where cells are growing and add 1 mL of cell detachment solution.

2. Lay the flask down to let the solution completely contact with the layer of cells and wait for about 5 min. Add 9 mL of NCCIT medium.

3. Repeatedly rinse the wall cells grow on for several times with the medium in the flask, detaching as many cells as possible from the wall of the flask. Ensure that most of the cells are at the bottom of the flask suspended in the medium.

4. Transfer the entire suspension into a 15-mL centrifuge tube.

5. Centrifuge the cultured cells at $1{,}000 \times g$ for 3 min and remove the supernatant.

6. Resuspend the pellet in 5 mL of fresh NCCIT medium.

7. Always count cell before harvesting and only harvest cells when they reach 10^6 cells/mL.

8. Before cell counting, make sure the medium and the cells are mixed well. Combine 10 μL cell suspension and 10 μL trypan blue stain in one well of a 96-well plate and mix well.

9. Inject 10 μL mixture into the cleft of the hemocytometer and count the number of both living cells (transparent) and dead cells (blue).

10. Determine the percentile of cell viability and calculate the approximate total cell number in the medium. If viability is >50%, replate at higher density.

11. Repeat step 5. To the pellet, add 10 mL of PBS buffer.

12. Gently mix the cells and buffer using a pipette. Centrifuge at $1{,}000 \times g$ for 3 min and rinse with PBS another two times, then collect the cell pellet after the last centrifugation minimum amount of PBS buffer (see Note 3).

3.2. Recovery and Purification of GAGs

Start with approximately 10^7 NCCIT cells prepared and counted using a hemocytometer as described in Subheading 3.1. Total GAG extraction as described previously (23) requires a multistep procedure (see Note 4).

1. Lyophilize cells by freezing the cell pellet from Subheading 3.1.2.12 at –60°C for 30 min. Place the frozen cell pellet into a tube in a freeze dryer bottle (Fisher Scientific) and attach to a lyophilizer. The sample is freeze-dried overnight under pressure of $<1.3 \times 10^{-8}$ bar at a collector temperature of –40°C.

2. Defatting involves the three-step washing of the cell pellet with 3 mL each of 2:1, 1:1, 1:2 (v/v) chloroform/methanol. Samples are placed on a shaker at a speed of 200 rpm at room temperature. Each step takes about 8–10 h.

3. Between the steps, leave the mixture to sediment. Remove the supernatant portion with a glass pipette before adding the new wash.

4. Perform a proteolysis step by incubating defatted cell pellets with actinase E proteolysis solution at 55°C overnight.

5. For GAGs extraction, add dry urea and dry CHAPS to the proteolyzed aqueous sample to obtain a final concentration of 2% (wt%) in CHAPS and 8 M in urea. Remove particles from the resulting solutions by either centrifuging at $4,000 \times g$ or by passing the samples through a 0.22-μm syringe filter.

6. To recover and purify GAG use a Vivapure Mini Q H spin column (see Note 5). Wash and preequilibrate spin columns with 200 μL denaturing solution by centrifuging at $2,000 \times g$. Load sample (approximately 0.5 mL) onto the wet spin column and run through the spin columns under $2,000 \times g$. Wash the spin column once with 200 μL denaturing solution at $2,000 \times g$.

7. Next, wash the spin column five times at $2,000 \times g$ with 200 μL prewash solution to remove nonspecific binding materials.

8. Elute HS and CS GAGs from column by washing three times at $2,000 \times g$ with 50 μL of collection solution.

9. Desalt GAGs with a Microcon® Centrifugal Filter Units YM-3 (3,000 MWCO) spin column (see Note 6). To do so, load 100 μL of NaOH to prewash the spin column and centrifuge at $12,000 \times g$.

10. Rinse the column five times with 400 μL of water to remove all the NaOH, centrifuging after each wash at $12,000 \times g$. Make sure the eluate is at pH 7 before proceeding further using pH strips.

11. Load GAG samples and centrifuge at $12,000 \times g$.

12. Wash the membrane five times with 400 μL of water to completely remove salts and other small molecules, centrifuging at $12,000 \times g$ after each wash.

13. The GAGs are recovered from the top layer of the filtration membrane by inverting the membrane and centrifuging at $1,000 \times g$.

14. Then rinse the surface of membrane three times, each time with 100 μL of water centrifuging at $1,000 \times g$ to obtain residual GAGs on membrane. Store the GAG-containing wash (approximately 350 μL) at 4°C or lyophilize for future use.

3.3. Analysis of Intact GAGs

1. Preparing the gel in a mini-gel electrophoresis system begins by washing all equipment and glass plates thoroughly with detergent before and rinsed extensively with distilled water before and after each use.

3.3.1. Preparing a Gel

2. Prepare a 0.75 mm thick, 22% gel by mixing 6 mL of front gel buffer, 36 μL of 10% (w/w) aqueous ammonium persulfate solution, and 6 μL of TEMED, and mix rapidly with a needle.

3. Inject the gel into the sandwich glass plates by syringe, leaving some space for a stacking gel.

4. Cover the upper layer of the gel with water. The gel should polymerize within about 30 min depending on the room temperature. After the polymerization is set, carefully remove the water.

5. Prepare the stacking gel by mixing 2 mL of stacking gel buffer with 60 µL of 10% (w/v) aqueous ammonium persulfate solution, and 2 µL of TEMED, mixing rapidly with a needle. Using a syringe, inject the stacking gel to the top of separating gel. Insert a comb, carefully avoiding any air bubbles. The stacking gel should polymerize within about 30 min at room temperature.

6. Assemble the inner core of gel system. Once the stacking gel polymerization is done, carefully remove the comb and pour in upper buffer. Make certain there is no leakage on the assembled inner core. Add the resolving buffer to the lower chambers of the gel system.

3.3.2. Loading and Running the Gel

1. Mix 5 µL of sample and 5 µL of 50% (w/v) sucrose.

2. Load 10 µL of each sample into a well. One well should contain a mixture of heparin oligosaccharide standards, such as oligosaccharides prepared by the partial enzymatic depolymerization of heparin, and one well should contain a purified heparin oligosaccharide, such as an octasaccharide. Phenol Red (0.5 µL) can be added to this well if a visible indicator if needed.

3. Complete the assembly of the gel system and connect to a power supply. Gel electrophoresis is performed at 200 V for 90 min or until the phenol red reaches the bottom of the plate.

3.3.3. Staining, Destaining, and Quantifying the Data

1. After electrophoresis is complete, carefully separate the gel from the glass plates.

2. Stain the gel with Alcian blue dye reagent for 30–60 min.

3. Completely destain the gel by shaking overnight in water.

4. Wrap the gel in clear plastic wrap and scan the gel with on a standard computer scanner (Fig. 2). The scan can then be digitized using UN-Scan-it software.

5. A standard curve, the log of the molecular weight of each heparin oligosaccharide band (disaccharide 665, tetrasaccharide 1330, hexasaccharide 1995, etc.) as a function of migration distance, is plotted.

6. The pure oligosaccharide is used to provide a counting frame by identifying the size of the oligosaccharides in the heparin oligosaccharide mixture.

7. The average molecular weights of the GAGs isolated from the cells are calculated based on this standard curve (24).

3.4. GAG
Depolymerization

1. Incubate intact GAGs recovered in Subheading 3.2 with the chondroitinase ABC and ACII enzymes at 37°C for 10 h.

2. Recover the products of the chondroitinase treatment using the Microcon® Centrifugal Filter Units YM-3 (3,000 MWCO). Refer to Subheading 3.2, steps 9–14 for detailed procedures.

3. In the current case, collect both the portion, which passes through the membrane, and the portion remaining above the membrane (the retentate). The CS/DS disaccharides have a molecular weight <3,000 and passing through the membrane should be collected in three washes, combined and lyophilized and used for further LC-MS analysis (Subheading 3.5.1). Continue with the retentate with the following step.

4. GAGs in the retentate, remaining on the top of the filtration membrane, are next incubated with the heparinase I, II, and III enzyme mixture at 37°C for 10 h. The HS disaccharides have a molecular weight <3,000 and passing through the Microcon® Centrifugal Filter Units YM-3 (3,000 MWCO) membrane should be collected in three washes, combined,

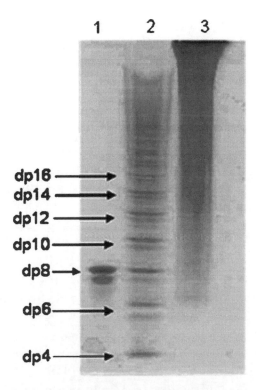

Fig. 2. PAGE analysis on GAGs isolated from NCCIT cells. *Lane 1* shows heparin oligosaccharide with degree of polymerization of eight (dp8). *Lane 2* shows heparin oligosaccharide standard where the degree of polymerization (dp) from four (tetrasaccharide) to 14 tetradecasaccharide is labeled. *Lane 3* shows the isolated intact GAG mixture. PAGE analysis with Alcian blue staining confirmed that GAGs were present by a broad band of expected polydispersity. After digitizing the gels using UN-Scan-it software, the average MW of GAGs were calculated based on the heparin oligosaccharide standards. The average molecular weight of GAGs from NCCIT cells is 15.53 kDa.

and lyophilized and used for further LC-MS analysis (see Subheading 3.5.2).

3.5. Disaccharide Analysis

This method has been optimized by our laboratory and has been found to work well (2, 25, 26). Two different eluent systems are required for the optimum resolution of the CS/DS and heparin/HS disaccharides.

3.5.1. CS/DS Disaccharide Analysis

1. Inject 8 µL of disaccharide standards containing 10 ng of each disaccharide or 8 µL of CS/DS disaccharide sample from Subheading 3.4) onto an ACQUITY UPLC™ BEH C18 column.

2. Use HPLC solution A and HPLC solution B to elute CS/DS disaccharides at 100 µL/min.

3. The elution conditions at 45°C are solution A for 10 min, followed by a linear gradient from 10 to 40 min of 0–50% solution B.

4. The column effluent enters the UV detector followed by the source of the electrospray ionization mass spectrometer for continuous detection. Set the electrospray interface in positive ionization mode with the skimmer potential 40.0 V, capillary exit 40.0 V and a source of temperature of 350°C to obtain maximum abundance of the ions in a full scan spectra (350–2,000 Da, ten full scans/s).

5. Use nitrogen as a drying (8 L/min) and nebulizing gas (40 p.s.i.).

6. Use UV detection at 232 nm with simultaneous extracted ion chromatogram (EIC) detection (see Note 7). The results of this analysis are presented in Fig. 3 and Table 1.

3.5.2. Heparin/HS Disaccharide Analysis

1. Use a Zorbax SB-C18 column (Agilent, 0.5 × 250 mm, 5 µm) and inject 8 µL of disaccharide standards containing 10 ng of each disaccharide or 8 µL of heparin/HS disaccharide sample from Subheading 3.4.

2. Use HPLC solution C and HPLC solution D to elute heparin/HS disaccharides at 10 µL/min.

3. The elution conditions at 20°C are solution C for 20 min, followed by a linear gradient from 20 to 45 min of 0–50% solution D.

4. The column effluent should enter the UV detector followed by the source of the electrospray ionization mass spectrometer for continuous detection.

5. Add another 5 µL/min of acetonitrile just after column and before MS to make the solvent and TrBA easy spray, and easy evaporate in the ion-source.

6. Set the electrospray interface in negative ionization mode with the skimmer potential –40.0 V, capillary exit –40.0 V

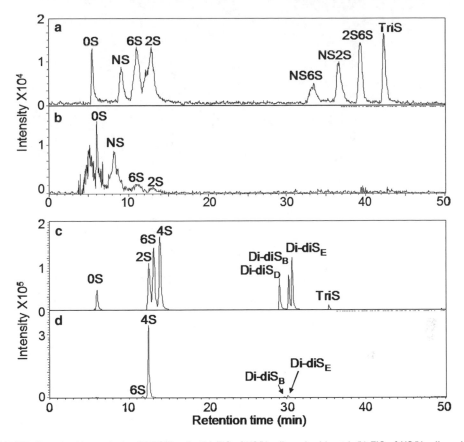

Fig. 3. LC-MS disaccharide analysis of NCCIT cells. (**a**) EIC of HS/Hp disaccharide std. (**b**) EIC of HS/Hp disaccharide of NCCIT cells. (**c**) EIC of CS/DS disaccharide std. (**d**) EIC of CS/DS disaccharide of NCCIT cells.

and a source of temperature of 325°C to obtain maximum abundance of the ions in a full scan spectra (150–1,500 Da, ten full scans/s).

7. Use nitrogen as a drying (5 L/min) and nebulizing gas (20 p.s.i.). Detect with UV at 232 nm with simultaneous extracted ion chromatogram (EIC) detection (see Note 8). The results of this analysis are presented in Fig. 3 and Table 1.

4. Notes

1. NCCIT cells were established by Shinichi Teshima (National Cancer Institute, Tokyo, Japan) in 1985 from a mediastinal mixed germ cell tumor (27). This pluripotent stem cell line is capable of somatic and extraembryonic differentiation. The undifferentiated cells are equivalent to a stage intermediate between seminoma and embryonal carcinoma. They will differentiate in response to retinoic acid (28).

Table 1
Disaccharide composition of NCCIT cells

HS/Hp	0S	NS	6S	2S	NS6S	NS2S	2S6S	TriS
	23.8%	47.6%	18.1%	10.5%	n.d.	n.d.	n.d.	n.d.

CS/DS	0S	2S	6S	4S	NS6S	NS2S	2S6S	TriS
					Di-diS$_D$	Di-diS$_B$	Di-diS$_E$	
	n.d.	n.d.	1.2%	94.3%	n.d.	0.3%	4.2%	n.d.

HS/Hp: 0S, ΔUA-GlcNAc; NS, ΔUA-GlcNS; 6S, ΔUA-GlcNAc6S; 2S, ΔUA-GlcNAc6S; TriS, ΔUA2S-GlcNS6S. CS/DS: 0S, ΔUA-GalNAc; 2S, ΔUA2S-GalNAc; 6S, ΔUA-GalNAc6S; 4S, ΔUA-GalNAc4S; Di-diS$_B$, ΔUA2S-GalNAc4S; Di-diS$_D$, ΔUA-GalNAc4S6S; TriS, ΔUA2S-GalNAc4S6S; n.d., not detected.

2. It is best to freshly prepare actinase E proteolysis enzyme solution from dry protein immediately before use.

3. We tested series of different number of cells for GAGs extraction, recovery and subsequent disaccharide analysis, we found that 1×10^6 is currently the minimum number of cells required for disaccharide quantification.

4. This method for total GAGs extraction and recovery was developed in our lab, involving in the use of a simple recovery and purification that relies on protease digestion and strong anion-exchange chromatography on a spin column followed by salt. Urea, a nonionic denaturant, is known to solubilize most proteins, and Chaps, a zwitterionic surfactant, is commonly used to solubilize hydrophobic molecules such as triglycerides. Approximately 90% of GAGs can be recovered using this method.

5. When Vivapure spin filters are used, make sure the centrifugal force is not $>2,000 \times g$. For each centrifugation, adjust time carefully so that there is always residual liquid above on the top of the membrane to avoid dryness and membrane cracking.

6. When Microcon® Centrifugal Filter Units YM-3 (3,000 MWCO) spin columns are used, make sure the centrifugal force is not $>14,000 \times g$. For each centrifugation, adjust time carefully so that there is always residual liquid above on the top of the membrane to avoid dryness and membrane cracking.

7. The sensitivity in this method improved to 0.2 ng/disaccharide of CS/DS.

8. The sensitivity in this method improved to 2 ng/disaccharide of HS/Hp.

Acknowledgment

Our laboratory acknowledges generous support from the New York State Department of Health and the Empire State Stem Cell Board in the form of grant number N08G-264.

References

1. Raman, R., Raguram, S., Venkataraman, G., Paulson, J. C., and Sasisekharan, R. (2005) Glycomics: an integrated systems approach to structure-function relationships of glycans. *Nat. Methods* **2**, 817–824.

2. Zhang, F., Zhang, Z., Thistle, R., McKeen, L., Hosoyama, S., Toida, T., et al. (2009) Structural characterization of glycosaminoglycans from zebrafish in different ages. *Glycoconj. J.* **26**, 211–218.

3. Park, Y., Yu, G., Gunay, N. S., and Linhardt, R. J. (1999) Purification and characterization of heparan sulphate proteoglycan from bovine brain. *Biochem. J.* **344(Pt 3)**, 723–730.

4. Warda, M., Toida, T., Zhang, F., Sun, P., Munoz, E., Xie, J., et al. (2006) Isolation and characterization of heparan sulfate from various murine tissues. *Glycoconj. J.* **23**, 555–563.

5. Nairn, A. V., Kinoshita-Toyoda, A., Toyoda, H., Xie, J., Harris, K., Dalton, S., et al. (2007)

Glycomics of proteoglycan biosynthesis in murine embryonic stem cell differentiation. *J. Proteome Res.* **6**, 4374–4387.

6. Linhardt, R. J., and Toida, T. (2004) Role of glycosaminoglycans in cellular communication. *Acc. Chem. Res.* **37**, 431–438.

7. Linhardt, R. J. (2003) 2003 Claude S. Hudson Award address in carbohydrate chemistry. Heparin: structure and activity. *J. Med. Chem.* **46**, 2551–2564.

8. Johnson, Z., Proudfoot, A. E., and Handel, T. M. (2005) Interaction of chemokines and glycosaminoglycans: a new twist in the regulation of chemokine function with opportunities for therapeutic intervention. *Cytokine Growth Factor Rev.* **16**, 625–636.

9. Capila, I., and Linhardt, R. J. (2002) Heparin-protein interactions. *Angew. Chem. Int. Ed. Engl.* **41**, 390–412.

10. Beenken, A., and Mohammadi, M. (2009) The FGF family: biology, pathophysiology and therapy. *Nat. Rev. Drug Discov.* **8**, 235–253.

11. Hoffman, L. M., and Carpenter, M. K. (2005) Characterization and culture of human embryonic stem cells. *Nat. Biotechnol.* **23**, 699–708.

12. Giacomini, M., Baylis, F., and Robert, J. (2007) Banking on it: public policy and the ethics of stem cell research and development. *Soc. Sci. Med.* **65**, 1490–1500.

13. Uygun, B. E., Stojsih, S., and Matthew, H. (2009) Effects of immobilized glycosaminoglycans influence proliferation and differentiation of mesenchymal stem cells. *Tissue Eng. Part A* **15(11)**, 3499–3512.

14. Kumarasuriyar, A., Murali, S., Nurcombe, V., and Cool, S. M. (2009) Glycosaminoglycan composition changes with MG-63 osteosarcoma osteogenesis in vitro and induces human mesenchymal stem cell aggregation. *J. Cell Physiol.* **218**, 501–511.

15. Dombrowski, C., Song, S. J., Chuan, P., Lim, X., Susanto, E., Sawyer, A. A., et al. (2009) Heparan sulfate mediates the proliferation and differentiation of rat mesenchymal stem cells. *Stem Cells Dev.* **18**, 661–670.

16. Carney, S. L., and Muir, H. (1988) The structure and function of cartilage proteoglycans. *Physiol. Rev.* **68**, 858–909.

17. Poole, A. R. (1986) Proteoglycans in health and disease: structures and functions. *Biochem. J.* **236**, 1–14.

18. Bayliss, M. T., Osborne, D., Woodhouse, S., and Davidson, C. (1999) Sulfation of chondroitin sulfate in human articular cartilage. The effect of age, topographical position, and

zone of cartilage on tissue composition. *J. Biol. Chem.* **274(22)**, 15892–15900.

19. Rizkalla, G., Reigner, A., Bogoch, E., and Poole, A. R. (1992) Studies of the articular cartilage proteoglycan aggrecan in health and osteoarthritis. Evidence for molecular heterogeneity and extensive molecular changes in disease. *J. Clin. Invest.* **90**, 2268–2277.

20. Plaas, A. H., Wong-Palms, S., Roughley, P. J. Midura, R. J., and Hascall, V. C. (1997) Chemical and immunological assay of the nonreducing terminal residues of chondroitin sulfate from human aggrecan. *J. Biol. Chem.* **272**, 20604–20610.

21. Hitchcock, A. M., Yates, K. E., Shortkroff, S., Costello, C. E., and Zaia, J. (2007) Optimized extraction of glycosaminoglycans from normal and osteoarthritic cartilage for glycomics profiling. *Glycobiology* **17(1)**, 25–35.

22. Pervin, A., Gallo, C., Jandik, K. A., Han, X.-J., and Linhardt, R. J. (1995) Preparation and structural characterization of large heparin-derived oligosaccharides. *Glycobiology* **5**, 83–95.

23. Zhang, F., Sun, P., Munoz, E., Chi, L., Sakai, S., Toida, T., et al. (2006) Microscale isolation and analysis of heparin from plasma using an anion-exchange spin column. *Anal. Biochem.* **353**, 284–286.

24. Edens, R. E., Al-Hakim, A., Weiler, J. M., Rethwisch, D. G., Fareed, J., and Linhardt, R. J. (1992) Gradient polyacrylamide gel electrophoresis for determination of the molecular weights of heparin preparations and low-molecular-weight heparin derivatives, *J. Pharm. Sci.* **81**, 823–827.

25. Zhang, Z., Park, Y., Kemp, M. M., Zhao, W., Im, A. R., Shaya, D., et al. (2009) Liquid chromatography-mass spectrometry to study chondroitin lyase action pattern. *Anal. Biochem.* **385**, 57–64.

26. Zhang, Z., Xie, J., Liu, H., Liu, J., and Linhardt, R. J. (2009) Quantification of heparan sulfate disaccharides using ion-pairing reversed-phase microflow high-performance liquid chromatography with electrospray ionization trap mass spectrometry. *Anal. Chem.* **81**, 4349–4355.

27. Teshima, S., Shimosato, Y., Hirohashi, S., Tome, Y., Hayashi, I., Kanazawa, H., Kakizoe, T. (1988) Four new human germ cell tumor cell lines. *Lab Invest.* 59, 328–336.

28. Damjanov, I., Horvat, B., Gibas, Z. (1993) Retinoic acid-induced differentiation of the developmentally pluripotent human germ cell tumor-derived cell line, NCCIT. *Lab Invest.* 68, 220–232.

Chapter 20

Drill Hole Defects: Induction, Imaging, and Analysis in the Rodent

Andre Obenaus and Pedro Hayes

Abstract

Advances in stem therapy, scaffolds, and therapeutic biomolecules are accelerating bone repair research, and model systems are required to test new methods and concepts. The drill hole defect is one such model and is used to study a variety of bone defects and potential therapies designed to repair these injuries. We detail the methodologies required to successfully generate and evaluate drill hole defects. Although performing a successful drill hole defect requires patience and dexterity, investing the time to perfect the technique will provide ample opportunity for the researcher to expand his/her particular research interests. Mastering this technique will allow testing of stem cell therapies, novel scaffold designs, and biomolecules that can be used for clinical translation.

Key words: Drill hole defect, Surgery, Stereotaxic, Magnetic resonance imaging, Computerized tomography

1. Introduction

Advances in stem cell therapy, scaffold design, and therapeutic biomolecules are rapidly advancing bone repair research. The bulk of the studies investigating stem cell therapy have been related to cardiovascular and neuronal repair therapies for disease. However, stem cell therapy for bone injury is now being used to repair bone defects (1, 2). Much of the recent research has focused on the implantation of mesenchymal stem cells (MSC) to evaluate their repair potential (1–3). Genetic modification of MSCs to produce active biomolecules is also an area of active research (4, 5). Primary MSCs can repair bone injury in numerous models, and a recent report demonstrates clinical potential (6), but additional work is required prior to their widespread use.

Nicole I. zur Nieden (ed.), *Embryonic Stem Cell Therapy for Osteo-Degenerative Diseases*, Methods in Molecular Biology, vol. 690, DOI 10.1007/978-1-60761-962-8_20, © Springer Science+Business Media, LLC 2011

Artificial constructs such as scaffolds have been used to support implanted cells and for the development and improvement of supportive structures to assist in bone repair. Optimization of scaffolds including shape, degrees of porosity, and novel compounds have demonstrated that bone repair can be facilitated by combinations of these factors (7–10). Abnormal and ectopic bone formation supports the notion that scaffold materials and design are important factors to consider in design repair strategies (11). The use of scaffold materials seeded with MSC has shown promise in a variety of bone injury models (5, 12, 13).

Imaging of implantation of MSC, scaffolds, and other supportive cells or materials is paramount for clinical translatability. X-ray and, more recently, computed tomography (CT) have been used to noninvasively evaluate bone loss and repair (14, 15). CT-derived data can provide useful information on vascularity, density, and progress of repair mechanisms. CT also provides the ability to undertake three-dimensional analysis that is more robust than two-dimensional methods (15). More recently, magnetic resonance imaging (MRI) has been developed clinically to evaluate bone injury (16). Quantitative MRI from T1-weighted imaging can provide diagnostic information about the bone structure and bone marrow (17). Dynamic contrast-enhanced MRI (DCE-MRI) can also be used to assess vascularity within bone structures to demonstrate improvement after injury and therapeutic interventions (18). However, invasive infusions of polymers into the vascular system can also be used for visualization by CT (14). Interestingly, ultrasound has also been used to improve the regenerative potential in bone restoration methods that are then assessed with MRI (19). Bioluminescent and positron emission tomography (PET) methods after cell transfection with various reporters or labeling with radioligands have also shown promise for noninvasive localization, but are not particularly useful for assessment of bone structure (2, 20). Thus, a range of noninvasive imaging methods can be used to assess the efficacy of therapeutic approaches to bone repair and restoration.

While numerous models of bone injury exist, drill hole defects are a simple model system to evaluate bone repair mechanisms, used in a broad range of research to investigate prospective therapeutic approaches (18, 21). Drill hole defects can be performed at a variety of sites within the body including the tibia, femur, and cranium (22, 23). These drill hole defects range is size and location dependent upon the research questions being answered. The detailed methods described in this chapter use the cranial approach due to the following reasons: (1) the cranium provides easy access, (2) the method can be performed by relative novices, (3) the cranium is easy to access for tissue and therapeutic implantation, and (4) the cranium provides an easy site for subsequent imaging to evaluate the effectiveness of treatment. The cranial surgical procedure primarily involves drilling or removing a portion of cranium.

To successfully complete this procedure, the researcher must have the capability to perform several tasks, including: (1) the induction of anesthesia, (2) surgical access to the bone, (3) drilling and removing a piece of the cranium, and (4) closing the incision via suturing techniques. In this chapter, we outline the steps required to generate a drill hole defect and then to assess the defect using clinically relevant noninvasive imaging modalities.

The drill hole defect can be readily produced in a variety of locations. Using the methods and procedures outlined in this chapter, the researcher can generate the defects in virtually any bone within the mammalian body. We have also outlined the general techniques for the noninvasive imaging of these defects, including CT and MRI, and described what these defects may look like and some rudimentary analysis methods.

2. Materials

2.1. Induction of Anesthesia

1. Oxygen gas tanks.
2. Isoflurane vaporizer and induction box.
3. Isoflurane.
4. Instrument used to perform the pedal reflex test (application of pressure on a rat's hind-limb ankle joint results in the foot being withdrawn; reflex is absent in deeply anesthetized animals).
5. Hair Clippers.

2.2. Drill Hole Defect

The materials required to perform a successful drill hole defect includes surgical tools/supplies, solutions, and other equipment. In preparation for the drill hole defect, it is wise to gather the items listed as follows (see Figs. 1 and 2):

1. Isofluorane vaporizer and induction box.
2. Isofluorane
3. Animal (Rodent) stereotaxic frame; For juvenile rat and mouse drill hole defect, a custom stereotaxic attachment is needed.
4. Betadine.
5. Q-tips.
6. Scalpel Handle (no. 4).
7. Scalpels (no. 21).
8. 2″ × 2″ Gauze Sponges.
9. Two Needle Retractors (fabricate retractors if performing drill hole defect in mice).
10. Hydrogen Peroxide.
11. Surgical Drill.

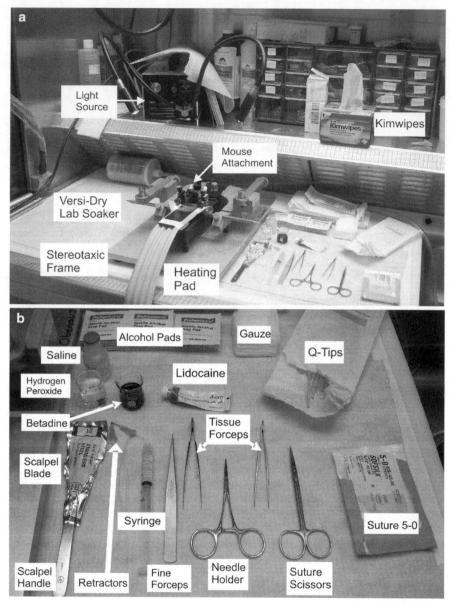

Fig. 1. Surgical station and supplies needed to perform a drill hole defect. (**a**) The surgical station setup including the stereotactic holder with a mouse attachment shown. The heating pad is important to maintain the temperature of the animals while under anesthesia. (**b**) A collection of surgical instruments and supplies required to undertake a drill hole defect surgery.

12. Trephine or Drill bit (utilize diameter for your research purposes).

13. 0.9% Saline.

14. Fine Forceps.

15. Kimwipes.

16. Suture Scissors.

17. Suture needle 5-0 (½, 17 mm, taper).

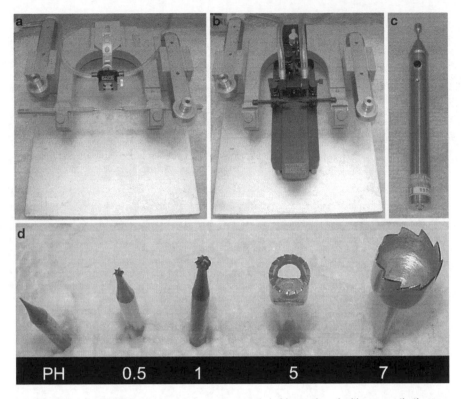

Fig. 2. Stereotactic holders and drills/trephines (**a**). A rat stereotactic holder equipped with an anesthetic nose cone. Note the ear bars (**b**). The same rat stereotactic holder with an attached mouse holder. Note smaller ear bars and more constricted working space. A surgical drill (**c**) and drills/trephines of varying sizes (**d**). Drill bits are primarily used in performing drill hole defect for smaller defect sizes. Trephines are used in performing drill hole defects in which a larger diameter hole is desired (sizes are in mm; PH denotes pinhole).

18. Lidocaine.

19. Tissue Forceps for surgical incisions.

20. Needle Holder for suturing.

21. Heating pad for postsurgical recovery.

22. Light Source (High Intensity Illuminator).

23. A surgical recovery chamber. See Note 1.

2.3. Imaging Equipment

To confirm a proper drill hole defect, computed tomography (CT) and/or magnetic resonance imaging (MRI) modalities can be employed. In addition, these modalities may pose a research benefit to the study depending on the focus of the project by the addition of quantitative imaging data. See Note 2.

2.3.1. MicroCT (Mouse or Rat)

Images can be acquired via the use of a micro X-ray computerized tomography (micro-CT) unit (for example, MicroCAT II®, ImTek Inc., Knoxville, TN) equipped with a 80 kVp X-ray source which effectively produces a reconstructed volume resolution of ~15 µm (Fig. 3a).

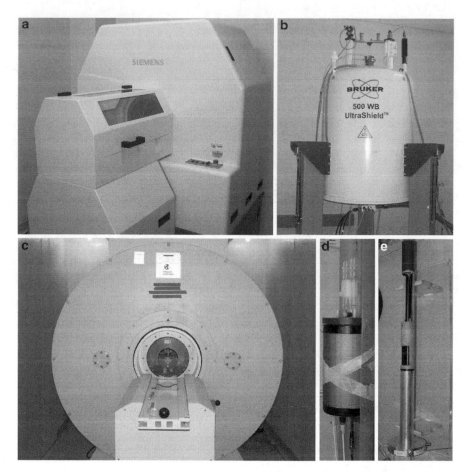

Fig. 3. Noninvasive imaging methods. The Siemens MicroCAT II MicroCT system (**a**) can accommodate large animals if desired. Our imaging suite has a Bruker Advance 11.7T MRI (**b**), and a Bruker 4.7T MRI Scanner (**c**) that can accommodate a range of experimental subjects (mice to dogs). A volume radiofrequency coil with animal handling including anesthetic nose cone (**d**) for the 4.7T MRI and for the 11.7T MRI scanner, a similar volume radiofrequency coil (**e**) is available for mice and small rodents.

2.3.2. MRI (Mouse)

When performing MRI on mice, one can use a Bruker Advance 11.7T MRI (8.9-cm bore) with a 3.0 cm (internal diameter) volume radiofrequency coil (Bruker Biospin, Billerica, MA) or similar scanner/coil with comparable capabilities (Fig. 3b, e). Imaging resolution of this instrument is on the order of ~100 μm/pixel.

2.3.3. MRI (Rat)

When performing MRI on rats, one can use a Bruker 4.7T 30 cm horizontal bore instrument equipped with 250 mT/m microgradients (slew rate 1,000 mT/s) and a 116 mm (internal diameter) quadrature receiver coil or similar scanner/coil with comparable capabilities (Fig. 3c, d). Imaging resolution of this instrument is on the order of ~150 μm/pixel.

2.4. Analysis Materials

Upon completion of the CT or MRI, 3D reconstruction of the drill hole defect can be generated, using advanced image analysis software (i.e. Amira, Mercury Computer Systems, Inc.). This generates not only publication-quality figures but also extractable quantification of data, such as volumes etc. See Note 3.

3. Methods

3.1. Induction of Anesthesia/Hair Trimming

1. Place animal in induction box. Be sure that you have an isoflurane vaporizer, filled with isoflurane, and oxygen supply attached to the box.

2. Begin induction of anesthesia by setting the isoflurane vaporizer to 3% and the oxygen flow rate to 2.5 L/min. At this time, several of the animal's vital signs will need to be monitored, which include respiration rate/pattern, heart rate, and animal color (skin tone should be pink). Monitoring of these vital signs should be continued throughout the procedure with adjustments in isoflurane level made accordingly to maintain animal health.

3. Once the animal is completely under anesthesia (~5 min), remove from the induction box and place on surface where hair trimming is to take place.

4. Prior to beginning the hair trimming, check to ensure that animal is completely under anesthesia by administering a pedal reflex test. If the animal exhibits a positive pedal reflex (i.e. limb withdrawal), then place it back in the induction box and readminister anesthesia. If the animal exhibits a negative pedal reflex, proceed to the hair trimming.

5. Using a standard hair clipper, remove the hair in the region where the drill hole defect is to be performed (typically area between the ear level and the eye). This is accomplished by stabilizing the animal's head between your index and middle finger and simply cutting the hair away (Fig. 4a, b), see Note 4.

6. Once the head has been shaved, place the animal back in the induction box and continue to administer anesthesia (3% isoflurane, 2.5 L/min oxygen) until the animal's respiratory rate is 0.5–1 breaths/s. Assessment can be done using manual counting methods or by use of a physiological monitoring system (i.e. Biopac Systems, Inc.)

3.2. Drill Hole Defect

1. With the animal's respiratory rate at 0.5–1 breath/s, begin the flow of anesthesia to the stereotactic holder (3% isoflurane, 2.5 L/min oxygen), see Note 5.

2. Remove animal from the induction box and place in the stereotactic holder. When placing the animal in the stereotactic holder, ensure that the front incisors are affixed around the bite bar and that the anesthetic gas cone completely covers the animal's nostrils (Figs. 1a and 2a, b).

3. Secure the head of the animal by placing the ear bars in the animal's ears. When doing this, place the pointed tip of the ear bar into the ear canal. When advancing the ear bar, lift the ear bar dorsally, with respect to the animal. There are also tick marks on the stereotactic holder that enable one to perfectly center the animal's head.

4. Advance the ear bars until resistance is met inside the ear canal. Ideally, this resistance should be the result of the ear bar meeting the caudal most part of the zygomatic arch. Check tick marks to ensure that the animal's head is centered within the stereotactic holder. Proper placement of the animal's head should result in an audible pop sound (Fig. 4c). See Note 6.

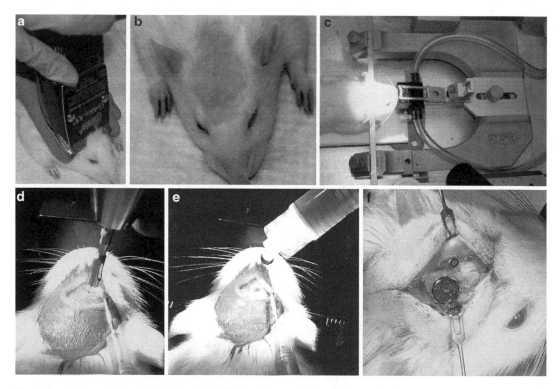

Fig. 4. Surgical procedures for initiation of a drill hole defect. (**a**) The rodent's head should be stabilized using the index and middle finger while the clippers are held in the other hand. Successful trimming results in a bald region between animal's eyes and ears (**b**). Placement of a rat in stereotactic holder where the head should be centered using the tick marks located on the ear bars (**c**). Retraction of the skin flap exposing the cranium (**d, e**). Drill placement and irrigation techniques (**d, e**). The drill should be parallel to the craniotomy site (**d**), and irrigation should take place every 4–7 s (**e**). Sample drill hole defects (3 and 1 mm) with the cranium removed demonstrating the intact dura (**f**) (*outlines*).

5. Sterilize the head of the animal in the area in which the incision is to be made by applying Betadine using a Q-tip as an applicator (Fig. 4d).

6. Using a no. 21 scalpel blade attached to a no. 4 scalpel handle; make a midline incision directly dorsal to the sagittal suture that extends from the region above the eyes to the animal's ears. Ensure that not too much pressure is applied to the scalpel blade as you do not want to cut into the underlying cranium. See Note 7.

7. Clean any blood that may appear as a result of making the incision, with a Q-tip or gauze.

8. Using a set of retractors, pull the skin flaps laterally from the incision to expose the cranium. Most drill hole defects will use bregma as a reference which should be visible at this time. Should the bregma be unidentifiable, a Q-tip with hydrogen peroxide on it can be used to clean the surface of the cranium so that the bregma becomes visible (Fig. 4d, f). See Note 8.

9. Using the coordinates deemed appropriate for the particular study, use a surgical drill and trephine to perform the drill hole defect, see Note 9.

10. After 4–7 s of drilling, irrigate the drill site with saline and dry with gauze. This will not only cool the drill site but also allow one to ascertain if more drilling is needed. If further drilling is required, proceed to drill in 4–7 s intervals with intermittent saline irrigation and drying. If the drill hole defect appears sufficient enough to remove the piece of cranium, proceed to attempt to remove the piece of cranium (Fig. 4d, e).

11. To remove the drilled piece of cranium, use fine forceps and try to lift the piece of bone out. When doing this, have the forceps as parallel to the piece of cranium as possible (to prevent touching the brain) and pull horizontally as opposed to vertically to prevent damage. When removing the bone fragment from the cranium, take care to ensure that the dura mater is not compromised. This will decrease bleeding and improve the integrity of the drill hole defect. Should any blood be present after removing the bone gentle application of kimwipes can be used to absorb residual amounts of blood, care being taken not to touch the brain. If after attempting to remove the bone it is determined that it is still attached to the rest of the skull, continue to drill and assess as indicated in steps 10 and 11.

12. Once the bone has been removed, a hole in the brain with intact dura should be visible, and we illustrate here a 3-mm and a 1-mm drill hole defect (Fig. 4f). See Note 10.

13. Once the drill hole defect is complete, employ a simple interrupted suture technique using a 5-0 silk suture to close the incision. The simple interrupted suture technique is a series of single sutures that bring the skin together and then tied off. The sutures should be evenly placed along the length of the incision (see http://emedicine.medscape.com/article/1128240-overview for more details on suturing).

14. After the incision has been sutured closed, apply lidocaine topically to the incision site. See Note 11.

3.3. Imaging Methods

3.3.1. Magnetic Resonance Imaging

After performing the drill hole defect and closing the incision, MRI can be employed to visualize the hole in a noninvasive manner (Fig. 5a, b). Under isoflurane anesthesia, the animal can be placed in either supine or prone position on a bed that can be inserted into the MRI. This bed and how the animal is attached will vary between users. In our own laboratory, we place the animals in a prone position, and the motion of the animal is minimized by the use of a Delron plastic stereotactic holder (see Fig. 3d). Anesthesia gases flow through a nose cone similar to surgical stereotactic holders. When imaging limbs, these are often immobilized by tape. To best visualize the drill hole defect, one should perform T2-weighted and proton-density (PD) sequences

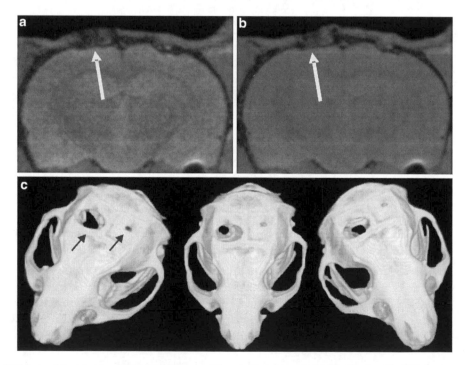

Fig. 5. Drill hole defects are easily identifiable on T2-weighted (**a**) and proton-density (**b**) MRI images (*arrows*). Reconstructed CT images illustrate the drill hole defect after a 3-mm (left side of skull) and 1-mm drill hole (right side of skull) (*arrows*) (**c**).

in both coronal and sagittal planes of view using the following parameters; see Note 12:

T2: TR/TE = 3,563.4 ms/20 ms, NEX = 2, FOV = 2.8 cm, Matrix = 256^2, Slices = 25, Slice Thickness = 1 mm, and Slice Interval = 1 mm.

PD: TR/TE=3,000 ms/20 ms, NEX = 2, FOV = 2.8 cm, Matrix = 256^2, Slices = 25, Slice Thickness = 1 mm, and Slice Interval = 1 mm.

These imaging sequence parameters can be altered as needed using the user interface on the MRI (or CT; see below).

3.3.2. Micro Computed Tomography

In addition to MRI, microCT can be used to visualize the drill hole defect in a noninvasive manner. As this methodology is sensitive to varying tissue density, the images produced from this modality will produce images in which the bone is in high contrast, allowing for easy identification of the hole produced as a result of the drill hole defect (Fig. 5c). Animal handling systems during the CT also vary as noted above, and we use a very similar system to that described in Subheading 3.3.1. When acquiring such images, the parameters should be used as follows:

Binning = 4 × 4, Exposure Time=125 ms, X-ray Voltage = 75.0 kVp, Anode Current = 1,000.0 µA, Rotational Steps = 360,

Rotation = 360°, Light: Dark Calibration Exposures = 25, and Matrix = 512^2.

These imaging sequence parameters can be altered as needed using the user interface on the MicroCT; see Note 13.

3.4. Analysis

Once imaging data has been acquired (CT or MRI), advanced imaging software, such as Amira, can be employed to generate 3D reconstructions of the 2D images (Fig. 5c) and extract quantifiable data pertaining to the particular study. Furthermore, density, length, volume, and other measurements can be readily obtained depending upon the research questions. See Note 14.

---

4. Notes

1. While the instruments shown in Fig. 1 and described in the detailed protocol are optimal variations, availability of supplies will dictate the actual mix of supplies and instruments. If available, one should always consult with a veterinarian to obtain guidance, as institutional rules and directives may vary.

2. Scheduling experiments and animal surgery may in some laboratories require coordination with user facilities that provide

access to these imaging instruments. Some of the user facilities may also charge a fee for imaging. It is recommended that the user meets with the imaging facility manager to discuss the timing, experimental design, and associated operational issues related to noninvasive imaging. It should also be noted that while in vivo imaging is optimal, ex vivo imaging after fixation can also provide valuable data.

3. While many laboratories use expensive three-dimensional reconstruction software programs such as Amira, free software, such as NIH image, can be used to extract and visualize imaging datasets. Often, however, these free programs are not powerful enough, and three-dimensional reconstruction is not optimal.

4. Take care not to cut the animal's whiskers or ears in the process. If the anesthetic plane is sufficiently deep, shaving the head is simply done outside the anesthesia induction box.

5. This will require an anesthesia splitter in which the flow of gas can be directed from the induction box to the stereotactic device.

6. Ensuring the proper placement of the ear bars is a critical element to obtain accurate and reproducible data. There are numerous resources on the internet that can provide additional guidance on correct placement of the ear bars. However, correct placement and insertion of the ear bars can be tested to ensure that there is no lateral (side-to-side) movement of the rodents' head, but the up and down movement is still unrestricted.

7. It is imperative that a sterile surgical approach is utilized. A clean and sterilized work environment (use 90% alcohol) is important. We strongly recommend sterilization of all surgical instruments, sutures, and supplies (e.g. gauze etc.) or at a minimum, the use of a hot bead sterilizer, taking care to let the instruments cool sufficiently prior to use.

8. Use of bregma or other surface landmarks on the cranium are important only if the researcher wants to localize the bone drill hole defect at the same precise location for each experiment. In some cases, this accuracy may not be necessary, and visual placement of the defect may be sufficient.

9. While drilling the hole, there should be no downward force applied so as to eliminate the trephine going completely through the cranium and into the brain. The drill should also remain as perpendicular to the cranium as possible to eliminate one side of the trephine cutting deeper than the other. If this occurs, the chances of hitting the brain will increase, and it will become increasingly difficult to remove the piece of cranium.

10. If a mistake is made that can be seen by excessive bleeding, it is recommended that the experimental animal be discarded and a new surgical attempt be made.

11. Long-term pain management is also recommended. Buprenorphine, a synthetic opiate, is often used in rodents to manage pain. Again, please consult with your designated veterinarian for guidance.

12. Other MR sequences can be utilized depending on the research questions. Sequences such as diffusion-weighted imaging and T1-weighted imaging can be useful.

13. An important caveat in designing your experiments is to note that CT imaging will deposit a radiation dose in the animal. Sensitive cellular constructs, such as stem cells, could be adversely affected, particularly with repeated CT imaging. For sample doses to various structures within the mouse and rat, please see Obenaus and Smith (24).

14. For examples of various types of bone analysis, the reader is referred to Willey et al. (25).

References

1. Kumar, S., and Ponnazhagan, S. (2007) Bone homing of mesenchymal stem cells by ectopic alpha 4 integrin expression. *FASEB J.* **21**, 3917–3927.

2. Lee, S.W., Padmanabhan, P., Ray, P., Gambhir, S.S., Doyle, T., Contag, C., et al. (2009) Stem cell-mediated accelerated bone healing observed with in vivo molecular and small animal imaging technologies in a model of skeletal injury. *J Orthop Res.* **27**(3), 295–302.

3. Hayashi, O., Katsube, Y., Hirose, M., Ohgushi, H., and Ito, H. (2008) Comparison of osteogenic ability of rat mesenchymal stem cells from bone marrow, periosteum, and adipose tissue. *Calcif Tissue Int.* **82**(3), 238–247.

4. Chang, S.C., Chuang, H., Chen, Y.R., Yang, L.C., Chen, J.K., Mardini, S., et al. (2004) Cranial repair using BMP-2 gene engineered bone marrow stromal cells. *J Surg Res.* **119**(1), 85–91.

5. Cui, L., Liu, B., Liu, G., Zhang, W., Cen, L., Sun, J., et al. (2007) Repair of cranial bone defects with adipose derived stem cells and coral scaffold in a canine model. *Biomaterials.* **28**(36), 5477–5486.

6. Centeno, C.J., Busse, D., Kisiday, J., Keohan, C., Freeman, M., and Karli, D. (2008) Increased knee cartilage volume in degenerative joint disease using percutaneously implanted, autologous mesenchymal stem cells. *Pain Physician.* **11**(3), 343–353.

7. Bikram, M., Fouletier-Dilling, C., Hipp, J.A., Gannon, F., Davis, A.R., Olmsted-Davis, E.A., et al. (2007) Endochondral bone formation from hydrogel carriers loaded with BMP2-transduced cells. *Ann Biomed Eng.* **35**(5), 796–807.

8. Ge, Z., Tian, X., Heng, B.C., Fan, V., Yeo, J.F., and Cao, T. (2009) Histological evaluation of osteogenesis of 3D-printed poly-lactic-co-glycolic acid (PLGA) scaffolds in a rabbit model. *Biomed Mater.* **4**(2), 21001.

9. Petersen, W., Zelle, S., and Zantop, T. (2008) Arthroscopic implantation of a three dimensional scaffold for autologous chondrocyte transplantation. *Arch Orthop Trauma Surg.* **128**(5), 505–508.

10. Williams, J.M., Adewunmi, A., Schek, R.M., Flanagan, C.L., Krebsbach, P.H., Feinberg, S.E., et al. (2005) Bone tissue engineering using polycaprolactone scaffolds fabricated via selective laser sintering. *Biomaterials.* **26**(23), 4817–4827.

11. Claase, M.B., de Bruijn, J.D., Grijpma, D.W., and Feijen, J. (2007) Ectopic bone formation in cell-seeded poly(ethylene oxide)/poly(butylene terephthalate) copolymer scaffolds of varying porosity. *J Mater Sci Mater Med.* **18**(7), 1299–1307.

12. Krupa, P., Krsek, P., Javornik, M., Dostál, O., Srnec, R., Usvald, D., et al. (2007) Use of 3D geometry modeling of osteochondrosis-like iatrogenic lesions as a template for press-and-fit scaffold seeded with mesenchymal stem cells. *Physiol Res.* **56(Suppl 1),** S107–S114.

13. Weinand, C., Pomerantseva, I., Neville, C.M., Gupta, R., Weinberg, E., Madisch, I., et al. (2006) Hydrogel-beta-TCP scaffolds and stem cells for tissue engineering bone. *Bone.* **38(4),** 555–563.

14. Bolland, B.J., Kanczler, J.M., Dunlop, D.G., and Oreffo, R.O. (2008) Development of in vivo muCT evaluation of neovascularisation in tissue engineered bone constructs. *Bone.* **43(1),** 195–202.

15. Park, C.H., Abramson, Z.R., Taba, M. Jr., Jin, Q., Chang, J., Kreider, J.M., et al. (2007) Three-dimensional micro-computed tomographic imaging of alveolar bone in experimental bone loss or repair. *J Periodontol.* **78(2),** 273–281.

16. Datir, A.P. (2007) Stress-related bone injuries with emphasis on MRI. *Clin Radiol.* **62(9),** 828–836.

17. Ecklund, K., Vajapeyam, S., Feldman, H.A., Buzney, C.D., Mulkern, R.V., Kleinman, P.K., et al. (2010) Bone marrow changes in adolescent girls with anorexia nervosa. *J Bone Miner Res.* **25(2),** 298–304.

18. Ehrhart, N., Kraft, S., Conover, D., Rosier, R.N., and Schwarz, E.M. (2008) Quantification of massive allograft healing with dynamic contrast enhanced-MRI and cone beam-CT: a pilot study. *Clin Orthop Relat Res.* **466(8),** 1897–1904.

19. Moinnes, J.J., Vidula, N., Halim, N., and Othman, S.F. (2006) Ultrasound accelerated bone tissue engineering monitored with magnetic resonance microscopy. *Conf Proc IEEE Eng Med Biol Soc.* **1,** 484–488.

20. Love, Z., Wang, F., Dennis, J., Awadallah, A., Salem, N., Lin, Y., et al. (2007) Imaging of mesenchymal stem cell transplant by bioluminescence and PET. *J Nucl Med.* **48(12),** 2011–2020.

21. Pereira, A.C., Fernandes, R.G., Carvalho, Y.R., Balducci, I., Faig-Leite, H. (2007) Bone healing in drill hole defects in spontaneously hypertensive male and female rats' femurs. A histological and histometric study. *Arq Bras Cardiol.* **88(1),** 104–109.

22. Katae, Y., Tanaka, S., Sakai, A., Nagashima, M., Hirasawa, H., and Nakamura, T. (2009) Elcatonin injections suppress systemic bone resorption without affecting cortical bone regeneration after drill-hole injuries in mice. *J Orthop Res.* **27(12),** 1652–1658.

23. Nagashima, M., Sakai, A., Uchida, S., Tanaka, S., Tanaka, M., and Nakamura, T. (2005) Bisphosphonate (YM529) delays the repair of cortical bone defect after drill-hole injury by reducing terminal differentiation of osteoblasts in the mouse femur. *Bone.* **36(3),** 502–511.

24. Obenaus, A., and Smith, A. (2004) Radiation dose in rodent tissues during micro-CT Imaging. *J X-Ray Sci Technol.* **12,** 241–249.

25. Willey, J.S., Grilly, L.G., Howard, S.H., Pecaut, M.J., Obenaus, A., Gridley, D.S., et al. (2008) Bone architectural and structural properties after 56Fe26+ radiation-induced changes in body mass. *Radiat Res.* **170(2),** 201–207.

Chapter 21

Measurement and Illustration of Immune Interaction After Stem Cell Transplantation

Stephan Fricke

Abstract

A variety of stem cells, including embryonic, mesenchymal, and hematopoietic stem cells, have been isolated to date, resulting in the current investigation of many therapeutic applications. These stem cells offer a high potential in cell replacement therapies or in the regeneration of organ damage. One current obstacle in using these stem cells in clinical applications are the unknown or unexplained mechanisms regarding the activation of immune responses as well as their given potential of immune activity, which can attack the host tissue. Similarly, the unknown immunological environment, which can benefit tumor growth, also restrains the rapid clinical implementation of stem cells.

We have shown that several techniques for measurement or illustration of immune responses in a hematopoietic murine $CD4^{k/o}$ mice transplantation model might be beneficial to get new insight into in vivo behavior of transplanted stem cells. Subjected to the transplantation setups (allogeneic, syngeneic, or xenogenic transplantation) different immune responses (enhancement of $CD4^+$ T cells, cytokine activity) as well as different effects of the transplanted cells on the host organs (organ destruction, toxicity) are detectable. The methods used to describe such immune responses will be presented here.

Key words: Immune response, Transgenic mice, Transplantation, Stem cell grafts, Cytometric Bead Array™, Flow cytometry, Immunohistochemistry, Histology

1. Introduction

Today stem cells from many sources [including bone marrow, adipose tissue, peripheral blood, human umbilical cord blood, and embryonic stem cells (ESCs)] are cultivated and used for experimental and therapeutic approaches. These cells offer an attractive potential in cell replacement and cell repair therapy due to their inherent plasticity and ability to self-renew. A variety of diseases, e.g. diabetes mellitus (1), neurologic diseases (2), and hematopoietic diseases (3) benefit from these characteristics, but

Nicole I. zur Nieden (ed.), *Embryonic Stem Cell Therapy for Osteo-Degenerative Diseases*, Methods in Molecular Biology, vol. 690, DOI 10.1007/978-1-60761-962-8_21, © Springer Science+Business Media, LLC 2011

application could be accompanied by severe complications. The lack of knowledge of immune interactions between transplanted cells and cells in the host tissue limits the translation into clinical use. The immune responses activated by graft cells are important to predict and characterize their therapeutic effects. For instance, the MHC class I molecule H2K(b) is expressed by ESCs, and without immune challenge they are able to escape immune recognition by H2K(b)-reactive CD8(+) T cells. Therefore, it has been suggested that ESCs and their terminally differentiated derivatives may possess a fragile immune privilege (4). In contrast, after hematopoietic stem cell transplantation (HSCT), graft versus host disease (GvHD) caused by the graft stem cells is still a major complication (5).

Measurement of cellular and humoral immunological responses in transplantation medicine might increase our understanding of in vivo behavior of transplanted stem cells regarding their mechanisms of graft rejections. Furthermore, these parameters may answer, if stem cells have inherent immunologic activity with regard to graft versus host reactions. Here, we will describe the measurement and illustrate immune responses using an examplatory hematopoietic transplantation model in transgenic, murine $CD4^{k/o}$ mice (6, 7). Since the murine CD4 antigen is knocked out in the host animals, murine stem cell engraftment can be characterized by the numbers of arising T cells expressing murine CD4. From the various available techniques for the measurement of immune responses and interactions, we will illustrate flow cytometry as one measurement of cell analysis. A Cytometric Bead Array™ will be used for the investigation of cytokine expression after transplantation, histological analysis for the characterization of toxicity of transplanted cells, and immunohistochemistry for the determination of donor organ chimerism. The described methods may certainly also be used for analyzing immune responses in other stem cell transplantation models, such as transplantations with ESCs.

2. Materials

2.1. Drawing Blood from Mice

1. Heparin-coated capillaries (Greiner Biochemica) for retroorbital blood taking. Store at room temperature.
2. Microcentrifuge tubes (e.g. Eppendorf).
3. Applicator for ring caps with capillary ejection (Hirschmann Laborgeräte GmbH & Co. KG). Store at room temperature.
4. 500 mL glass box (Duran Group) for anesthesia. Store at room temperature.
5. Swabs (Promedia) for the absorption of ether in Falcon tube. Store at room temperature.

6. Pulp (Zellstoffvertriebs GmbH) to avoid ether contact. Store at room temperature.

7. Fence cartridge (self-made) for placing the mice into diethyl ether without direct contact. Store at room temperature.

8. 50-mL Falcon tube (Greiner Bio-One).

9. Oxytetracycline (Jenapharm) eye ointment for prevention of eye infection. Store at 4°C.

10. Diethyl ether (Otto Fischer GmbH & Co. KG) for anesthesia. Do not use under an extractor hood or in an oxygen-rich environment. Store at room temperature in a dark environment.

11. Heparin (heparin–sodium 25,000, Ratiopharm) containing 10 μL microcentrifuge tubes (Eppendorf) for the collection of blood material. Store at 4°C.

2.2. Flow Cytometry and Hematology

1. For flow cytometry samples, use BD Falcon™ round-bottom tube (BD Biosciences), 12×75 mm, polystyrene (tube), sterile. Store at room temperature.

2. Animal blood counter (SCIL) must be calibrated for mouse blood (see Note 1).

3. SCIL animal ABC pack (SCIL). Reagents for 160 measurements. Handle with care and store at room temperature.

4. BD FACS lysing solution (BD Biosciences). Store at 2–25°C.

5. BD FACSCantoII™ flow cytometer with BD FACSDIVA™ software (both BD Biosciences).

6. Conjugated monoclonal antibodies: murine CD4 – PE-Cy™7; MHC-II (I-A[d], MHC class II alloantigen) – PE; MHC-II (I-A[b], MHC class II alloantigen) – FITC (all BD Biosciences). Store at 4°C (see Note 2).

7. Phosphate-buffered saline (10× PBS): 140 mM NaCl, 2.7 mM KCl, 10 mM Na_2HPO_4, and 1.8 mM KH_2PO_4. Adjust volume with Aqua destillata and the pH to 7.4. Filter sterilize or autoclave. Store at room temperature. Make a solution of 1× PBS containing 1% FBS (Gibco/Invitrogen) for washing samples. Store at 4°C.

2.3. Mouse Inflammation Cytometric Bead Array

1. BD™ Cytometric Bead Array (CBA) mouse inflammation cytokine kit (BD Biosciences) containing the following: mouse IL-6 capture beads, mouse IL-10 capture beads, mouse MCP-1 capture beads, mouse IFN-γ capture beads, mouse TNF capture beads, and mouse IL12-p70 capture beads; one vial each a 0.8 mL. One vial of mouse inflammation PE detection reagent (4 mL), two vials of mouse inflammation standards (0.2 mL, lyophilized), one vial cytometer setup beads (1.5 mL), one vial PE positive control detector

(0.5 mL), and one vial FITC positive control detector (0.5 mL). One bottle of ready to use wash buffer containing 130 mL is also included for washing samples and resuspending serum samples. Store at everything at 4°C (see Note 3).

2. Assay diluent (ready to use): one bottle containing 30 mL for washing and resuspending mouse inflammation standards, for dilution of serum samples and for use as a negative control. Store at 4°C.

3. Serum samples from mice. Store at –20°C (see Note 4).

4. BD FACSCalibur™ with BD CellQuest™ Pro software (BD Biosciences) (see Note 5).

5. 12×75 mm sample acquisition tubes for a flow cytometer (BD Falcon™) (BD Biosciences). Store at room temperature.

6. 15-mL Polypropylene tube (Greiner). Store at room temperature.

7. Autoclaved Eppendorf tubes (Eppendorf). Store at room temperature.

2.4. Histological Analysis by Hematoxylin–Eosin and Kernechtrot–Aniline Blue–Orange G Staining

1. Mayer's hemalaun solution (Merck) for hematoxylin–eosin (H&E) staining procedure. Store at room temperature.

2. Eosin Y (Sigma-Aldrich) for H&E staining procedure. Irritant agent, dispose of waste according to lab safety regulations. To prepare Eosin Y solution, add 1 g of Eosin Y to 50 mL distilled water. Next, prepare a 1:1 solution with 100% w/v ethanol (Merck). Store at room temperature.

3. Tap water as bluing reagent for H&E staining procedure. Needs to be partially softened to 8 dH water hardness.

4. Nuclear red (VWR, Prolabo) for Kernechtrot–Aniline Blue–Orange G (KAO) staining procedure: Make a 0.1% w/v nuclear red staining solution containing 1% w/v acetic acid by adding 0.1 g nuclear red to 100 mL of distilled water and 1 mL of acetic acid. Afterwards, filter the solution with a Folden Filter. Store at room temperature.

5. Aniline blue–Orange G (Halmi–Konecny solution) (8): 0.1% w/v aniline blue (VWR, Prolabo), 0.3% w/v Orange G (Sigma-Aldrich), 0.5% w/v wolfram phosphoric acid hydrate (Merck), and 1% w/v acetic acid (VWR, Prolabo) for KAO staining procedure. To prepare the KAO Halmi–Konecny solution, add 0.1 g Aniline blue, 0.3 g Orange G, 0.5 g wolfram phosphoric acid hydrate, and 1 mL glacial acetic acid in 100 mL distilled water. Boil the solution to 100°C for 1 min and let it cool to room temperature. Afterwards, filter the solution with a Folden Filter.

6. Isopropanol for KAO staining procedure. Highly flammable, store under fireproof conditions at room temperature.

7. Alcohol series (40–100% w/v): For the preparation of samples prior to embedding, ethanol methylated with 1% ethyl methyl ketone (Merck). Store at room temperature (see Note 6).

8. ddH$_2$O (Millipore) for staining procedure (H&E and KAO). Store at room temperature.

9. 35% w/v Formaldehyde (Carl Roth GmbH & Co. KG) for fixation of organs and tissues. Formaldehyde is toxic, store in a dark environment at room temperature, and only handle under an extractor hood (see Note 7).

10. Xylene (Rotipuran®, product of high purity), 99.9% p.a. (Carl Roth GmbH & Co. KG), for dehydration of organs and tissues. Xylene is harmful and highly flammable. Only handle under an extractor hood and store under fireproof conditions at room temperature.

11. Paraffin, Histosec pastilles (Merck) for embedding organs and tissues. Store at room temperature.

12. Methyl benzoate (Merck) for dehydration of organs and tissues. Methyl benzoate is harmful. Only handle under an extractor hood and store at room temperature.

13. Entellan (Merck) for embedding organs and tissues. Entellan is harmful. Only handle under an extractor hood store at room temperature.

14. Osteosoft®, decalcification reagent (Merck) to decalcify animal bones. Store at room temperature (see Note 8).

15. Whatman® ø 125 mm Folden Filter (Whatman GmbH).

16. Microscope cover glasses (Menzel-Gläser, Menzel GmbH & Co. KG) for embedding organs and tissues. Store at room temperature.

17. Polysine slides (Thermo Scientific, Menzel GmbH & Co. KG) for staining procedure. Store at room temperature.

18. Microtome blades, low profile – type 819 (Leica Biosystems), for cutting of embedded organs and tissues. Store at room temperature.

19. 33×24×12 mm Stainless steel embedding moulds (Bio-Optica). Store at room temperature.

20. Tissue embedding cassettes, Medim Uni-Safe (MDS-Group), for embedding of organs, especially intestines. Store at room temperature.

21. Leica-brush (Leica Biosystems) for cleaning and optimizing the preparations on the objective slide.

22. Incubator (Memmert GmbH & Co. KG) for keeping the paraffin in a liquid consistence.

23. Paraffin embedding station, Leica EG1150 H (Leica Biosystems), for manual embedding of organs and tissues.

24. Tissue processor, for example Type TP 1020 (Leica Biosystems), for automatic embedding of organs and tissues.

25. Rotary microtome, Leica RM2255 (Leica Biosystems), to cut organs and tissues for histological analysis. To cut osseous containing material for histological analysis use a sliding microtome, e.g. Leica SM2000R (Leica Biosystems).

26. Water bath, such as type-22721 (MDS-Group), for aiding in the preparation of the specimens.

27. Microscope, Axioskop 40 (Zeiss), to analyze the objective slides.

28. Pioneer balance (Ohaus), weighing machine for adjustment, and standardization of sample material.

2.5. Immunohistology

1. Cryostat, e.g. Leica CM3050 S (Leica Biosystems) and microtome blades, low profile – type 819 (Leica Biosystems), for cutting embedded organs and tissues. Store at room temperature.

2. Leica-brush (Leica Biosystems) for cleaning and optimizing the preparations on the objective slides.

3. 2-Methylbutane (isopentane, Carl Roth GmbH & Co. KG) for keeping the organs and tissues in a continuous frozen condition. This substance is harmful and highly flammable, only handle under an extractor hood and store under fireproof conditions.

4. Disposable base molds, $15 \times 15 \times 5$ mm or $24 \times 24 \times 5$ mm (Simport). Store at room temperature.

5. ddH$_2$O (Millipore).

6. Wet chamber (Carl Roth GmbH & Co. KG).

7. Phosphate-buffered saline (10× PBS): 140 mM NaCl, 2.7 mM KCl, 10 mM Na$_2$HPO$_4$, and 1.8 mM KH$_2$PO$_4$, with a final pH of 7.4. Filter sterilize or autoclave and store at room temperature.

8. Alcohol series (40–100% w/v); ethanol methylated with 1% ethyl methyl ketone (Merck) for dehydration of objective slides. Store at room temperature.

9. Xylene (Rotipuran®) ′99.9% p.a. (Carl Roth GmbH & Co. KG) for dehydration of organs and tissues. This substance is harmful and highly flammable. Only handle under an extractor hood and store under fireproof conditions at room temperature.

10. Entellan (Merck) for embedding organs and tissues. Entellan is harmful, only handle under an extractor hood and store at room temperature.

11. Acetone (Merck) to fix the organs and tissues. Store at room temperature.

12. SuperFrost Plus microscope slides (Thermo Scientific, Menzel GmbH & Co. KG) and microscope cover glasses (Menzel-Gläser, Menzel GmbH & Co. KG). Store at room temperature.

13. Tissue-Tek, O.C.T. compound (Sakura Finetek) for cryopreservation of organs and supporting medium for cutting procedure. Store at room temperature.

14. Dako pen wax crayon (Dako) for surrounding objects on the slides. Store at room temperature.

15. H_2O_2 solution, 30% w/v (Merck), for blocking endogenous peroxidase activity. Store at 4°C. Dilute to 0.3% w/v working concentration with in PBS.

16. Biotin blocking system consisting of avidin and biotin solutions and containing sodium azide (Dako). Used for blocking endogenous biotin (vitamin H). Store at 4°C.

17. Mayer's hemalaun solution (Merck), for counterstaining. Store at room temperature.

18. Antibodies for staining: rat anti-mouse CD4 $IgG_{2a}\kappa$, clone: RM4-5 (BD Biosciences). As secondary antibody use biotin-conjugated goat anti-rat $IgG_{2a}\kappa$ (BD Biosciences). As negative control, you may include an isotype control, such as rat polyclonal $IgG_{2a}\kappa$ (BD Biosciences). Store all antibodies at 4°C. Dilute antibodies to working concentration (1:100) with antibody diluent for immunohistochemistry (BD Biosciences).

19. Prediluted streptavidin-horseradish peroxidase (BD Biosciences), detection enzyme for the substrate of the secondary antibody. Store at 4°C.

20. 3,3′-Diaminobenzidine (DAB) substrate kit (BD Biosciences). Toxic, carcinogen. Store in a dark environment at 4°C. Prepare a DAB dilution by adding 40 µL of the DAB chromogen to 1 mL of DAB buffer. DAB dilution must be used within 6 h of preparation and at during this time stored at room temperature.

21. Liquid nitrogen (Linde GmbH). Handle with care and store in an insulated container.

22. Blocking buffer (10% w/v rat serum diluted in PBS with 10% w/v FBS) for blocking unspecific binding. Store at 4°C.

23. Immunofluorescence microscope, i.e., Eclipse TE 2000-E, Nikon, or other.

3. Methods

Immune reactions after transplantation of stem cells can be measured by several methods. These methods can detect whole cells, a variety of biologically soluble or stationary structures including proteins, cytokines, and nucleic acids and can also detect the change in organ structure after transplantation or application of stem cells in a host. Useful methods for illustration and measurement of immune responses include conventional ELISA (9), flow cytometry and histology (10), immunohistology (11), and Cytometric Bead Array (12). To receive reproducible and meaningful results by the several techniques, it is important to handle all samples and reagents accurately. Most of the methods used for the detection of stem cells or soluble molecules use monoclonal detection antibodies, which are highly sensitive and target even low amounts of an appropriate formation.

Flow cytometry is a common and almost standardized technique, which allows the detection of extracellular and intracellular molecules. Blood, bone marrow, and spleen cells from mice can be investigated by monitoring the expression or secretion of molecules. The measurement of cytokines and chemokines with the Cytometric Bead Array is advantageous and uses a series of particles with discrete fluorescence intensities (13). This allows detection and quantification of soluble analytes in a particle-based immunoassay. Compared to a conventional enzyme-linked immunosorbent assay (ELISA), the investigation of multiple analytes in small volume samples is now possible and less time consuming. The method can be used for the measurement of specific cytokines for a broad variety of samples (containing tissue culture supernatants, EDTA plasma, and serum samples). In addition, immunofluorescence techniques (e.g. microscopy or immunohistochemistry) are able to show organ distribution of transplanted cells (14). Attention should also be given to conventional histological analysis, which allows the investigation of organ structures after transplantation and gives valuable information about toxicity and damage. Flow cytometry and CBA analyses as well as histology and immunohistology techniques are demonstrated utilizing the triple transgenic transplantation model of $CD4^{k/o}$ mice as described above, but may be transferred into other animal models.

3.1. Blood Taking from Mice

1. Put the fence cartridge in the glass box and add 5 mL of diethyl ether. Afterwards, place the pulp over the cartridge to avoid direct contact between diethyl ether and the animal (see Note 9).

2. Take the mouse out of the cage, place it on the fence cartridge, and close the top cover of the glass box.

3. Take the nonsentient mouse out of the glass box and place one eye between thumb and index finger to pave the way for the heparin-coated capillaries (see Note 10).

4. Put the heparin-coated capillary under the eyeball and push carefully forward to the retrobulbar venous plexus as if throwing the capillary. Wait until the capillary is filled with blood. Repeat the step on the other capillary of the same animal for a maximum of two times.

5. Blow out the capillaries with the applicator for ring caps in prepared microcentrifuge tubes containing heparin. Centrifuge each microcentrifuge tube for 5 s at $1,000 \times g$.

6. Put oxytetracycline eye ointment on the manipulated eye and wait at least 1 min before placing the mice back into the cage.

3.2. Flow Cytometry and Hematology

1. Take 150 µL of blood from each mouse and incubate the blood with 10 µL of heparin (25,000 IE/mL). Preparation of samples for flow cytometry should be done within 3 h after blood taking.

2. Adjust the animal blood counter for measuring mouse blood according to manufacturer's instructions. Use the same samples that you will take into further flow cytometry analysis.

3. Provide BD flow cytometric tubes™ according to the amount of mouse blood samples with 100 µL of PBS/1% FBS and add 5 µL each of murine CD4-PE-Cy™7, MHC-II-PE, and MHC-II-FITC. Vortex each tube for 5 s.

4. Add 150 µL of mouse blood to each tube and mix immediately for 5 s. Incubate all tubes for 20 min at room temperature in a dark environment.

5. Using 2 mL of BD FACS™ lysing solution, make a dilution of 1:10 of the blood. Vortex carefully after adding the solution and incubate for an additional 10 min at room temperature. Afterwards, centrifuge all tubes for 5 min at $250 \times g$ and discard the supernatant (see Note 11).

6. Wash the pellet by carefully adding 3 mL of PBS/1% FBS. Mix gently to avoid spouting of the samples and centrifuge for 5 min at $250 \times g$. Discard the supernatant carefully and prepare the samples for measurement with the flow cytometer by adding 300 µL of PBS/1% FBS (see Note 12). Example results are shown in Fig. 1.

3.3. Mouse Inflammation Cytometric Bead Array

1. Prepare the mouse inflammation standards. For each run, a single standard curve must be created. Take one vial (0.2 mL) and transfer the lyophilized standard spheres to a 15 mL polypropylene tube. Add 2 mL of assay diluent buffer to get a master or top standard (concentration of 5 µg/mL) (see Note 13).

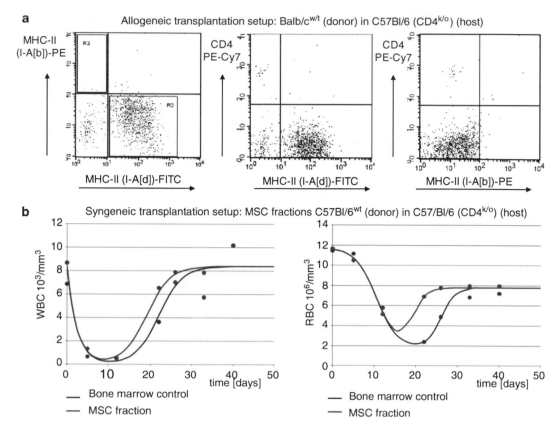

Fig. 1. Examples of flow cytometric analysis and blood counts. (**a**) Flow cytometric analysis of murine CD4 – PE-Cy™7, MHC-II (I-A[d], MHC class II alloantigen of Balb/c mice) – PE; MHC-II (I-A[b], MHC class II alloantigen of C57Bl/6 CD4$^{k/o}$ mice) – FITC from peripheral blood (gated for lymphocytes) on day 28 after transplantation. Lethally X-ray irradiated (8 Gy) C57Bl/6 CD4$^{k/o}$ mice received 5×10^6 allogeneic bone marrow stem cells of Balb/c^{wt} mice. Dot plots obtained from a flow cytometer show engraftment of bone marrow stem cells in the hosts represented by MHC-II (I-A[d]) and CD4 recovery. (**b**) Determination of white blood cells (WBCs) and red blood cells (RBCs) from peripheral blood by a SCIL animal blood counter was done over a period of 50 days (blood taking once per week) after transplantation. Lethally X-ray irradiated (8 Gy) C57Bl/6 CD4$^{k/o}$ mice received 5×10^6 syngeneic bone marrow stem cells of C57Bl/6^{wt} mice (enriched with 5×10^6 syngeneic splenocytes, bone marrow control) or 2×10^6 of syngeneic MSC fractions (MSC fraction, culture procedure not shown). Recovery rates of WBCs and RBCs provide evidence of a therapeutic effect with regard to reconstitution of hematopoesis.

Prepare eight microcentrifuge tubes with 100 μL of assay diluent buffer and label them with dilution 1:2, 1:4, 1:8, 1:16, 1:32, 1:64, 1:128, and 1:256. Create a standard dilution by transferring 100 μL of the top standard into the 1:2 microcentrifuge tube and mix well. Create the other dilutions by transferring 100 μL from the 1:2 tube to the 1:4 tube and so forth to the last one. Mix well in between steps.

2. Prepare the mixed mouse inflammation capture beads. First, determine the number of samples which should be investigated. Vortex the capture beads thoroughly for 5 s before they are used.

3. For each assay tube (BD Falcon™ round-bottom tube) to be analyzed, take 5 μL of each capture bead and mix all capture beads together in one single tube (e.g. in our protocol for MCP-1, capture beads and 20 100-μL tubes are required). Afterwards, vortex carefully for 15 s (see Note 14).

4. Subsequently, vortex the capture beads and add 25 μL of the mixture into each assay tube, including the unknown samples, negative controls, positive controls, as well as the mouse inflammation cytokine standard dilutions.

5. Pipette 25 μL of mouse inflammation PE detection reagent (ready to use) to all assay tubes. Then add 25 μL of mouse inflammation standard dilutions to the control assay tubes and 25 μL of each serum sample to other assay tubes. Vortex gently and incubate the assay tubes for 3 h in a dark chamber. Use the incubation time to perform the cytometer setup according to manufacturer's instructions (see Note 15).

6. Prepare the cytometer setup and the data acquisition and analysis software on the flow cytometer. Add 50 μL of the cytometer setup beads to three assay tubes (BD Falcon™ round-bottom tube) and label them from A to C. To assay tube B, pipette 50 μL of positive control detector linked with FITC, and to assay tube C, 50 μL of positive control detector linked with PE. Incubate tubes A, B, and C for 30 min at room temperature in a dark environment.

7. Add 500 μL of the wash buffer to each assay tube and centrifuge at $200 \times g$ for 5 min at room temperature. Carefully pipette and discard the supernatant.

8. Add 150 μL of the wash buffer again and resuspend the pellet immediately. Analyze the samples at once by flow cytometry to reduce background noise and to increase sensitivity.

9. Add 400 μL of wash buffer to assay tube B and C, and 450 μL to assay tube A. Follow the instructions for the "instrument setup with the cytometer setup beads" according the manual (see Note 5). Analyze the samples by flow cytometry. Example results are shown in Fig. 2.

3.4. Histological Analysis by H&E and KAO Staining

1. Kill mice according to the standards as per the American Veterinary Medical Association using procedures consistent with national animal protection law.

2. Prepare the organs and tissues (see additional procedure for osseous containing material in step 2) of the animals and put into 4% w/v formaldehyde. The preparation of all organs should be done immediately after death of the mice. Samples should not be fixed longer than 24 h to receive an optimal histological result. Until further handling, the formalin boxes must be kept under dark conditions to prevent formalin precipitation.

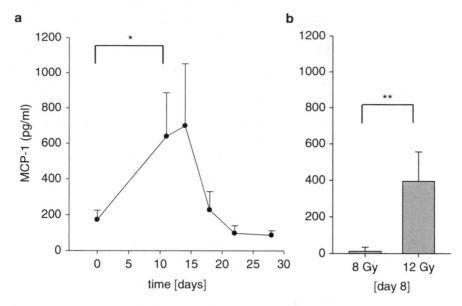

Fig. 2. MCP-1 levels in transplanted mice. (**a**) Determination of monocyte chemotactic protein-1 (MCP-1) over a period of 28 days after transplantation from sera of C57Bl/6 CD4^{k/o} mice by Cytometric Bead Array™. Lethally irradiated (8 Gy) C57Bl/6 CD4^{k/o} mice received 1×10^7 human PBMCs. Blood was taken by retroorbital bleeding once a week. After transplantation, there were increased serum levels of MCP-1. (**b**) Lethal X-ray irradiation (8 Gy) without transplantation does not lead to increased levels of MCP-1 in this xenogenic transplantation setup. However, 12 Gy X-ray irradiation leads to severe organ damage, which finally increases MCP-1 serum levels.

3. Fix osseous containing samples into 4% w/v formaldehyde for 24 h. Put the sample in a previously marked tissue embedding cassette. Before decalcification, flush the samples for 2 h with tap water and incubate the cassette in a glass beaker filled with Osteosoft® (see Notes 16 and 17).

4. You can use a paraffin embedding automate or you can do the following work manually. First, flush the fixed tissue for 2 h with running tap water to remove the residual formaldehyde. Then, pass the organs and tissues through an ascending alcohol series (remember to pass through each concentration twice) from 70 to 100% w/v, each for 1 h. Then, incubate the samples with isopropanol for 1 h and with methyl benzoate for 24 h.

5. Embed organs and tissues in paraffin for 48 h with the help of a paraffin embedding station. Fill embedding molds with 60°C prewarmed paraffin. Place the tissues and organs in molds and cover them with paraffin. Cool the samples in the molds to room temperature on the cold plate of the embedding station.

6. Remove the samples from the mold and store the blocks at room temperature (the blocks may be stored for a long time)

until preparation of histological slides. Cut paraffin blocks at the desired thickness (5 μm) on a microtome and float the sections on a 40°C water bath containing distilled water.

7. Transfer the sections onto a polysine slide. Allow the slides to dry overnight at 37°C and store slides at room temperature until ready for staining with H&E or KAO.

8. Prepare the slides for H&E staining. Remove the paraffin from the objective slides by incubating them twice with xylene for 5 min at room temperature.

9. Transfer the objective slides to a descending alcohol series (100, 90, 80, 70, and 50% w/v), 5 min for each incubation step, then transfer them to ddH$_2$O at room temperature.

10. Now place the objective slides in Mayer's hemalaun solution for 5 min at room temperature. Afterwards, wash them in running tap water for 10 min to reach a blue staining.

11. Add 1% w/v Eosin Y solution to the objective slides for 5 min at room temperature.

12. Afterwards, put the object slides in 40% w/v alcohol and pass them through an ascending (70–100% w/v) alcohol series (10 s for each incubation step).

13. Complete the process by incubating the objective slides with xylene for 5 min at room temperature and covering them with Entellan and cover glasses. Analyze the objective slides under the microscope.

14. Prepare the objective slides containing osseous substances for KAO staining according to Halmi–Konecny (8).

15. Remove the paraffin from the objective slides by incubating them twice with xylene for 5 min at room temperature.

16. Transfer the objective slides (5 min for each incubation step) to a descending alcohol series (100, 90, 80, 70, and 50% w/v), and then transfer them to ddH$_2$O at room temperature.

17. Incubate the objective slides for 5 min with the nuclear red staining solution and rinse them with ddH$_2$O for 5 s at room temperature.

18. Afterwards, incubate the objective slides for 5 min with a 5% w/v wolfram phosphoric acid solution and flush them with ddH$_2$O for 5 s at room temperature.

19. Now put the slides for 8 min in the previously prepared KAO Halmi–Konecny solution and rinse them with ddH$_2$O for 5 s at room temperature.

20. Put all objective slides in 96% w/v alcohol for 10 s at room temperature.

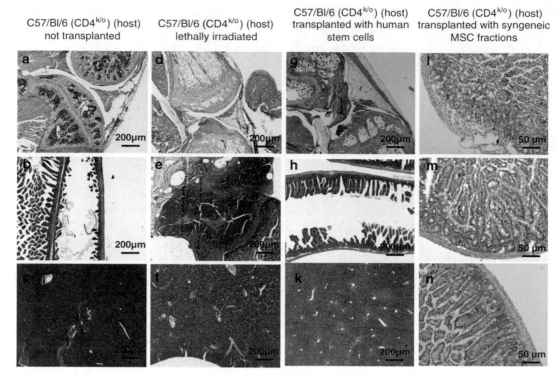

C57/Bl/6 (CD4$^{k/o}$) (host) not transplanted

C57/Bl/6 (CD4$^{k/o}$) (host) lethally irradiated

C57/Bl/6 (CD4$^{k/o}$) (host) transplanted with human stem cells

C57/Bl/6 (CD4$^{k/o}$) (host) transplanted with syngeneic MSC fractions

Fig. 3. Histology (KAO and H&E staining) of the bone marrow cavities, the gastrointestinal tract system, and the liver of irradiated and transplanted C57Bl/6 CD4$^{k/o}$ mice, as well as immunohistological staining of the gut. Knee joint (**a**) (KAO staining) of an untransplanted C57Bl/6 CD4$^{k/o}$ mouse shows hematopoietic islets with a prevalent form of erythropoiesis. Intestine (**b**) and liver (**c**) (both H&E stained) were normal with regard to histological structure and cell density. After lethal X-ray irradiation (8 Gy), a reduced cellularity with replacement of bone marrow cells by marrow adipose cells in the knee joint (**d**) could be observed. The gastrointestinal tract system shows a severe damage of all layers (**e**). The cell density in the liver (**f**) was found to be low. For the determination of possible therapeutic and repair effects of human umbilical cord blood stem cells (USCs), C57Bl/6 CD4$^{k/o}$ mice were lethally X-ray irradiated and transplanted with 2×10^6 USCs. Knee joints of several mice (**g**) show low amounts of hematopoietic islets and a recovery of the intestine (**h**) and partial recovery of the liver to a normal organ structure (**k**). After transplantation of syngeneic MSC fractions (cell culture procedure not shown), murine CD4 was detectable in the intestine (**l**, **m**) in previous murine CD4$^{k/o}$ mice as a sign of engraftment by immunohistology (streptavidin peroxidase technique), isotype control (**n**).

21. Dehydrate the samples for 30 s in isopropanol at room temperature and finish the work by incubating the slides with xylene for 5 min, and finally, by covering them with Entellan and cover glasses. Analyze the objective slides under the microscope. Example results are shown in Fig. 3.

3.5. Immunohistology

1. Prepare a stainless steel beaker with 2-methylbutane. Place the beaker in liquid nitrogen and allow it to cool for at least 10 min.

2. Prepare the organs and tissues of the animals and cut them to appropriate pieces that will fit into the base molds. The preparation of all organs should be done immediately.

3. Place the organs and tissues in prelabeled base molds filled with Tissue Tek at room temperature. Afterwards, quickly immerse the base mold with the fresh tissue into the steel beaker with cold 2-methylbutane for 5 min.

4. Remove the base molds from the 2-methylbutane and store them at −80°C until ready for sectioning process (see Note 18).

5. For sectioning, place the tissue block on the cryostat specimen disk and prepare sections of 6 μm thickness. Afterwards, place the sections on a Fisher superfrost slide and store them immediately at −80°C in a sealed slide box until further analysis.

6. Transfer the objective slides from −80 to −20°C, then fix the slides in cooled acetone for 2 min. Afterwards, dry the slides at room temperature.

7. Encircle the fixed preparations on the objective slides with a wax crayon. Wash the objective slides three times with PBS to remove the residual Tissue Tek.

8. For inhibition of endogenous peroxidase activity, incubate the objective slides with 0.3% w/v H_2O_2 (dissolved in PBS) for 10 min in a wet chamber, then wash them three times with PBS (2 min for each washing step).

9. For inhibition of unspecific binding, treat the preparations with 10% FBS w/v (dissolved in PBS) for 30–60 min at room temperature.

10. Quickly wash the objective slides with PBS, incubate them in avidin solution for 10 min, and finally wash them with PBS. Furthermore, incubate the preparations in biotin solution for 10 min and finally wash them with PBS.

11. Incubate the objective slides with the primary antibody, for 1 h at room temperature, then wash them three times with PBS, and wash the control slides with the isotype control (2 min for each washing step).

12. Afterwards, incubate the objective slides with the biotinylated secondary antibody, for 30 min at room temperature, then wash them three times with PBS (2 min for each washing step).

13. Cover the objective slides with the ready to use streptavidin-horseradish peroxidase for 30 min at room temperature, then wash them three times with PBS (2 min for each washing step). Afterwards, completely remove the PBS.

14. Incubate the objective slides with DAB dilution for 5 min until an obvious color intensity is achieved, then wash the slides three times with ddH_2O (1 min for each washing step).

15. Cover the objective slides with Mayer's hemalaun solution for a maximum incubation time of 1 min and then wash the slides with tap water for 10 min to visualize the blue staining.

16. Pass the slides through the ascending alcohol series (40–100% w/v) (1 min for each step), incubate with xylene for 5 min, and finally cover with Entellan and cover glasses. Analyze the objective slides under the microscope. Example results are shown in Fig. 3.

4. Notes

1. Before measuring the hematological parameters, it is essential to clean the blood counter according to the manufacturer's instructions to receive reproducible results and to avoid blockage of the equipment.

2. We only show an example of analyzing MHC-II antigens of Balb/c and C57Bl/6 mice. The technique can be used for detection of a broad range of other expressed molecules. We have found that the used antibodies work excellent with the noted concentrations. A variety of other antibodies are available, which often require determination of the optimal concentration before starting the experiments.

3. The described basic protocol can be used for the determination of a broad range of other cytokines from other available bead array kits. Modifications to the manufacturer's instructions may be necessary.

4. Mouse serum samples should be prepared accurately. Contamination of the serum with hemoglobin will decrease the sensitivity of determining soluble molecules. Serum samples should be frozen within 3 h after taking blood. Avoid repeated freeze–thaw cycles. If the serum samples are to be stored for over 1 month, freeze them at −80°C.

5. You can use other flow cytometric setups. We found that the used protocols for flow cytometry and Cytometric Bead Array™ are efficient and decrease hard work in optimizing and adaption to other systems.

6. For each staining procedure, prepare a new alcohol series. Repeated use of the same series degrades the quality of the histological slides. Use one alcohol series for no more than 50 objective slides.

7. You should be very careful when using formaldehyde. Using this substance outside of an extractor hood and inhaling, it will lead to streaming eyes and may exacerbate respiration.

Therefore, do not put your organs and tissues in formaldehyde outside of an extractor hood.

8. Each Osteosoft® solution should only be used for one procedure. Prepare a new solution after finishing one assay. A loss of color intensity does not influence activity of the reagent. It is possible to decalcify for longer than 7 days.

9. The method assures only temporary anesthesia for approximately 25 s. Be careful with the ether dose and always check the breathing of each mouse. Anesthesia by the inhalation of ether for mice is regulated by animal protection law and has to be applied for at the Regional Board of Animal Care. Alternative anesthesia forms (e.g. isoflurane anesthesia for mice) are possible.

10. To extend the effects of the anesthesia for over 25 s, add diethyl ether on a swab and put it into a Falcon tube. Place the tube on the nose of the mouse when the anesthesia begins to wear off.

11. Do not incubate the samples with the lysing solution for more than 10 min because a longer incubation time negatively influences the previously achieved binding of the labeled antibodies.

12. If it is not possible to measure the samples immediately, they can be fixed with a solution of PBS and 1% w/v of formaldehyde. The samples should be stored in cool conditions at 4°C, and then measured within 3 days by flow cytometry.

13. While establishing your method, you should follow manufacturer's instructions accurately. Dissolve the standard spheres for at least 15 min. Be careful during the procedure and only mix by pipetting. Do not mix vigorously.

14. Prepare the capture beads solution for the samples with an excess of three tubes to compensate for a probable loss of volume (pipetting error). However, the mixed capture beads cannot be stored and thus must be calculated exactly.

15. The standard curve only indicates a defined range from 20 to 5,000 pg/mL. The serum samples should be diluted with assay diluent 1:4 to make sure that the results are within the appropriate range.

16. As a guideline for the amount of Osteosoft® needed, a $15 \times 9 \times 3$ mm iliac crest should be covered with 50 mL of Osteosoft®.

17. Cover the Osteosoft® containing beaker with parafilm to avoid evaporation. For testing of the softness of osseous containing samples use a pin.

18. Before the cutting procedure starts, allow the samples to reach a temperature of −20°C. Do not directly cut blocks stored at −80°C.

References

1. Zulewski, H. (2006) Stem cells with potential to generate insulin producing cells in man. *Swiss. Med. Wkly.* **136**, 647–654.

2. Leker, R. R., and McKay, R. D. (2004) Using endogenous neural stem cells to enhance recovery from ischemic brain injury. *Curr. Neurovasc. Res.* **1**, 421–427.

3. Ball, L. M., Bernardo, M. E., Roelofs, H., Lankester, A., Cometa, A., Egeler, R.M., et al. (2007) Cotransplantation of ex vivo expanded mesenchymal stem cells accelerates lymphocyte recovery and may reduce the risk of graft failure in haploidentical hematopoietic stem-cell transplantation. *Blood* **110**, 2764–2767.

4. Wu, D., Boyd, A. S., and Wood, K. J. (2008) Embryonic stem cells and their differentiated derivatives have a fragile immune privilege, but still represent novel targets of immune attack. *Stem Cells* **26**, 1939–1950.

5. Shlomchik, W. D. (2007) Graft-versus-host disease. *Nat. Rev. Immunol.* **7**, 340–352.

6. Laub, R., Dorsch, M., Wenk, K., and Emmrich, F. (2001) Induction of immunologic tolerance to tetanus toxoid by anti-human CD4 in HLA-DR3(+)/human CD4(+)/murine CD4(–) multiple transgenic mice. *Transplant. Proc.* **33**, 2182–2183.

7. Laub, R., Dorsch, M., Meyer, D., Ermann, J., Hedrich, H. J., and Emmrich, F. (2000) A multiple transgenic mouse model with a partially humanized activation pathway for helper T cell responses. *J. Immunol. Methods.* **246**, 37–50.

8. Werksschrift, C. (1962). Ausgewählte Färbemethoden für Botanik, Parasitologie, Zoologie. Chroma-Ges. Schmid & Co., Stuttgart-Untertürkheim, p. 57.

9. De Simone, E. A., Saccodossi, N., Ferrari, A., and Leoni, J. (2008) Development of ELISAs for the measurement of IgM and IgG subclasses in sera from llamas (Lama glama) and assessment of the humoral immune response against different antigens. *Vet. Immunol. Immunopathol.* **126**, 64–73.

10. Tung, T. H., Mohanakumar, T., and Mackinnon, S. E. (2005) TH1/TH2 cytokine profile of the immune response in limb component transplantation. *Plast. Reconstr. Surg.* **116**, 557–566.

11. Ramanayake, T., Simon, D. A., Frelinger, J. G., Lord, E. M., and Robert, J. (2007) In vivo study of T-cell responses to skin alloantigens in Xenopus using a novel whole-mount immunohistology method. *Transplantation* **83**, 159–166.

12. Jimenez, R., Ramirez, R., Carracedo, J., Aguera, M., Navarro, D., Santamaría, R. et al. (2005) Cytometric bead array (CBA) for the measurement of cytokines in urine and plasma of patients undergoing renal rejection. *Cytokine* **32**, 45–50.

13. Carson, R. T., and Vignali, D. A. (1999) Simultaneous quantitation of 15 cytokines using a multiplexed flow cytometric assay. *J. Immunol. Methods.* **227**, 41–52.

14. Beckmann, J. H., Yan, S., Lührs, H., Heid, B., Skubich, S., Förster, R., et al. (2004) Prolongation of allograft survival in ccr7-deficient mice. *Transplantation* **77**, 1809–1814.

INDEX

A

Accutase 37, 51, 53, 55, 123, 125, 130, 132
Actin ring 274, 276, 280–283
Activin..58, 60, 61, 63, 64
Adult stem cell...2, 4, 9, 175
Aggrecan 5, 8, 96, 197, 256, 286
Alizarin Red S staining 186
Alkaline phosphatase
    alkaline phosphatase activity...................45, 48, 73, 208
    alkaline phosphatase assay 226, 260–261
All-trans retinoic acid.. 196, 199
Anesthesia 303, 304, 307, 310, 312, 316, 317, 331
Artificial chromosome 96
Ascorbic acid 11, 12, 140, 186, 190, 196, 199, 200, 257

B

Beta-catenin 12, 58, 197, 234
Beta-glycerophosphate 12, 139, 140, 199, 257
Biglycan.. 256, 257
Bioreacor
    coating 140
    inoculation.. 139, 141
    probe calibration... 147
    sterilization... 140
Blastocyst... 9, 32, 68, 107, 217
Bone morphogenetic protein.........................122, 196, 199
Bone remodeling 218, 239
Bone sialoprotein... 5, 8, 12
Bovine serum albumin.........................58–60, 86, 185, 200,
    222, 261

C

Cadherin
    cadherin–11 ....20, 196, 197
    E-cadherin...83, 84, 87
    N-cadherin 83–85, 87, 89, 93
Calcification 6, 262
Calcitonin receptor... 6, 274
Calcium
    assay... 266
    deposit 264
Calmodulin.. 258
Carbonic anhydrase II 274, 276, 278, 282
Cathepsin K...6, 9, 274, 276

cDNA synthesis.. 71, 209
Cell-cell adhesion 83
Cell count... 73, 77, 78, 116, 124,
    126–127, 141, 180, 200, 205, 226, 233, 292
Cell surface marker................. 65, 82–83, 90, 165, 167, 196
Chemically defined medium.. 61
Chondrocyte......................... 8, 11–14, 197, 200, 208, 256
Chondrogenesis...12, 14, 197–198
Chondrogenic differentiation 197–201, 207–208
Chondroitin/dermatan sulfate 285
Chromatin 10
Coating
    fetal bovine serum....................................34, 61, 62, 82,
    85, 97, 109, 122, 139, 155, 177, 185, 197, 219, 242
    gelatin 133
    matrigel.. 72
    sigmacote 140
Co-culture
    limb bud progenitor cells.. 197
    OP9, 247, 250
    primary chondrocytes 5
    ST2, 241, 248
Collagen
    collagen type I 5, 198, 219, 222,
    229–230, 259, 275, 278
    collagen type II 5, 13, 14
Collagenase IV 37, 51, 53, 55, 61,
    63, 65, 97, 98, 100, 109, 111, 212
Computed tomography 302, 305, 311
Conditioned medium 68, 98, 116, 244
Core binding factor alpha 1, 5
Crude osteoclast preparation 275–279
Cryosectioning 329
Cytometric bead array 316–318, 322–326, 330

D

Decorin...256, 257, 286
Delta-delta-Ct method 210
Dentine slice culture.................................276, 279–280
Dexamethasone 11, 12, 16, 21, 186,
    190, 196, 197, 220, 241, 242, 244, 257
Dicer...108
Dispase 63, 65, 71, 76, 79, 185,
    188, 212
DMMB assay 259–262

Nicole I. zur Nieden (ed.), *Embryonic Stem Cell Therapy for Osteo-Degenerative Diseases*, Methods in Molecular Biology, vol. 690,
DOI 10.1007/978-1-60761-962-8, © Springer Science+Business Media, LLC 2011

DNA delivery

chemical-based transfection...96

electroporation..96, 98

knock-in approach..96

nucleofection...96, 98

viral transduction.........................96, 98, 116, 118, 165

DNase treatment...90, 91

Drawing blood...316–317

Drill hole defect...301–313

E

EB assay...69, 73

EB formation

analysis..145

bioreactor...143

hanging drop...143, 152

microwell patterning..138

static suspension..152

suspension..152

Embryoid body.....................71, 73–75, 103, 104, 135–148

Embryonic stem cell

derivation...2, 9, 82, 176

passaging

cut & paste...52–53

EDTA split...98, 101, 102

enzymatically...288, 294

Glass beads...38, 51–52

mechanically..53

therapies...2, 175

three dimensional culture...............................32, 225

Encapsulation

alginate...198

poly-(D, L)-lactic acid...16

polyethylene glycol hydrogel...198

self-assembling peptide scaffold...................................217

Endochondral ossification.............6, 7, 13, 20, 21, 218, 286

Epithelial-mesenchymal transition...................83, 184, 191

Expansion

feeder-free..72

large-scale..121–133

Extracellular matrix.....................................4–8, 60–61, 136, 198, 218, 229, 256–258, 270, 285, 286

F

Feeder layer...9, 31, 32, 35–36, 40–44, 54, 62–63, 82, 89, 93, 157, 165, 176, 184, 187

Fetal bovine serum.....................................34, 61, 62, 82, 85, 97, 109, 122, 139, 155, 177, 185, 197, 219, 242, 244, 275, 287

Fibroblast growth factor

basic.............................37, 58, 61, 82, 97, 176, 177, 185

Fibronectin..5, 32, 60–64

Fluorescence activated cell sorting (FACS).........................
82, 85–87, 90, 93, 96, 124, 125, 165, 169, 177, 178, 180, 181, 317, 318, 323

G

Gelatin coating...53, 133

Gel electrophoresis..........................92, 222, 288, 293, 294

Gene expression..................................12, 14, 74, 79, 96, 98, 103, 107, 108, 136, 165, 202, 208, 211

Genetic manipulation...96

Germ layer..104, 117, 121, 135, 151

Glycomics...285–299

Glycosaminoglycan

depolymerization.....................................288, 294–296

molecular weight analysis.................................287–288

purification...287, 292, 299

Graft *versus* host disease..168, 316

H

Hanging drop protocol...................136–138, 143, 144, 158

Hematopoietic cells................8, 11, 218, 240, 241, 250, 273

Hematoxylin-eosin staining...318

Hemocytometer...70, 73, 77, 78, 86, 99, 123, 124, 126, 127, 139, 141, 200, 205, 292

Heparin'heparan sulfate...285, 288

Host...4, 96, 98, 108, 167–169, 316, 322

Hyaluronan..285

Hydroxyapatite..5, 20, 233, 256, 259

Hypertrophic chondrocyte....................................6–8, 218

I

IMAGE J...211, 212

Imaging...201, 301–313

Immune interaction..315–333

Immune response...316, 322

Immunofluorescence.......................221, 226–228, 321, 322

Induced pluripotent stem cell...............16, 33, 65, 107, 184

Inner cell mass...9, 10, 82, 107, 185, 217

Insulin-like growth factor 1..197

Integrin...258, 274

Interleukin...273

Intramembraneous ossification.....................6, 7, 13, 14, 20

K

Karyotyping..62–64

Keratan sulfate...256, 285

Kernechtrot-Aniline Blue-Orange G
(KAO) staining...318–320

L

Laminin...32, 60

Lentivirus

production of recombinant particles.........110, 114–116

transduction...................................96, 98, 116, 118, 165

Leukemia inhibitory factor..............................9, 36, 82, 86, 122, 135, 157, 217, 244

Lithium chloride 58
Lowry assay 261, 262, 266–268

M

Macrophage colony-stimulating factor.................. 240, 273
Magnetic resonance imaging.........................60, 69, 72, 73,
    76–78, 97, 100, 101, 103, 104, 110, 116
Matrix metalloproteinase 9.. 274
Mesenchymal stem cells
    criteria for multipotency 164
    derivation.. 175–182
    immunomodulatory properties 168–169
    isolation 226
    passaging 14, 34, 35, 38, 40,
        45–47, 51–53, 62, 65, 79, 126, 132, 169, 170,
        180–181, 290
    surface antigen markers 83
Mesoderm 5, 6, 9–12, 21,
    73, 75, 76, 78, 121, 136, 137, 145, 164, 167, 184,
    217, 223, 231, 240
Microarray 12, 109, 111, 112, 117, 196, 197
Microcarrier
    releasing cells 132
    seeding.. 159
Microwell 138, 151–161
Microwell patterning
    EB formation................................ 69, 73, 79, 136, 138,
        139, 141, 148, 152, 153
    photomask production... 156
    wafer 156, 158, 159
3D microenvironment 198
Mineralization...6, 12–14, 139,
    140, 196, 210–212, 218, 219, 226–227, 257–259
Morphometric image analysis 210, 212
Mouse embryonic fibroblast
    freezing.. 35
    inactivation
        irradiation 42–43
        mitomycin C.. 40, 43–44
    Passaging 35
mTeSR 32, 33, 37, 50–53, 69,
    72, 73, 76–79
Mycoplasma 33, 147, 205

N

Nanog................................ 9, 65, 75, 82, 124, 129–131, 145
Neural commitment 122, 125, 130–131, 133
Neural crest 5, 7, 20, 166, 221
Nodal..58
Non-destructive cell assay.. 83

O

Oct–3/4 ....10, 124, 221
Oil Red O staining.. 187, 190

OP 9 cells
    maintenance... 246
    preparation.. 242–244
    thawing.. 245–246
Osteoblast..7, 8, 11–14, 96,
    145, 196, 197, 207, 218, 219, 255–259, 282
Osteocalcin....................5, 8, 12, 13, 96, 145, 196, 207, 258
Osteoclast...5, 7, 8, 11–14, 240–242,
    244, 250, 273–280, 282
Osteogenesis... 1, 2, 8, 12, 13, 168,
    183, 195–197, 257, 258, 264, 265
Osteogenic differentiation......................... 12, 20, 139, 140,
    145, 148, 186, 190, 195–197, 199–200, 206–207,
    210, 212, 218, 223–226, 230, 233, 234
Osteoporosis...1, 2, 258

P

Parathyroid hormone......................................258, 259, 273
Pit-formation assay................................274–276, 278–280,
    282, 283
Plasmid DNA.. 102–104
Platelet-derived growth factor................................ 176, 177
Pluripotency 4, 9–12, 15, 32, 45, 48, 55, 58,
    60, 61, 65, 67–80, 82–83, 95, 107, 108, 111, 124
Polymerase chain reaction 87–88, 92, 201
Positron emission tomography 302
Pre-cartilage condensations.. 198
Primitive streak..12, 13, 20, 83
Proliferation..8, 12, 15, 32,
    58, 63, 64, 81, 105, 136, 147, 168, 183, 184, 192,
    230, 257, 258, 286
Promotor 7, 8, 10–13, 20, 32, 96, 98,
    103, 104, 113, 117, 118, 152, 158, 168–170, 176,
    197, 198, 218, 219, 235
Prostaglandin..273
Proteoglycan 5, 13, 14, 208, 256–257,
    259, 261, 262, 285

Q

Quantitative PCR 71, 75–78, 201, 208–211

R

Raclure method 165, 183–192
RAD16-I...219, 220, 225, 231
Receptor activator of nuclear factor
    (NF)-kB ligand.. 5, 274
Rejection 2, 4, 15, 16, 170, 316
RNA-induced silencing complex.................................... 108
RNA interference 108
RNA isolation 208–209, 223

S

Scaffold.......................5, 16, 20, 21, 198, 217–234, 301, 302
SCID mice 15, 16, 71, 76

Self-renewal..........................15, 58, 72, 81, 82, 96, 121, 163
Serum-free...33, 57–65, 122, 123,
        127, 128, 130–133, 191
Serum replacer...58
Shear stress 122
Somatic cell nuclear transfer...4
Sox2..10
Sox9..8, 13, 14
Spinner flask...122–125
Stage-specific embryonic antigen
        SSEA–1...68, 82, 83, 129, 130
        SSEA–3/4 ....60, 64, 82, 83
Static culture..122, 142–144
ST2 cells
        maintenance...243–244, 247
        preparation...243–244, 247
        thawing..247
Sterile technique...33, 34
Stirred culture 123–124
Surgical drill 303, 305, 309
Suspension culture bioreactor...138
Suturing..303, 305, 310

T

Tartrate-resistant acidic phosphatase
        Staining 6, 9, 242, 249, 275
T-Brachyury 12, 137, 145
Teratoma
        teratoma formation assay.........................15, 16, 76, 166
Transfection
        antibiotic selection...109
        identification of stable clones......................................96

kill curve 34, 35, 101, 106, 168, 325
        picking clones 103
Transforming growth factor 197, 200
Transgenic approach...96
Transplantation..4, 15, 16, 68, 69,
        166–169, 196, 315–331
Trimethylation marks 10
Trophoectoderm 10
Trypan blue 78, 123, 124, 126,
        127, 139, 141, 180, 200, 205, 247, 287, 292
Trypsin 34–43, 47, 49, 54,
        62–64, 71, 74, 78, 86, 89, 93, 97, 114, 123–127,
        130, 132, 133, 139, 141, 143, 156, 157, 177–181,
        185, 188, 200, 205, 206, 208, 213, 220, 223, 225,
        226, 230, 240, 243, 244, 247, 248, 259, 278, 287

V

Viability..............................20, 124, 126–127, 166, 287, 292
Viral vector
        adenovirus... 108
        lentivirus...110, 116, 118
Vitamin D3 .... 139, 145, 148, 199, 212, 215
Vitronectin
        vitronectin receptor.......................................6, 274, 278
von Kossa staining..................................221–223, 279–280

W

Wnt/beta-catenin..12, 58, 234

Y

Y–27632 ....69, 72, 73, 78, 79